# Phage Display of Peptides and Proteins

## A Laboratory Manual

# Phage Display of Peptides and Proteins

## A Laboratory Manual

*Edited by*

**Brian K. Kay**
**Jill Winter**
**John McCafferty**

**Academic Press**

San Diego   London   Boston   New York   Sydney   Tokyo   Toronto

**Find Us on the Web! http://www.apnet.com**

This book is printed on acid-free paper. ∞

Academic Press, Inc.
A Division of Harcourt Brace & Company
525 B Street, Suite 1900, San Diego, California 92101-4495

*United Kingdom Edition published by*
Academic Press Limited
24-28 Oval Road, London NW1 7DX

Library of Congress Cataloging-in-Publication Data

Phage display of peptides and proteins : a laboratory manual / edited
    by Brian K. Kay, Jill Winter, John McCafferty.
            p.          cm.
    Includes bibliographical references and index.
    ISBN 0-12-402380-0 (alk. paper)
    1. Bacteriophages--Laboratory manuals.    2. Microbial
biotechnology--Laboratory manuals.    3. Viral proteins--Laboratory
manuals.    4. Affinity chromatography--Laboratory manuals.
5. Peptides--Biotechnology--Laboratory manuals.    6. Proteins-
-Biotechnology--Laboratory manuals.    I. Kay, Brian K.    II. Winter,
Jill.    III. McCafferty, John, Dr.
QR342.P456    1996
576'.6482--dc20                                      96-13576
                                                     CIP

PRINTED IN THE UNITED STATES OF AMERICA
96   97   98   99   00   01   EB   9   8   7   6   5   4   3   2   1

## 3 Vectors for Phage Display

Norris Armstrong, Nils B. Adey, Stephen J. McConnell, and Brian K. Kay

## 4 Microbiological Methods

James E. Rider, Andrew B. Sparks, Nils B. Adey, and Brian K. Kay

## 5 Construction of Random Peptide Libraries in Bacteriophage M13

Nils B. Adey, Andrew B. Sparks, Jim Beasley, and Brian K. Kay

## 6 Construction and Screening of Antibody Display Libraries

John McCafferty and Kevin S. Johnson

## 7 Phagemid-Displayed Peptide Libraries

Diane Dottavio

## 8  Multiple Display of Foreign Peptide Epitopes on Filamentous Bacteriophage Virions

**Pratap Malik, Tamsin D. Terry, and Richard Perham**

## 9  Phage Libraries Displaying Random Peptides Derived from a Target Sequence

**Dion H. du Plessis and Frances Jordaan**

## 10 Display and Selection of Proteins on Genetic Packages

**Robert Charles Ladner**

## 11 A cDNA Cloning System Based on Filamentous Phage: Selection and Enrichment of Functional Gene Products by Protein/Ligand Interactions Made Possible by Linkage of Recognition and Replication Functions

**Mark Suter, Maria Foti, Mathias Ackermann, and Reto Crameri**

## 12 Multicombinatorial Libraries and Combinatorial Infection: Practical Considerations for Vector Design

**Régis Sodoyer, Luc Aujame, Frédérique Geoffroy, Corinne Pion, Isabelle Peubez, Bernard Montègue, Paul Jacquemot, and Jocelyne Dubayle**

## 13 Screening Phage-Displayed Random Peptide Libraries

**Andrew B. Sparks, Nils B. Adey, Steve Cwirla, and Brian K. Kay**

## 14 Substrate Phage

**David J. Matthews**

## 15 Phage Display: Factors Affecting Panning Efficiency

**John McCafferty**

## 16 Preparation of Second-Generation Phage Libraries

**Nils B. Adey, Willem P. C. Stemmer, and Brian K. Kay**

## 17 Nonradioactive Sequencing of Random Peptide Recombinant Phage

**Barbara L. Masecar, Christopher A. Dadd, George A. Baumbach, and David J. Hammond**

# 18  Measurement of Peptide Binding Affinities Using Fluorescence Polarization

**Thomas Burke, Randall Bolger, William Checovich, and Robert Lowery**

*Numbers in parentheses indicate the pages on which the authors' contributions begin.*

**Mathias Ackermann** (195) Institute for Virology, Faculty of Veterinary Medicine, University of Zurich, 8057 Zurich, Switzerland

**Nils B. Adey** (35, 55, 67, 227, 277) Myriad Genetics, Salt Lake City, Utah 84108

**Norris Armstrong** (35) University of California, Berkeley, Berkeley, California 94720

**Luc Aujame** (215) Combinatorial Technology, Pasteur Mérieux S.V., 69280 Marcy L'Étoile, France

**George A. Baumbach** (293) Bayer Corporation, Clayton, North Carolina 27520

**Jim Beasley** (67) Department of Chemistry, University of North Carolina at Chapel Hill, Chapel Hill, North Carolina 27599

**Randall Bolger** (305) PanVera Corporation, Madison, Wisconsin 53705

**Thomas Burke** (305) PanVera Corporation, Madison, Wisconsin 53705

**William Checovich** (305) PanVera Corporation, Madison, Wisconsin 53705

**Reto Crameri** (195) Swiss Institute of Allergy and Asthma Research, 7270 Davos, Switzerland

**Steve Cwirla** (227) Affymax Research Institute, Palo Alto, California 94304

**Christopher A. Dadd** (293) Bayer Corporation, Clayton, North Carolina 27520

**Diane Dottavio** (113) Leukosite, Inc., Cambridge, Massachusetts 02142

**Jocelyne Dubayle** (215) Combinatorial Technology, Pasteur Mérieux S.V., 69280 Marcy L'Étoile, France

**Dion H. du Plessis** (141) Biochemistry Division, Onderstepoort Veterinary Institute, Onderstepoort 0110, Republic of South Africa

**Maria Foti** (195) Institute for Virology, Faculty of Veterinary Medicine, University of Zurich, 8057 Zurich, Switzerland

**Frédérique Geoffroy** (215) Combinatorial Technology, Pasteur Mérieux S.V., 69280 Marcy L'Étoile, France

**David J. Hammond** (293) Bayer Corporation, Clayton, North Carolina 27520

**Ronald H. Hoess** (21) DuPont Merck Pharmaceutical Co., Wilmington, Delaware 19880

**Paul Jacquemot** (215) Combinatorial Technology, Pasteur Mérieux S.V., 69280 Marcy L'Étoile, France

**Kevin Johnson** (79) Cambridge Antibody Technology, The Science Park, Melbourn, Cambridgeshire SG8 6JJ, England

**Frances Jordaan** (141) Biochemistry Division, Onderstepoort Veterinary Institute, Onderstepoort 0110, Republic of South Africa

**Brian K. Kay** (21, 35, 55, 67, 227, 277) Department of Biology, University of North Carolina at Chapel Hill, Chapel Hill, North Carolina 27599

**Robert Charles Ladner** (151) Protein Engineering Corporation, Cambridge, Massachusetts 02138

**Robert Lowery** (305) PanVera Corporation, Madison, Wisconsin 53705

**Pratap Malik** (127) Cambridge Centre for Molecular Recognition, Department of Biochemistry, University of Cambridge, Cambridge, CB2 1QW, United Kingdom

**Barbara L. Masecar** (293) Bayer Corporation, Clayton, North Carolina 27520

**David Matthews** (255) Arris Pharmaceutical Corporation, South San Francisco, California 94080

**John McCafferty** (79, 261) Cambridge Antibody Technology, The Science Park, Melbourn, Cambridgeshire SG8 6JJ, England

**Stephen J. McConnell*** (35) Cytogen Corporation, Princeton, New Jersey 08540

**Bernard Montègue** (215) Combinatorial Technology, Pasteur Mérieux S.V., 69280 Marcy L'Étoile, France

*Present address: Chugai Biopharmaceutical, Inc., San Diego, California 92121

**Richard Perham** (127) Cambridge Centre for Molecular Recognition, Department of Biochemistry, University of Cambridge, Cambridge, CB2 1QW, United Kingdom

**Isabelle Peubez** (215) Combinatorial Technology, Pasteur Mérieux S.V., 69280 Marcy L'Étoile, France

**Corinne Pion** (215) Combinatorial Technology, Pasteur Mérieux S.V., 69280 Marcy L'Étoile, France

**James E. Rider** (55) Department of Biology, University of North Carolina at Chapel Hill, Chapel Hill, North Carolina 27599

**Régis Sodoyer** (215) Combinatorial Technology, Pasteur Mérieux S.V., 69280 Marcy L'Étoile, France

**Andrew B. Sparks** (55, 67, 227) Curriculum in Genetics and Molecular Biology, University of North Carolina at Chapel Hill, Chapel Hill, North Carolina 27599

**Willem P. C. Stemmer** (277) Affymax Research Institute, Palo Alto, California 94304

**Mark Suter** (195) Institute for Virology, Faculty of Veterinary Medicine, University of Zurich, 8057 Zurich, Switzerland

**Tamsin D. Terry** (127) Cambridge Centre for Molecular Recognition, Department of Biochemistry, University of Cambridge, Cambridge, CB2 1QW, United Kingdom

**Robert Webster** (1) Department of Biochemistry, Duke University Medical Center, Durham, North Carolina 27710

## 1988—A Year of Discovery

Lunch on January 19, 1988, was an epiphany. Not because of the food, I hasten to add, which was indifferent deli fare, but because of the ideas that arose during the mealtime conversation with Vidal de la Cruz and Tom McCutchan of the Malaria Division at the National Institute of Allergy and Infectious Diseases.

I had just given a talk to the malaria group about what has now come to be called "phage display." Phage display had its genesis in May 1984, while I was working on filamentous bacteriophage during my sabbatical in Bob Webster's lab in the Biochemistry Department at Duke University. I'd been experimenting with pIII, one of the coat proteins, and it occurred to me that because of its domain structure, its surface location at one tip of the virion, and its apparent flexibility as evidenced in electron micrographs, the protein might tolerate being fused to foreign polypeptides without losing its function. It would be an unusual fusion protein, I realized, because it would be associated with the infectious particle itself. Conventional phage expression vectors like λgt11, in contrast, encode fusion proteins that aren't incorporated into the virion.

If the fused foreign polypeptide were indeed displayed on the surface of the filamentous phage, it probably would be accessible to antibodies. This raised the possibility that a large library of pIII fusions could be surveyed for polypeptides that bind a specific antibody using a powerful selection technique: affinity purification. First, the antibody would be tethered to a solid support; the phage library would then be passed over the support, allowing those few particles displaying an antibody-binding polypeptide to react with the antibody and thereby become attached to the support;

after nonbinding particles had been washed away, the phage that remained—highly enriched for particles displaying antibody-binding polypeptides—would be eluted under conditions that didn't destroy the phage's infectivity. The eluted phage could then be cloned and propagated indefinitely by infecting fresh bacterial host cells. Affinity purification would make it easy to survey billions of clones *en masse,* whereas dozens of plaque-lifts had to be laboriously processed to survey just a million λgt11 clones. Paul Modrich, down the hall from Webster's lab, had just what I needed to test the idea: the gene for *Eco*RI restriction enzyme (a source of clonable gene fragments), antibody against the enzyme (to affinity-select clones displaying fragments of the enzyme), and vast quantities of biochemically pure enzyme (a competitive inhibitor to check the specificity of the antibody-phage reaction). It worked! I reported the first results at the Cold Spring Harbor phage meeting in August 1984 and published a 1985 paper in *Science.* By the time I gave my talk to the NIAID malaria group three and a half years later, Steve Parmley, a grad student in my lab, had developed a practical display vector and a very effective affinity selection method.

I still thought of phage display mostly as an alternative to λgt11 and other conventional expression vectors, but a passing remark of McCutchan's at that lunch utterly changed my outlook. He said (more or less): "You know, you could use phage to out-Geysen Geysen." I didn't know who he was talking about, but I soon found out. Mario Geysen had devised simple methods for synthesizing large numbers of peptides on pins in microtiter format. His "pepscan" technology is used widely for mapping the epitopes recognized by antiprotein antibodies. More importantly, he'd devised a clever way of searching through all possible octapeptide sequences to delineate an epitope for an antibody without any usable knowledge of its specificity. If synthetic oligonucleotides coding for all possible short peptides (hexamers, for instance) could be cloned into a display vector, you'd have a universal "epitope library" that might contain peptide ligands for almost any antibody. It would be easy to identify a ligand for any given antibody: simply use it to affinity-select the relevant clones by Parmley's technique. I saw that by making degenerate oligonucleotides, I could produce all the necessary coding sequences in a single run of the chemical synthesizer. It should be easy to make the universal library.

By the time I submitted my next NIH proposal five weeks later, the epitope library dominated my plans for the future. I imagined numerous immunological problems that might be approached with the new technology (if I succeeded in developing it). As it happened, Hannah Alexander, who had done much of the key work on protein epitopes in Richard Lerner's lab at Scripps, now worked in Columbia. She helped me through a crash course in the field, and forged an alliance with the Scripps group that was to prove invaluable when it came time to test the epitope library concept. The paper Parmley and I submitted to *Gene* in June 1988 outlined the concept, and in August, Shannon Flynn, an undergraduate, started synthesizing a miniature peptide library that contained (among many other sequences) the epitope for antibodies of known specificity from Lerner's group.

In October 1988, while preparing to resubmit my NIH proposal (the previous one had failed), I happened to read the *Science* article by Bird *et al.* on single-chain antibodies. I was inspired by a new vision. I imagined a universal library of single-chain antibodies displayed on phage. The antibodies would have one or a few framework sequences, but their complementarity-determining regions would be randomized to generate a vast diversity of binding specificities. The user could use any antigen of interest to affinity-purify clones whose displayed antibodies happened to bind that antigen. In this way, monoclonal antibodies to almost any antigen might be obtained by simple microbiological means, without the need for animals or animal cells in culture. I thought these artificial antibodies would eventually replace conventional and monoclonal antibodies for many purposes.

The eventful year of 1988 closed appropriately with yet another broadening of my horizons. On December 15, I was sitting in Jim Larrick's office at Genelabs, Inc. "Drugs!" he said cryptically. He explained that if the epitope library were used to find new peptide ligands for receptors other than antibodies, these ligands would represent a new and broadly applicable route to drug candidates. It was so obvious! Yet, up to then my plans had been exclusively immunological—reflecting, I suppose, my doctoral and postdoctoral training in that field. From that day on, drug discovery was to become a new element in my concept of phage-display technology.

Work on the epitope library did not begin in earnest until the summer of 1989, both because I was occupied with other projects and because my lab was not funded between May 1988 and July 1989, when my new NIH grant—finally successful!—started. Two people came to the lab then; Jamie Scott, a postdoctorate student who constructed and tested our first epitope library and has been my close collaborator ever since; and a new technician, Robert Davis, who carried out many of the key experiments and has managed my lab now for six years.

When I look back on my year of discovery, I'm struck—and a little embarrassed—at how parochial my vision of phage display was at the beginning, before my education at the hands of my fellow scientists. Phage-display technology, as we now understand it, and as is exemplified in this book, has been very much a communal invention. There's little sense in trying to apportion credit for such advances to individuals—for patent purposes or any other reason. But there is much sense, I think, in maintaining a vigorous, public community of scientists, talking freely with one another, whose ideas, excitement, and results are selected, amplified, recombined, tested, and reselected much as are the phage libraries.

George P. Smith
*Division of Biological Sciences*
*University of Missouri*

## General Discussion of Phage Display

The ability to display peptide and proteins on the surface of bacteriophage M13 particles has had a major impact on the fields of immunology, cell biology, protein engineering, physiology, and pharmacology. Phage display is a rapidly growing technology that we feel fortunate to have been a part of. In writing this book, we have attempted to archive useful protocols, information, and thoughts in one useful place.

We intend this book to be a laboratory manual, meant to help not only the experienced user but the neophyte as well. Each chapter contains an introduction to the research problem or technique featured, step-by-step protocols, troubleshooting tips, sample data, and references. To assist experimenters, we have collected reagent recipes, general protocols, and sequences in the appendix.

Despite our best efforts, typos or oversights will inevitably have crept into some parts of this book. We welcome any comments you might have that would allow us to correct any such inconveniences in future editions or instantly on the world wide web page.

Please note that there is a redundancy of some protocols from chapter to chapter. Yet within each seemingly redundant protocol (ligation, electroporation, etc.) there are subtle differences reflecting the authors' preferences, optimizations, and budgets. We have allowed this redundancy in order for this variety to be accessible to the reader.

## Establishment of an Internet Home Page

To make this book a "living" document, we have established a World Wide Web page for the purpose of having an up-to-date listing of corrections or modifications to the book and to phage display references. The address is http://www .unc.edu/depts/biology/bkay/phagedisplay.html, and the home page can be accessed through the computer programs Mosaic, Netscape, or MacWeb.

## Acknowledgments

We thank our fellow book contributors and our colleagues for their help on this project. J. W. thanks Steve Rosenberg and Mike Doyle for helpful and interesting discussions. We dedicate this book to our family members: Helen, Emily, Allison, Marco, Nicole, Michelle, Pat, Eamonn, and Ciara.

*The editors,*
*Brian, Jill, & John*

# Biology of the Filamentous Bacteriophage

## *Robert E. Webster*

## INTRODUCTION TO THE LIFE CYCLE OF THE BACTERIOPHAGE

A number of filamentous phage have been identified which are able to infect a variety of gram negative bacteria. They have a single-stranded, covalently closed DNA genome which is encased in a long cylinder approximately 7 nm wide by 900 to 2000 nm in length. The best characterized of these phage are M13, fl, and fd, which infect *Escherichia coli* containing the F conjugative plasmid. The genomes of these three bacteriophage have been completely sequenced and are 98% homologous (Van Wezenbeek *et al.,* 1980; Beck and Zink, 1981; Hill and Petersen, 1982). Because of their similarity and their dependence on the F plasmid for infection, M13, fl, and fd are collectively referred to as the Ff phage.

Infection of *E. coli* by the Ff phage is initiated by the specific interaction of one end of the phage with the tip of the F pilus. This pilus is encoded by genes in the *tra* operon on the F conjugative plasmid (Fig. 1). The F pilus is required for conjugal transfer of the F plasmid DNA or chromosomal DNA containing the integrated plasmid DNA into recipient bacteria lacking the plasmid DNA (Willetts and Skurray, 1987; Ippen-Ihler and Maneewannekul, 1991; Frost *et al.,* 1994). It consists of a protein tube which is assembled and disassembled by a polymerization and depolymerization process from pilin subunits in the bacterial inner membrane (Frost,

*Phage Display of Peptides and Proteins*
Copyright © 1996 by Academic Press, Inc. All rights of reproduction in any form reserved.

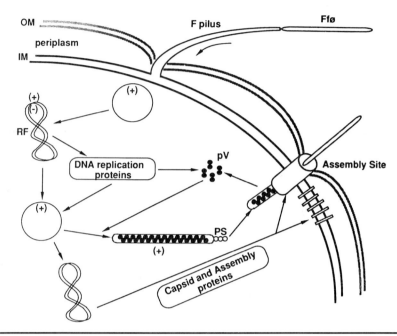

**FIGURE I**    Schematic representation of the life cycle of the Ff bacteriophage. OM, outer membrane; IM, inner membrane; (+), the bacteriophage single-stranded DNA; (-), the complementary DNA strand; RF, the double stranded replicative form; PS, the bacteriophage packaging signal and pV, the product of the bacteriophage gene V.

1993). Conjugation is thought to be initiated by interaction of the tip of the F pilus and the envelope of the recipient bacteria. Pilus retraction, by depolymerization of pilin subunits into the inner membrane, then draws the donor (F plasmid containing) and recipient bacteria together to facilitate the processes required for DNA transfer (Ippen-Ihler and Manneewannekul, 1991).

In the case of phage infection, it also is thought that the end of the phage attached to the pilus tip is brought to the membrane surface by depolymerization of the F pilus. There, the major capsid proteins integrate into the membrane and the phage DNA is translocated into the cytoplasm in a process which requires the bacterial TolQ, R, and A proteins. In the cytoplasm, bacterial enzymes synthesize the complementary strand and convert the infecting phage DNA into a supercoiled, double-stranded replicative form (RF) molecule (Fig. 1). This molecule serves as a template for transcription and translation from which all of the phage proteins are synthesized. Some of the phage products, in concert with bacterial enzymes, direct the synthesis of single-stranded phage DNA which is converted to additional RF molecules. The production of phage proteins increases with the accumulation of these RF molecules. Capsid proteins and other phage proteins involved in the assembly of the particle become integrated into the cell envelope. Proteins involved in DNA replication remain in the cytoplasm. When the phage specific single-stranded DNA binding protein, pV, reaches the proper concentration, it sequesters the newly synthesized phage single-stranded DNA into a complex. The DNA in this pV–DNA complex is not converted to RF but rather is assembled into new phage particles.

Assembly occurs at the bacterial envelope, at a site where the inner and outer membranes are in close contact. During the assembly process, the molecules of the pV single-stranded binding protein are displaced and the capsid proteins assemble around the DNA as it is extruded through the envelope. The transfer of the capsid proteins from their integral cytoplasmic membrane location to the assembling phage particle is directed by three phage-specific noncapsid assembly proteins and the bacterial protein thioredoxin. Assembly continues until the end of the DNA is reached and the phage is released into the media. The assembly process is tolerated quite well by the host, as the infected bacteria continue to grow and divide with a generation time approximately 50% longer than that for an uninfected bacteria. About 1000 phage particles are produced during the first generation following infection, after which the host produces approximately 100–200 phage per generation. This continues for as many generations as has been measured so the curing rate is very low (Model and Russel, 1988). Plaques are quite turbid and of varying size and contain approximately $10^8$ infective particles.

As a consequence of their structure and life cycle, the Ff phage have served as valuable tools for biological research. Since replication of DNA or assembly of the phage is not constrained by the size of the DNA, the Ff phage are excellent cloning vehicles (see Smith, 1988, for review). The result of an insertion of foreign DNA into a nonessential region merely results in a longer phage particle. Thus, large quantities of single-stranded DNA containing foreign DNA inserts ranging from a few to thousands of nucleotides can be easily obtained. These can be used for sequence analysis as well as other manipulations, such as the creation of substrates to study DNA mismatch repair (Su *et al.*, 1988) or reactions involved in recombination (Bianchi and Radding, 1983).

By combining the best features of plasmids and the phage, new cloning vectors called phagemids have been constructed (see Mead and Kemper, 1988, for review). They contain the replication origin and packaging signal (PS) of the filamentous phage together with the plasmid origin of replication and gene expression systems of the chosen plasmid. Phagemids can maintain themselves as plasmids directing the expression of the desired proteins in the bacteria. Infection with a filamentous helper phage activates the phage origin of replication and the resulting phagemid single-stranded DNA is encapsulated into phage-like particles using helper phage proteins.

In recent years, vectors have been developed that allow the display of foreign peptides on the surface of the filamentous phage particle (Cesareni, 1992). By insertion of specific oligonucleotides or entire coding regions into genes encoding specific phage capsid proteins, chimeric proteins can be produced which are able to be assembled into phage particles (Hoess, 1993; Smith and Scott, 1993; Winter *et al.*, 1994). This results in the display of the foreign protein on the surface of the phage particle. It is this use of the Ff phage that is the subject of this book. To aid in the understanding of this powerful technique, this chapter will describe aspects of the

**TABLE I**    Genes and Gene Products of f1 Bacteriophage

| Gene | Amino acids[a] | Molecular weight[a] | Function |
|------|-----------|------------------|----------|
| I | 348 | 39,502 | Assembly |
| II | 410 | 46,137 | DNA replication |
| III | 406 | 42,522 | Minor capsid protein |
| IV | 405 | 43,476 | Assembly |
| V | 87 | 9,682 | Binding ssDNA |
| VI | 112 | 12,342 | Minor capsid protein |
| VII | 33 | 3,599 | Minor capsid protein |
| VIII | 50 | 5,235 | Major capsid protein |
| IX | 32 | 3,650 | Minor capsid protein |
| X | 111 | 12,672 | DNA replication |
| XI | 108 | 12,424 | Assembly |

[a] The number of amino acids and the molecular weight are for the mature proteins. The initiating methionine is included in proteins which do not contain an amino terminal signal sequence

biology of the Ff phage. It will concentrate on the present knowledge of the structure and location of the phage proteins in both the infected bacteria and the mature virion. An attempt will be made to relate this "static" picture of these proteins to the interactions which might occur between them during the dynamic processes of phage assembly and infection. Both of these processes occur at the bacterial membranes, making detailed biochemical analysis difficult. Consequently, we have only a crude picture of the mechanism of phage assembly and disassembly based on molecular biological and genetic analysis. Still, such knowledge should aid in designing better display vectors which, in turn, may further our understanding of the biology of these interesting bacteriophage. The excellent review by Model and Russel (1988) contains a complete compendium of references on the subject up to 1987. Therefore, references cited in this chapter will primarily be to papers published subsequent to that review.

## THE GENOME AND ITS PRODUCTS

The genome of the Ff phage is a single-stranded covalently closed DNA molecule. It encodes 11 genes, the products of which are listed in Table 1. Two of these genes, X and XI, overlap and are in-frame with the larger genes II and I (Model and Russel, 1988; Rapoza and Webster, 1995). The genes are grouped on the DNA according to their functions in the life cycle of the bacteriophage (Fig. 2). One group encodes proteins required for DNA replication, the second group encodes the proteins which make up the capsid and the third group encodes three proteins involved in the membrane associated assembly of the phage. There also is the "Intergenic Region," a short stretch of DNA which encodes no proteins. It contains the sites of origin for the synthesis of the (+) strand (phage DNA) or (-) strand as well as a hairpin region which is the site of initiation for the assembly of the phage particles (packaging signal).

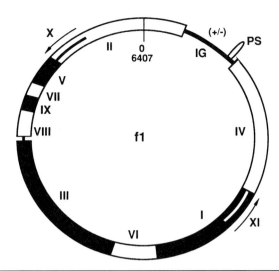

**FIGURE 2**     The genome of the f1 bacteriophage. The single-stranded DNA contains 11 genes. It has 6407 nucleotides which are numbered from the unique *Hind*II site located in gene II (represented by 0). IG, Intergenic Region; PS, the packaging signal; (+/-), the position of the origin of replication of the viral and complementary DNA strands; arrows indicate the location of the overlapping genes X and XI. The figure is adapted from Model and Russel (1988).

## DNA Replication Proteins

Genes II, V, and X encode cytoplasmic protein products which are involved in the replication of the phage DNA (see Model and Russel, 1988, for review). The infecting single-stranded genome is converted to a supercoiled double-strand or RF DNA by bacterial enzymes. pII binds and introduces a specific cleavage on the (+) strand of the RF DNA allowing the host enzymes to use the resulting 3′ end as a primer for synthesis of a new (+) strand. This rolling circle replication continues for one round and the resulting (+) strand is circularized by pII. The new (+) strand either is converted to another replicative form DNA by host enzymes or is bound by the phage pV subsequent to packaging into phage particles.

Dimers of pV bind single-strand DNA in a highly cooperative manner. It is this binding that prevents the conversion of the newly synthesized single-stranded DNA to RF and thus sequesters the (+) strand DNA for assembly into the phage particle. The result of this interaction is a pV-DNA structure approximately 880 nm long and 8 nm in diameter containing one (+) strand and approximately 800 pV dimers (Gray, 1989; Skinner *et al.*, 1994; Olah *et al.*, 1995). The structure appears to be wound in a left-handed helix with the pV dimers binding two strands of single-stranded DNA running in opposite directions along the helix. The PS is at one end of this rod-like structure, allowing it to be available for the initiation of phage assembly (Bauer and Smith, 1988).

pX is the result of a translational initiation event at methionine codon 300 in gene

II and therefore has the same sequence as the carboxyl-terminal third of pII. It is essential for the proper replication of the phage DNA and appears to function, in part, as an inhibitor of pII function (Fulford and Model, 1984).

## Capsid Proteins

The products of the next group of genes, III, VI, VII, VIII, and IX, constitute the capsid proteins of the phage particle. All reside in the cytoplasmic membrane (Fig. 3) until they are assembled into a phage particle. pVIII is the most abundant capsid protein, as approximately 2700 molecules are needed to form the protein cylinder around the DNA. It is synthesized as a 73-amino acid precursor containing a 23-residue amino-terminal signal sequence which is removed after the protein is inserted into the membrane in a manner independent of the bacterial Sec system (Kuhn and Troschel, 1992). The mature 50-amino acid pVIII resides in the cytoplasmic membrane via its hydrophobic region (residues 20–40) with its amino terminus in the periplasm (Wickner, 1975) and its positively charged carboxyl-terminal 10 amino acids in the cytoplasm (Ohkawa and Webster, 1981). Recent evidence suggests that the membrane associated pVIII adopts an alpha helical conformation with the carboxyl-terminal 30 residues traversing the membrane and extending into the cytoplasm in an orientation perpendicular to the surface of the membrane (McDonnell *et al.,* 1993).

The remaining four minor capsid proteins, which make up the ends of the phage particle, are also located in the membrane prior to assembly into mature bacteriophage

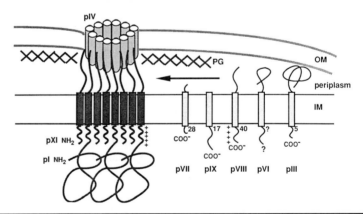

**FIGURE 3**          A schematic representation of the known topologies of the membrane-associated phage capsid and assembly proteins. The interactions between the pI, pIV, and pXI assembly proteins shown are based on the evidence cited in the text. The numbers on the cytoplasmic side of the inner membrane refer to the residue at the end of the membrane-spanning region for that protein. OM, outer membrane; IM, inner membrane; PG, peptidoglycan layer. The series of + symbols represents the positive side chains of the amphiphilic helices of pI, pXI, and pVIII adjacent to the cytoplasmic face of the inner membrane. The arrow represents the hypothetical movement of the capsid proteins into the site formed by pI, pXI, and pIV during assembly

(Endemann and Model, 1995). pIII is synthesized with an 18-residue amino-terminal signal sequence and requires the bacterial Sec system for insertion into the membrane (Rapoza and Webster, 1993). After removal of the signal sequence, the mature 406 residue pIII spans the cytoplasmic membrane one time with only the final five carboxyl-terminal residues in the cytoplasm (Davis et al., 1985). This places the amino-terminal 379 residues in the periplasm. Evidence suggests that the first 356 amino-terminal residues are responsible for phage adsorption and DNA penetration during infection when pIII is part of the mature phage particle (see The Infection Process). This region contains 8 cysteines which are involved in intramolecular disulfide bonds and buried in the interior of the folded portion of pIII (Kremser and Rasched, 1994). There also are two regions of glycine rich repeats of EGGGS (residues 68–86) and GGGS (residues 218–256), the function of which has not been determined (see Model and Russel, 1988).

pVI, pVII, and pIX are not synthesized with signal sequences and the mechanism responsible for their insertion into the membrane is not known. Both pVII and pIX are presumed to span the membrane with their amino-terminal portion exposed to the periplasm. This topology is based only on the observation that these molecules retain their amino-terminal formyl group after membrane insertion (Simons et al., 1981). In the case of pIX, such an orientation would put the positively charged carboxyl-terminal region in the cytoplasm as expected. The sequence of the 112-amino acid pVI predicts one hydrophobic membrane-spanning region (residues 36–59). It has been shown that pVI is an integral membrane protein (Endemann and Model, 1995) but the topology of this protein in the membrane is unknown.

## Assembly Proteins

Gene products pI, pIV, and pXI are required for phage assembly but are not part of the phage particle. pI is synthesized without an amino-terminal signal sequence but its insertion into the membrane appears to be dependent on the Sec system of the host (Rapoza and Webster, 1993). It spans the membrane once with its amino-terminal 253 residues in the cytoplasm and its carboxyl-terminal 75 amino acids exposed to the periplasm (Guy-Caffey et al., 1992). Production of moderate amounts of pI from plasmid-encoded gene I results in cessation of bacterial growth and loss of membrane potential (Horabin and Webster, 1988). This suggests that pI monomers may interact in the membrane to allow the transmembrane regions to form a channel through the cytoplasmic membrane. Consistent with this hypothesis is the observation that chimeric proteins which contain the pI transmembrane region in the middle of the chimeric molecule also are able to span the membrane and cause cessation of cell growth when present in moderate amounts (Horabin and Webster, 1988; Guy-Caffey and Webster, 1993).

pXI was initially designated pI* since it is the result of a translation initiation event at codon 241 of gene I (Guy-Caffey et al., 1992; Rapoza and Webster, 1995). The resulting 108-residue pXI is homologous to the carboxyl-terminal portion of pI. Like pI, pXI spans the membrane once with its carboxyl-terminal 75 amino acids in the periplasm (Fig. 3). Production of moderate amounts of pXI from a plasmid does not stop cell growth, suggesting that pXI cannot interact to form a channel in a manner similar to pI (Rapoza and Webster, 1995). However, like pI, pXI has been shown to be essential for phage assembly.

The product of gene IV is synthesized with a 21-residue amino-terminal signal sequence. It is secreted across the cytoplasmic membrane in a manner dependent on the Sec system of the host (Brissette and Russel, 1990; Russel and Kazmierczak, 1993; Rapoza and Webster, 1993). pIV then integrates into the outer membrane where it is present as an oligomer of 10–12 pIV monomers (Kazmierczak *et al.,* 1994). The amino-terminal half of pIV is exposed to the periplasm while the carboxyl-terminal half appears responsible for forming the portion of the pIV oligomer embedded in the outer membrane. The carboxyl-terminal region of pIV shows significant sequence homology to a number of bacterial proteins which are involved in the export of certain bacterial factors into the medium (Russel, 1994a). All these observations suggest that the pIV oligomer may form a large exit port in the outer membrane through which the assembled phage particle can pass. However, this port must be gated in some way since bacteria containing pIV oligomers in the outer membrane remain impervious to compounds which normally do not pass through the outer membrane (Russel, 1994b).

Two observations suggest a direct interaction between the gene IV and gene I/XI products. Russel (1993) isolated a mutant in the periplasmic portion of pI/pXI which can compensate for a mutation in the periplasmic portion of pIV. The second observation was that pI/pXI and pIV from fl and the related phage Ike were not functionally interchangeable. However, if the assembly proteins were from the same phage, some heterologous phage could be assembled (Russel, 1993). That is, pI, pIV, and pXI from fl phage are able to support, at low efficiency, the growth of an Ike phage lacking its own gene I and IV. Periplasmic interactions between these assembly proteins could form a channel connecting the inner and outer membranes to allow extrusion of the assembling phage particle (Fig. 3). It is proposed that the outer membrane-associated pIV oligomer acts as the gate in this channel to maintain the integrity of the membrane in the absence of phage assembly (Russel, 1994b). The existence of such a channel is consistent with the earlier observation that assembly of the bacteriophage occurs at sites where the inner and outer membrane appear to be in close contact (Lopez and Webster, 1985).

There is evidence suggesting possible interactions between the capsid and assembly proteins within the inner membrane. As outlined above, direct interaction of the carboxyl-terminal regions of pI and pXI with the pIV oligomers would suggest an inner membrane oligomer of pI and pXI, as depicted in Fig. 3. Genetic evidence also suggest a possible interaction between the transmembrane regions of pI/pXI and the major capsid protein pVIII (Russel, 1993; Rapoza and Webster, 1995). Mutational studies are consistant with interactions between the transmembrane helices of pVIII itself (Deber *et al.,* 1993). Data from immunoprecipitation experiments point to possible association of pIII and pVI with the major coat protein, pVIII, within the membrane (Endemann and Model, 1995). Gailus *et al.,* (1994) suggest that pIII and pVI may interact in the membrane prior to assembly into phage but this could not be confirmed by immunological techniques (Endemann and

Model, 1995). It should be pointed out that the evidence for any membrane associated interactions between these phage proteins is indirect, and thus any conclusion is preliminary.

## THE Ff BACTERIOPHAGE PARTICLE

The wild-type Ff phage particle is approximately 6.5 nm in diameter and 930 nm in length (Fig. 4). The DNA is encased in a somewhat flexible cylinder composed of approximately 2700 pVIII monomer units. At one end of the particle, there are about 5 molecules each of pVII and pIX. The other end contains approximately 5 molecules each of pIII and pVI. The DNA is oriented within the virion with the hairpin PS located at the end of the particle containing the pVII and pIX proteins.

X-ray diffraction and other physical studies have given a fairly complete description of the pVIII cylinder portion of the virion (Opella *et al.*, 1987; Glucksman *et al.*, 1992; Marvin *et al.*, 1994; Overman and Thomas, 1995; Williams *et al.*, 1995). Except for the amino-terminal five amino acids, pVIII appears to be present in an uninterrupted alpha helical conformation. The pVIII monomers are arranged in an overlapping manner so that the carboxyl-terminal 10–13 residues form the inside wall of the particle. This region contains four positively charged lysine residues which are arranged to form the positive face of an amphiphilic helix. The structure studies predict that these positive charges interact with the sugar phosphate backbone

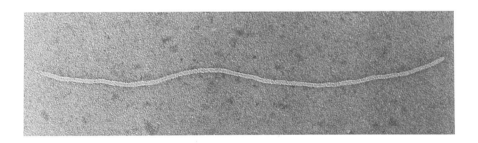

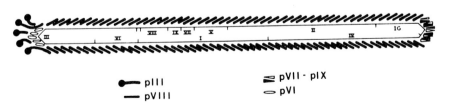

| **FIGURE 4** | The Ff bacteriophage particle. The top is an electron micrograph of a negatively stained particle with the pVII–pIX end at the right. The bottom is a schematic representation of the phage particle showing the location of the capsid proteins and the orientation of the DNA. The figure is adapted from Webster and Lopez (1985). |

of the DNA. Mutational studies suggest that these lysine residues do have some electrostatic interaction with the DNA (Hunter *et al.*, 1987; Greenwood *et al.*, 1991). For example, changing one of these lysine residues to glutamine results in the production of phage particles which are approximately 35% longer than wild-type bacteriophage. The additional mutant pVIII molecules in the longer particle presumably are required to maintain the appropriate ratio of positive charges of the capsid to negative charges of the DNA backbone.

The amino-terminal portion of pVIII, with its acidic residues, is exposed on the outside of the particle. The residues connecting the amino and carboxyl regions of pVIII interact with each other to form the inner core of the protein cylinder (Williams *et al.*, 1995). Much of this middle portion of pVIII is part of the membrane spanning region of the protein before it is assembled into the particle (Fig. 3). The axis of the helical pVIII monomer in the phage particle is tilted approximately 20° to the long axis of the particle, gently wrapping around the long axis of the virus in a right-handed way.

One end of the particle contains approximately five molecules each of the small hydrophobic pVII and pIX proteins. The arrangement of these proteins in the capsid and their interaction with pVIII is not known. A model for the structure of pVII/pIX at the end of the particle has been proposed and suggests that one of these proteins must be buried close to the DNA while the other is exposed at the surface (Makowski, 1992). The observation that antibodies to pIX but not pVII are able to interact with one end of the phage particle would suggest that only pIX is exposed and pVII is buried (Endemann and Model, 1995). Immunoprecipitation experiments on detergent-disrupted phage suggest an interaction between pVII and pVIII major capsid protein in the intact particle.

Approximately 5 molecules each of pIII and pVI make up the other end of the particle and account for about 10–16 nm of the phage length (Specthrie *et al.*, 1992). A major portion of the amino-terminal portion of pIII can be removed from the phage particle by protease treatment (Gray *et al.*, 1981; Armstrong *et al.*, 1981). Under certain conditions, this portion of pIII can be visualized by electron microscopy as a knob-like structure at one end of the phage (Gray *et al.*, 1981; see Fig. 4). It is this amino-terminal portion which is required for the infection process. The carboxyl-terminal 150 residues of pIII, together with pVI, presumably interact with pVIII to form one of the ends of the particle (Crissman and Smith, 1984; Kremser and Rasched, 1994). Proteins pVI and pIII appear to specifically recognize each other as they remain associated following disruption of the phage particle with certain detergents (Gailus and Rasched, 1994; Endemann and Model, 1995). It is proposed that the hydrophobic amino terminus of pVI is buried within the particle (Makowski, 1992). Presumably, the rather hydrophilic carboxyl-terminal region of this protein is near the surface, although antibody against the carboxyl-terminal 13 amino residues of pVI did not react with phage (Endemann and Model, 1995). However, the observation that phage particles containing foreign proteins fused to the carboxyl end of pVI can be produced would argue that the carboxyl terminus of pVI is very near the surface of the phage particle (Jespers *et al.*, 1995).

## THE INFECTION PROCESS

Infection of *E. coli* by the Ff bacteriophage is at least a two-step process (see Model and Russel, 1988). The first involves the interaction of the pIII end of the phage particle with the tip of the F conjugative pilus (Fig. 1). Retraction of the pilus, presumably by depolymerization of the pilin subunits into the inner membrane, brings the tip of the phage to the membrane surface. It is not known whether this event is a result of the normal polymerization–depolymerization cycles inherent to the pili or whether attachment of the phage can trigger pilus retraction (Frost, 1993).

The second step involves the integration of the pVIII major capsid proteins and perhaps the other capsid proteins into the inner membrane together with the transloca- tion of the DNA into the cytoplasm. This translocation event requires the products of the bacterial *tolQRA* genes (Russel *et al.*, 1988; Webster, 1991). Mutations in any one of these genes block the uptake of the phage DNA into the cytoplasm. However, the phage can still adsorb to the tip of the F pilus and conjugation is normal in these mutant bacteria (Sun and Webster, 1987). The TolQRA proteins also are involved in the translocation of the group A colicins (A, E1, E2, E3, K, L, N, S4) to their site of action in the bacteria (reviewed in Webster, 1991). These bacteriocins are able to bind to their outer membrane receptor in the *tol* mutant bacteria but such binding does not lead to cell death. Since both the phage and the colicins can bind to their respective receptors but do not infect or kill bacteria containing mutations in the *tolQRA* genes, these mutant bacteria have been designated "tolerant" (tol) to filamentous phage or the group A colicins.

The TolQRA proteins are located in the inner membrane. TolQ spans the membrane three times with the bulk of the protein in the cytoplasm (Kampfenkel and Braun, 1993; Vianney *et al.*, 1994). Both TolR and A span the membrane one time via a region near their amino terminus, leaving major portions of each protein exposed in the periplasm (Levengood *et al.*, 1991; Muller *et al.*, 1993). The structure and length of the periplasmic portion of TolA suggest that a region near the carboxyl-terminal end may interact with the outer membrane. There is some evidence that the transmem- brane regions of these three Tol proteins interact with each other (Lazzaroni *et al.*, 1995). All these observations suggest that the TolQRA proteins form some type of complex which can communicate between the inner and outer membranes. One of the normal functions of these Tol proteins may be to maintain the integrity of the outer membrane as bacteria containing mutants in these *tol* genes freely leak periplasmic proteins into the media (see Webster, 1991).

Recognition of the receptor (pilus tip) and the appropriate Tol proteins appears to be a function of the amino-terminal region of pIII. Removal of the amino-terminal 36 kDa of pIII by subtilisin renders the phage particle noninfectious (Gray *et al.*, 1981; Armstrong *et al.*, 1981). Infection of the bacteria by the phage is inhibited by the addition of this proteolytic fragment, presumably because the fragment competes with the phage for the pilus tip or the TolQRA proteins. This is consistent with the observation that production of pIII or its amino-terminal half confers resistance to infection by the Ff phages (Boeke *et al.*, 1982). It is probably the production of pIII that confers resistance to both superinfection by the Ff phage and the effect of the group A colicins in Ff-infected bacteria.

Analysis of a set of deletion mutants in gene III suggests that separate regions of the 406-residue pIII protein are involved in the pilus adsorption and TolQRA-dependent penetration steps (Stengele *et al.*, 1990). The pilus recognition site is present in residues 99–196 while residues 53–107 are involved in the penetration of the DNA into the bacterium. Presumably it is this latter region which specifically interacts with the TolQRA inner membrane proteins. The colicins also have a specific amino-terminal domain which appears to recognize the TolQRA proteins (Webster, 1991). This region of colicins E1 and A has been shown to interact with the carboxyl-terminal portion of TolA (Benedetti *et al.*, 1991) suggesting a similar interaction between TolA and the same region of pIII. This would explain why the overproduction of a fragment containing the first 96 residues of gene III is able to confer tolerance (resistance) to the effect of some group A colicins (Boeke *et al.*, 1982). Comparison of the amino-terminal domains of pIII with the E and A colicins reveal that all contain regions of glycine-rich repeats. Whether these regions are involved in any interactions with TolA or any of the other Tol proteins is unclear.

It is not known whether the absorption and penetration domains must be contiguous for pIII to function in the infection process. Also unclear is the relation of these two domains to the carboxyl-terminal third domain of pIII which interacts with the phage particle. The observation that a chimeric protein containing the amino-terminal 85% of pIII fused to the enzymatic carboxyl-terminal domain of colicin E3 is active only against bacteria expressing F pili suggests that some of these domains can act independently (Jakes *et al.*, 1988). Until a clear picture of the function of these regions of pIII emerges, it is best to continue to fuse foreign proteins at or near the amino terminus of pIII if one wishes to create an infectious particle.

However, it is possible to create infectious hybrid phage containing a mixture of wild-type and chimeric pIII (for example, see Lowman *et al.*, 1991). All that is needed is to have the foreign peptide fused to the carboxyl-terminal region of pIII so it will assemble with wild-type pIII into particles. Recently, attempts have been made to design a type of "two-hybrid system" in which active phage will be produced based on the interaction between two different peptides (Duenas and Borrebaeck, 1994; Gramatikoff *et al.*, 1994). In this case, one of the interacting peptides is fused to the carboxyl terminus of the amino-terminal adsorption and penetration domain of pIII and the other interacting species is attached to the amino-terminal portion of the carboxyl-terminal part of pIII responsible for phage assembly (see Chapter 3 for description of these display techniques).

The fate of pIII following the TolQRA-mediated translocation of the DNA into the cytoplasm is unknown. Some experiments suggest that it may be cleaved during the infection process but this has not been confirmed. There is no evidence that intact pIII integrates into the membrane in a state competent to be incorporated into a newly assembled phage particle. This is in contrast to the pVIII major capsid protein of the infecting bacteriophage, which has been shown to join the pool of newly synthesized pVIII and be assembled into newly formed particles (see Model and Russel, 1988). It also is proposed that pVII and pIX may enter the membrane and be used for the synthesis of new particles. This is based on the observation that infection of nonsuppressing bacteria with gene VII or IX amber mutant phage results in the production of one to three infectious polyphage per cell (Webster and Lopez, 1985). Polyphage are multilength phage particles containing many unit length molecules of the single-stranded phage DNA.

## THE ASSEMBLY PROCESS

Assembly of the Ff filamentous phage is a membrane-associated process which takes place at specific assembly sites where the inner and outer membrane are in close contact (Lopez and Webster, 1985). These assembly sites may be the result of specific interactions between pI, pIV, and pXI as previously described (Fig. 3). However, at some stage in the assembly process the capsid proteins must become part of the assembly site since they are all present as integral membrane proteins until assembled into the phage particle. Interaction of the pV–phage DNA complex with proteins in the assembly site presumably is the initiating event for assembly of the particle. This is followed by a progressive process in which the pV dimers are displaced from, and the appropriate capsid protein added to the DNA as it is extruded through the bacterial envelope into the media. Throughout this whole process, the membrane must retain its integrity, as the bacteria continues to grow and divide. The assembly process is conveniently divided into three parts, initiation, elongation and termination, reflecting the events required for packaging the different ends and the long cylinder of the phage (reviewed in Russel, 1991).

### Initiation

The substrate for the assembly process is the pV phage DNA complex. Genetic evidence suggests that an interaction between the DNA PS and the cytoplasmic domain of pI initiates assembly (Russel and Model, 1989). Similar genetic evidence suggests an interaction of PS with membrane-associated pVII and pIX. Presently, there is no direct biochemical evidence for any of these specific PS–protein interactions. There also is no evidence for the order of addition of pVII, pIX, and even pVIII to the PS during the initiation. One possibility is that pI is able to direct an orderly assembly of pVII, pIX, and pVIII around the PS to form the tip of the particle. It is also possible that these three proteins might interact to form the complete tip of the phage in the membrane which then recognizes the PS in a pI-directed reaction. With regard to this latter hypothesis, Endemann and Model (1995) present some evidence for the association of pVII with pVIII in the membrane, although they were unable to detect any membrane-associated pVII/pIX complex.

The pI, pIV, and pXI assembly proteins may be able to interact with each other before the start of the formation of the pVII/pIX end of the particle (Fig. 3). This is based on the observation that assembly sites are able to form in bacteria infected with phage containing amber mutants in genes VII or IX (Lopez and Webster, 1985). Assembly sites are defined by electron microscopy as regions of contact between

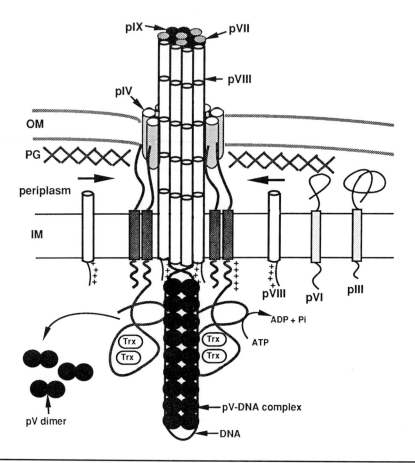

A schematic representation of the elongation process in phage assembly. The assembling phage is depicted extruding through an assembly site formed by pI, pIV, and pXI. Trx represents molecules of thioredoxin required for assembly. Both Trx and ATP are proposed to react with pI during assembly. The arrows in the periplasm suggest that capsid proteins are able to enter the assembly site from the adjacent membrane area. The + and other symbols are described in the legend to Fig. 3. The proteins are represented by the same shapes shown in Fig. 3 except that the periplasmic and transmembrane portions of pVIII are depicted as a tube in both the membrane and the phage particle. This figure is only meant to depict interactions that occur during assembly and does not represent any specific structure for the proteins or DNA involved in the process.

the inner and outer bacterial membranes from which phage emerge in bacteria infected with wild-type phage. In the case of mutant-infected bacteria, they are defined as the increase in the number of membrane contact regions over the number found in uninfected bacteria (Lopez and Webster, 1985). The assembly sites seen in the gene VII or IX mutant-infected bacteria may be defective as bacteria quickly cease to grow and divide after infection with these mutant phage (see Model and Russel, 1988).

## Elongation

Following the initiation event, the phage is elongated by the processive set of reactions which removes the pV dimers from the phage DNA and replaces them with pVIII as the DNA is extruded through the membrane (Fig. 5). These reactions are probably catalyzed by the cytoplasmic portion of pI and may require the hydrolysis of ATP (Russel, 1991). The host protein, thioredoxin, also is required and genetic evidence suggests that it interacts with the cytoplasmic domains of pI (Russel and Model, 1983, 1985; Lim *et al.*, 1985). Assembly does not require the redox activity of thioredoxin but only that thioredoxin be present in its reduced conformation. It is suggested that thioredoxin confers processivity on the elongation process (see Model and Russel, 1988).

Removal of the pV dimers by pI would result in two antiparallel strands of naked DNA being present near the membrane surface of the assembly site (Fig. 5). Since the structure in the DNA is different than that in the phage particle, some alteration would be necessary to the DNA to allow the carboxyl portion of pVIII to interact with it. The 12 amino acids of both pI and pXI adjacent to the cytoplasmic face of the membrane can form a positively charged amphiphilic helix similar to that present in the carboxyl 10 residues of the pVIII protein. Perhaps these regions of pI and/or pXI interact with the DNA just stripped of pV to help the DNA obtain the proper configuration to interact with the pVIII protein (Rapoza and Webster, 1995). In this regard, these is some evidence for an interaction between the transmembrane regions of pVIII, pI, and pXI (Russel, 1993).

Whether the proposed interaction between the periplasmic portions of the pI, pIV, and pXI assembly proteins is maintained during elongation is unknown. Once the assembling phage particle is being extruded through the pIV oligomer in the outer membrane, the presence of a continuous channel between the pI proteins and the pIV oligomer might not be needed. Such an alteration of the pI–pIV channel would allow the periplasmic portion of the pVIII, pVI, and pIII capsid easier access to the assembling particle (as depicted by arrows in Fig. 5). This is consistent with the observation that phage can be assembled with pVIII carrying an amino-terminal extension containing up to six additional amino acids (Kishchenko *et al.*, 1994; Iannolo *et al.*, 1995). Chimeric pVIII containing larger inserts can be assembled into phage particles only in the presence of normal pVIII (Greenwood *et al.*, 1991b; Kang *et al.*, 1991; Markland *et al.*, 1991). The amount of these larger chimeric proteins present in the mature phage particle is inversely proportional to the size of the insert, usually accounting for 1–20% of the major pVIII coat protein in the phage. Since these large foreign proteins are present in the amino terminus of pVIII, they would be in the periplasm. Any tight channel between the gene I proteins and pIV might hinder these large periplasmic inserts from assembling into phage.

## Termination

When the end of the DNA is reached, assembly is terminated by the addition of pVI and pIII. In the absence of either of these proteins, assembly goes on with pVIII continuing to encapsulate another DNA. In this way polyphage containing multiple copies of the genome are produced. Polyphage produced in the absence of pVI are very unstable compared to polyphage produced in the absence of pIII. This observation suggests that the addition of pVI stabilizes the phage structure and provides a structure for the binding of pIII. This does not preclude the possibility of an interaction of the carboxyl region of pIII with pVI in the membrane, as Endemann and Model (1995) observed that pIII could be immunoprecipitated from membranes of infected bacteria using anti-pVI antisera.

The final step in assembly is the release of the completed phage particle from the bacterial envelope. Presumably, this requires the extrusion of the completed phage through the outer membrane. It is not known whether an assembly site can be used for the synthesis of more than one phage particle. It is possible that the addition of the pVI and pIII during termination results in a breakdown of at least the inner membrane portion of the assembly site. In this case, a new site, capable of initiating assembly, would have to be formed. If this were the case, then initiation might be expected to be one of the rate-limiting steps in phage assembly.

It is the membrane-associated assembly that has made the filamentous phage such a valuable tool in biological research. Since the capsids are assembled around the DNA as it is extruded through the membrane-associated assembly site, there are no constraints, other than a possible shear constraint, on the length of DNA packaged. This property led to its use as a cloning vehicle. Similarly, the membrane-associated assembly properties of the capsid proteins allow the packaging of chimeric proteins into the phage particle. Most of the interactions involved in the assembly process appear to occur near the portion of the capsid proteins which initially span the membrane. Thus foreign proteins fused to the periplasmic portions of the pVIII or pIII capsid proteins have little, if any, influence on the assembly of the capsid proteins around the DNA. Theoretically, any protein has a good chance of being packaged into a phage particle when fused to the periplasmic portion of pVIII, pIII, or perhaps the pVI capsid protein, providing that at least two criteria can be satisfied. The foreign protein must be translocated efficiently across the inner membrane into the periplasm so that enough of the protein is available for assembly into particles. Once in the periplasm, it must be able to enter the assembly site and not interfere with the processes that occur at this site during assembly. The apparent flexibility of the assembly process has led to an impressive array of applications for the use of phage display, as discussed in the next chapter. Further knowledge about the assembly mechanism of the filamentous phage should lead to the design of better display vectors.

## References

Armstrong, J., Perham, R. N., and Walker, J. E. (1981). Domain structure of bacteriophage fd adsorption protein. *FEBS Lett.* **135,** 167–172.

Bauer, M., and Smith, G. P. (1988). Filamentous phage morphogenetic signal sequence and orientation of DNA in the virion and gene V protein complex. *Virology* **167,** 166–175.

Beck, E., and Zink, B. (1981). Nucleotide sequence and genome organization of filamentous bacteriophages fl and fd. *Gene* **16,** 35–58.

Benedetti, H., Lazdunski, C. and Lloubes, R. (1991). Protein import into *Escherichia coli:* Colicins A and E1 interact with a component of their translocation system. *EMBO J.* **10,** 1989–1995.

Bianchi, M. E., and Radding, C. M. (1983). Insertions, deletions and mismatches in heteroduplex DNA made by RecA protein. *Cell (Cambridge, Mass.)* **35**, 511–520.

Boeke, J. D., Model, P., and Zinder, N. D. (1982). Effects of bacteriophage f1 gene III protein on the host cell membrane. *Mol. Gen. Genet* **186**, 185–192.

Brissette, J. L., and Russel, M. (1990). Secretion and membrane integration of a filamentous phage-encoded morphogenetic protein. *J. Mol. Biol.* **211**, 565–580.

Cesareni, G. (1992). Peptide display on filamentous phage capsids. A powerful new tool to study protein-ligand interaction. *FEBS Lett.* **307**, 66–70.

Crissman, J. W., and Smith, G. P. (1984). Gene 3 protein of filamentous phages: Evidence for a carboxyl-terminal domain with a role in morphogenesis. *Virology* **132**, 445–455.

Davis, N. G., Boeke, J. D., and Model, P. (1985). Fine structure of a membrane anchor domain. *J. Mol Biol.* **181**, 111–121.

Deber, C. M., Khan, A. R., Li, Z., Joensson, C., Glibowicka, M. and Wang, J. (1993). Val - Ala mutations selectively alter helix-helix packing in the transmembrane segment of phage M13 coat protein. *Proc. Natl. Acad. Sci. USA* **90**, 11648–11652.

Duenas, M., and Borrebaeck, C. A. K. (1994). Clonal selection and amplification of phage displayed antibodies by linking antigen recognition and phage replication. *Bio Technology* **12**, 999–1002.

Endemann, H., and Model, P. (1995). Location of filamentous phage minor coat proteins in phage and in infected bacteria. *J. Mol. Biol.* **250**, 496–506.

Frost, L. S. (1993). Conjugative pili and pilus-specific phages. *In* "Bacterial Conjugation" (D. B. Clewell, ed.), pp. 189–221. Plenum, New York.

Frost, L. S., Ippen-Ihler, K., and Skurray, R. A. (1994). Analysis of the sequence and gene products of the transfer region of the F factor. *Microbiol. Rev.* **58**, 162–210.

Fulford, W., and Model, P. (1984). Gene X of bacteriophage f1 is required for phage DNA synthesis: Mutagenesis of in-frame overlapping genes. *J. Mol. Biol.* **178**, 137–153.

Gailus, V., and Rasched, I. (1994). The adsorption protein of bacteriophage fd and its neighbour minor coat protein build a structural entity. *Eur. J. Biochem.* **222**, 927–931.

Gailus, V., Ramsperger, U., Johner, C., Kramer, H., and Rasched, I. (1994). The role of the adsorption complex in the termination of filamentous phage assembly. *Res. Microbiol.* **145**, 699–709.

Glucksman, M. J., Bhattacharjee, S., and Makowski, L. (1992). Three-dimensional structure of a cloning vector. X-ray diffraction studies of filamentous bacteriophage M13 at 7A resolution. *J. Mol. Biol.* **226**, 455–470.

Gramatikoff, K., Georgiev, O., and Shaffner, W. (1994). Direct interaction rescue, a novel filamentous phage technique to study protein-protein interactions. *Nucleic Acids Res.* **22**, 5761–5762.

Gray, C. W. (1989). Three-dimensional structure of complexes of single-stranded DNA-binding proteins with DNA. IKe and fd gene 5 proteins from left-handed helices with single-stranded DNA. *J. Mol. Biol.* **208**, 57–64.

Gray, C. W., Brown, R. S., and Marvin, D. A. (1981). Adsorption complex of filamentous fd virus. *J. Mol. Biol.* **146**, 621–627.

Greenwood, J., Hunter, G. J., and Perham, R. N. (1991). Regulation of filamentous bacteriophage length by modification of electrostatic interactions between coat protein and DNA. *J. Mol. Biol.* **217**, 223–227.

Greenwood, J., Willis, A. E., and Perham, R. N. (1991b). Multiple display of foreign peptides on a filamentous bacteriophage. Peptides from *Plasmodium falciparum* circumsporozoite protein as antigens. *J. Mol. Biol.* **220**, 821–827.

Guy-Caffey, J. K., and Webster, R. E. (1993). The membrane domain of a bacteriophage assembly protein: Membrane insertion and growth inhibition. *J. Biol. Chem.* **268**, 5496–5503.

Guy-Caffey, J. K., Rapoza, M. P., Jolley, K. A., and Webster, R. E. (1992). Membrane localization and topology of a viral assembly protein. *J. Bacteriol.* **174**, 2460–2465.

Hill, D. F., and Petersen, G. B. (1982). Nucleotide sequence of bacteriophage f1 DNA. *J. Virol.* **44**, 32–46.

Hoess, R. H. (1993). Phage display of peptides and protein domains. *Curr. Opin. Struct. Biol.* **3**, 572–579.

Horabin, J. L., and Webster, R. E. (1988). An amino acid sequence which directs membrane insertion causes loss of membrane potential. *J. Biol. Chem.* **263**, 11575–11583.

Hunter, G. J., Powitch, D. H., and Perham, R. N. (1987). Interactions between DNA and coat protein in the structure and assembly of filamentous bacteriophage fd. *Nature (London)* **327**, 252–254.

Iannolo, G., Minenkova, O., Petruzzelli, R. and Cesareni, G. (1995). Modifying filamentous phage capsid: Limits in the size of the major capsid protein. *J. Mol. Biol.* **248,** 835–844.

Ippen-Ihler, K., and Maneewannekul, S. (1991). Conjugation among enteric bacteria: Mating systems dependent on expression of pili. *In* "Microbial Cell-cell Interactions" (M. Dworkin, ed.), pp. 35–69. American Society for Microbiology, Washington, DC.

Jakes, K. S., Davis, N.G., and Zinder, N. D. (1988). A hybrid toxin from bacteriophage f1 attachment protein and colicin E3 has altered cell receptor activity. *J. Bacteriol.* **170,** 4231–4238.

Jespers, L. S., Messens, J.H., Keyser, A. D., Eeckhout, D., Brande, I. V. D., Gansemans, Y. G., Lauwereys, M. J., Viasuk, G. P., and Stassens, P. E. (1995). Surface expression and ligand-based selection of cDNAs fused to filamentous phage gene VI. *Bio Technology* **13,** 378–382.

Kampfenkel, K., and Braun, V. (1993). Membrane topologies of the TolQ and TolR proteins of *Escherichia coli:* Inactivation of TolQ by a missense mutation in the proposed first transmembrane segment. *J. Bacteriol.* **175,** 4485–4491.

Kang, A. S., Barbas, C. F., Janda, K. D., Benkovic, S. J., and Lerner, R. A. (1991). Linkage of recognition and replication functions by assembling combinatorial antibody Fab libraries along phage surfaces. *Proc. Natl. Acad. Sci. U.S.A.* **88,** 4363–4366.

Kazmierczak, B., Mielke, D. L., Russel, M., and Model, P. (1994). Filamentous phage pIV forms a multimer that mediates phage export across the bacterial cell envelope. *J. Mol. Biol.* **238,** 187–198.

Kishchenko, G., Batlwala, H. and Makowski, L. (1994). Structure of a foreign peptide displayed on the surface of bacteriophage M13. *J. Mol. Biol.* **241,** 208–213.

Kremser, A., and Rasched, I. (1994). The adsorption protein of filamentous phage fd: Assignment of its disulfide bridges and identification of the domain incorporated in the coat. *Biochemistry* **33,** 13954–13958.

Kuhn, A., and Troschel, D. (1992). Distinct steps in the insertion pathway of bacteriophage coat proteins. *In* "Membrane Biogenesis and Protein Targeting" (W. Newport, and R. Lill, eds.). pp. 33–47. Elsevier, New York.

Lazzaroni, J. C., Vianny, A., Popot, J. L., Benedetti, H., Samatey, F., Lazdunski, C., Portalier, R., and Geli, V. (1995). Transmembrane alpha-helix interactions are required for the functional assembly of the *Escherichia coli* Tol complex. *J. Mol. Biol.* **246,** 1–7.

Levengood, S. K., Beyer, W. F., Jr, and Webster, R. E. (1991). Tol A: A membrane protein involved in colicin uptake contains an extended helical region. *Proc. Natl. Acad. Sci. U.S.A.* **88,** 5939–5943.

Lim, C. J., Haller, B., and Fuchs, J. A. (1985). Thioredoxin is the bacterial protein encoded by fip that is required for filamentous bacteriophage f1 assembly. *J. Bacteriol.* **161,** 799–802.

Lopez, J., and Webster, R. E. (1985). Assembly site of bacteriophage f1 corresponds to adhesion zones between the inner and outer membranes of the host cell. *J. Bacteriol.* **163,** 1270–1274.

Lowman, H. B., Bass, S. H., Simpson, N., and Wells, J. A. (1991). Selecting high-affinity binding proteins by monovalent phage display. *Biochemistry* **30,** 10832–10838.

Makowski, L. (1992). Terminating a macromolecular helix. Structural model for the minor proteins of bacteriophage M13. *J. Mol. Biol.* **228,** 885–892.

Markland, W., Roberts, B. L., Saxena, M. J., Guterman, S. K., and Ladner, R. C. (1991). Design, construction and function of a multicopy display vector using fusions to the major coat protein of bacteriophage M13. *Gene* **109,** 13–19.

Marvin, D. A., Hale, R. D., Nave, C., and Helmer Citterich, M. (1994). Molecular models and structural comparisons of native and mutant class I filamentous bacteriophages. *J. Mol. Biol.* **235,** 260–286.

McDonnell, P. A., Shon, K., Kim, Y., and Opella, S. J. (1993). fd coat protein structure in membrane environments. *J. Mol. Biol.* **233,** 447–463.

Mead, D. A., and Kemper, B. (1988). Chimeric single-stranded DNA phage-plasmid cloning vectors. *In* "Vectors: A Survey of Molecular Cloning Vectors and Their Uses" (R. L. Rodriquez, and D. T. Denhardt, eds.), pp. 85–111. Butterworth, Boston.

Model, P., and Russel, M. (1988). Filamentous bacteriophage. *In* "The Bacteriophages" (R. Calendar, ed.), Vol. 2, pp. 375–456. Plenum, New York.

Muller, M. M., Vianney, A., Lazzaroni, J. C., Webster, R.E., and Portalier, R. (1993). Membrane topology of the *Escherichia coli* Tol R protein required for cell envelope integrity. *J. Bacteriol.* **175,** 6059–6061.

Ohkawa, I., and Webster, R. E. (1981). The orientation of the major coat protein of bacteriophage f1 in the cytoplasmic membrane of *Escherichia coli*. *J. Biol. Chem.* **256,** 9951–9958.

Olah, G. A., Gray, D. M., Gray, C. W., Kergil, D. L., Sosnick, T. R., Mark, B. L., Vaughan, M. R. and Trewhella, J. (1995). Structures of fd gene 5 protein-nucleic acid complexes: A combined solution scattering and electron microscopy study. *J. Mol. Biol.* **249,** 576–594.

Opella, S. J., Stewart, P. L., and Valentine, K. G. (1987). Protein structure by solid-state NMR spectroscopy. *Q. Rev. Biophys.* **19,** 7–49.

Overman, S. A., and Thomas, G. J., Jr. (1995). Raman spectroscopy of the filamentous virus Ff (fd, fl, M13): Structural interpretation for coat protein aromatics. *Biochemistry* **34,** 5440–5451.

Rapoza, M. P., and Webster, R. E. (1993). The filamentous bacteriophage assembly proteins require the bacterial SecA protein for correct localization to the membrane. *J. Bacteriol.* **175,** 1856–1859.

Rapoza, M. P., and Webster, R.E. (1995). The products of gene I and the overlapping in-frame gene XI are required for filamentous phage assembly *J. Mol. Biol.* **248,** 627–638.

Russel, M. (1991). Filamentous phage assembly. *Mol. Microbiol.* **5,** 1607–1613.

Russel, M. (1993). Protein-protein interactions during filamentous phage assembly. *J. Mol. Biol.* **231,** 689–687.

Russel, M. (1994a). Phage assembly: A paradigm for bacterial virulence factor export. *Science* **265,** 612–614.

Russel, M. (1994b). Mutants at conserved positions in gene IV, a gene required for assembly and secretion of filamentous phage. *Mol. Microbiol.* **14,** 357–369.

Russel, M., and Kazmierczak, B. (1993). Analysis of the structure and subcellular location of filamentous phage pIV. *J. Bacteriol.* **175,** 3998–4007.

Russel, M., and Model, P. (1983). A bacterial gene, *fip,* required for filamentous bacteriophage fl assembly. *J. Bacteriol.* **154,** 1064–1076.

Russel, M., and Model, P. (1985). Thioredoxin is required for filamentous phage assembly. *Proc. Natl. Acad. Sci. U.S.A.* **82,** 29–33.

Russel, M., and Model, P. (1989). Genetic analysis of the filamentous bacteriophage packaging signal and of the proteins that interact with it. *J. Virol.,* **63,** 3284–3295.

Russel, M., Whirlow, H., Sun, T. P., and Webster, R. E. (1988). Low-frequency infection of F-bacteria by transducing particles of filamentous bacteriophages. *J. Bacteriol.* **170,** 5312–5316.

Simons, G. F. M., Konings, R. N. H., and Schoenmakers, J. G. G. (1981). Genes *VI, VII,* and genes *IX* of phage M13 code for minor capsid proteins of the virion. *Proc. Natl. Acad. Sci. U.S.A.* **78,** 4194–4198.

Skinner, M. M., Zhang, H., Leschnitzer, D. H., Guan, Y., Bellamy, H., Sweet, R. M., Gray, C. W., Konings, R. N., Wang, A. H., and Terwilliger, T. C. (1994). Structure of the gene V protein of bacteriophage fl determined by multiwavelength x-ray diffraction on the selenomethionyl protein. *Proc. Natl. Acad. Sci. U.S.A.* **91,** 2071–2075.

Smith, G. P. (1988). Filamentous phages as cloning vectors. *In* "Vectors: A Survey of Molecular Cloning Vectors and Their Uses" (R. L. Rodriquez, and D. T. Denhardt, eds.), pp. 61–84. Butterworth, Boston.

Smith, G. P., and Scott, J. K. (1993). Libraries of peptides and proteins displayed on filamentous phage. *In* "Methods in Enzymology" (R. Wu., ed.), Vol. 217, pp. 228–257. Academic Press, San Diego, CA.

Specthrie, L., Bullitt, E., Horiuchi, K., Model, P., Russel, M., and Makowski, L. (1992). Construction of a microphage variant of filamentous bacteriophage. *J. Mol. Biol.* **228,** 720–724.

Stengele, I., Bross, P., Garces, X., Giray, J., and Rasched, I. (1990). Dissection of functional domains in phage fd adsorption protein-discrimination between attachment and penetration sites. *J. Mol. Biol.* **212,** 143–149.

Su, S.-S., Lahue, R. S., Au, K. G., and Modrich, P. (1988). Mispair specificity of methyl-directed DNA mismatch correction *in vitro. J. Biol. Chem.* **263,** 6829–6835.

Sun, T. P., and Webster, R. E. (1987). Nucleotide sequence of a gene cluster involved in entry of E colicins and single-stranded DNA of infecting filamentous bacteriophages into *Escherichia coli. J. Bacteriol.* **169,** 2667–2674.

Van Wezenbeek, P.M.G.F., Hulsebos, T.J.M., and Schoenmakers, J.G.G. (1980). Nucleotide sequence of the filamentous bacteriophage M13 DNA genome: Comparison with phage fd. *Gene* **11,** 129–148.

Vianney, A., Lewin, T. M., Beyer, W. F., Jr., Lazzaroni, J. C., Portalier, R., and Webster, R.E. (1994). Membrane topology and mutational analysis of the TolQ protein of *Escherichia coli* required for the uptake of macromolecules and cell envelope integrity. *J. Bacteriol.* **176,** 822–829.

Webster, R. E. (1991). The *tol* gene products and the import of macromolecules into *Escherichia coli*. *Mol. Microbiol.* **5,** 1005–1011.

Webster, R. E., and Lopez, J. (1985). Structure and assembly of the class 1 filamentous bacteriophage. *In* "Virus Structure and Assembly" (S. Casjens, ed.). pp 235–268, Jones & Bartlett, Boston.

Wickner, W. (1975). Asymmetric orientation of a phage coat protein in cytoplasmic membrane of *Escherichia coli*. *Proc. Natl. Acad. Sci. U.S.A.* **72,** 4749–4753.

Willetts, N. S., and Skurray, R. (1987). Structure and function of the F factor and mechanism of conjugation. *In* "*Escherichia coli* and *Salmonella typhimurium*" (F. C. Neidhardt, J. L. Ingraham, K. B. Low, B. Magasanik, M. Schaechter, and H. E. Umbarger, eds.), pp. 1110–1133. American Society for Microbiology, Washington, D.C.

Williams, K. A., Glibowicka, M. Li, Z., Li, H., Khan, A. R., Chen, Y. M. Y., Wang, J., Marvin, D. A., and Deber, C. M. (1995). Packing of coat protein amphipathic and transmembrane helices in filamentous bacteriophage M13: Role of smallresidues in protein oligomerization. *J. Mol. Biol.* **252,** 6–14.

Winter, G., Griffith, A. D., Hawkins, R. E., and Hoogenboom, H. R. (1994). Making antibodies by phage display technique. *Annu. Rev. Immunol.* **12,** 433–455.

# Principles and Applications of Phage Display

## *Brian K. Kay and Ronald H. Hoess*

## INTRODUCTION

The display of peptides and proteins on the surface of bacteriophage represents a powerful new methodology for carrying out molecular evolution in the laboratory. The ability to construct libraries of enormous molecular diversity and to select for molecules with predetermined properties has made this technology applicable to a wide range of problems. The origins of phage display date to the mid-1980s when George Smith, on sabbatical in Bob Webster's laboratory at Duke University, first expressed a foreign segment of a protein on the surface of bacteriophage M13 virus particles. As a test case he fused a portion of the gene encoding the *Eco*RI endonuclease to the minor capsid protein pIII (Smith, 1985). Using a polyclonal antibody specific for the endonuclease, Smith demonstrated that phage containing the *Eco*RI–gIII fusion could be enriched more than 1000-fold from a mixture containing wild-type (nonbinding) phage with an immobilized polyclonal antibody. From these first experiments emerged two important concepts. First, using recombinant DNA technology, it should be possible to build large libraries (i.e., $10^8$) wherein each phage displays a unique random peptide. Second, the methodology provides a direct physical link between phenotype and genotype. That is, every displayed molecule has an addressable tag via the DNA encoding that molecule. Because of the ease and rapidity of DNA sequence analysis, selected molecules can be identified quickly. It is interesting to note that the idea of addressable tags is also now being adopted in some combinatorial chemical libraries (Needels *et al.*, 1993; Ohlmeyer *et al.*, 1993). Within a few years of George Smith's experiments the first phage-displayed random peptide libraries were assembled (Cwirla *et al.*, 1990; Devlin *et al.*, 1990; Scott and Smith, 1990), accompanied by reports that properly folded and functional proteins could also be displayed on the surface of M13 (Bass *et al.*, 1990; McCafferty *et al.*, 1990).

Perhaps one of the most impressive aspects of phage display is the variety of uses for the technology. A few of the many applications are listed and discussed below.

> *Phage Display of Natural Peptides*
> a. Mapping epitopes of monoclonal and polyclonal antibodies
> b. Generating immunogens
>
> *Phage Display of Random Peptides*
> a. Mapping epitopes of monoclonal and polyclonal antibodies
> b. Identifying peptide ligands
> c. Mapping substrate sites for proteases and kinases
>
> *Phage Display of Protein and Protein Domains*
> a. Directed evolution of proteins
> b. Isolation of high-affinity antibodies
> c. cDNA expression screening

## PHAGE DISPLAY OF NATURAL PEPTIDES

### Mapping Epitopes of Monoclonal and Polyclonal Antibodies

One of the motivations behind George Smith's original experiments was to use phage display as a means of identifying clones that were expressing epitopes recognized by a given antibody (Smith, 1985; de la Cruz *et al.*, 1988). Traditionally, epitope mapping of protein antigens has relied heavily on physical chemical analysis. These approaches have included: (1) fragmenting the purified antigen (protein) with various proteases, identifying reactive fragments, and sequencing them; (2) chemical modification experiments in which residues interacting with the antibody are protected from modification; (3) synthesizing a series of peptides corresponding to the primary structure of the antigen; and (4) direct physical characterization using NMR or X-ray crystallography. All of these methods are labor intensive and in general are not amenable to high-throughput analysis. As an alternative to these methods, phage display can be used to localize the antigenic epitope relatively quickly. Fragments of DNA that encode portions of the protein antigen are fused to the gene encoding capsid protein III. Phage can then be tested with the antibody to determine which displayed fragments react with the antibody. This has been done successfully for the *gag* gene of HIV (Tsunetsugu-Yokota *et al.*, 1991) and more recently for the outer capsid protein VP5 of the bluetongue virus (L.-F. Wang *et al.*, 1995). The obvious limitation of this approach is that each antigen represents the construction of a new phage library, albeit small, in comparison to most random peptide libraries. Furthermore, the starting point, DNA encoding the antigen, must encode the epitope.

## Generating Immunogens

Phage display can also be used for the purpose of generating immunogens. Short segments of various proteins have been displayed on M13 virus particles for the purposes of eliciting antibodies against the coat proteins of certain parasites and viruses. The repeat units of the circumsporozoite protein of the human malaria parasite *Plasmodium falciparum* have been displayed and led to the successful production of anti-circumsporozoite antibodies (de la Cruz *et al.,* 1988; Greenwood *et al.,* 1991). Various segments of different coat proteins of the human immunodeficiency virus (HIV) have been displayed for the purpose of generating a vaccine (Tsunetsugu-Yokota *et al.,* 1991; Minnenkova *et al.,* 1993; Veronese *et al.,* 1994). The immunological response to injected M13 viral particles is T-cell dependent and does not require adjuvants (Willis *et al.,* 1993). Thus, the use of recombinant M13 particles in generating antibodies is of great potential value to the academic scientist and clinician alike.

## PHAGE DISPLAY OF RANDOM PEPTIDES

One important application of phage display has been to construct combinatorial peptide libraries. Synthetic oligonucleotides, fixed in length but with unspecified codons, can be cloned as fusions to genes III or VIII of M13 where they are expressed as a plurality of peptide:capsid fusion proteins. The libraries, often referred to as random peptide libraries, can then be tested for binding to target molecules of interest. This is most often done using a form of affinity selection known as "biopanning" (Parmley and Smith, 1988). The library is first incubated with a target molecule followed by the capture of the target molecule with bound phage. The phage recovered in this manner can then be amplified and again selected for binding to the target molecule, thus enriching for those phage that bind the target molecule. Usually, three to four rounds of selection can be accomplished in 1 week's time, leading to the isolation of one to hundreds of binding phage. Thus, rare phage that bind can easily be selected from greater than $10^8$ different individuals in one experiment. The primary structure of the binding peptides is then deduced by nucleotide sequencing of individual clones. Random peptide libraries have successfully yielded peptides that bind to the combining site of antibodies, cell surface receptors, cytosolic receptors, extracellular and intracellular proteins, DNA, and many other targets (Kay, 1995).

As mentioned previously sequences coding for random peptides are fused to the amino terminus of either pIII or pVIII. Sequences fused to pIII have coded for peptides ranging in size from 6 amino acids (Scott and Smith, 1990) to 38 amino acids (Kay *et al.,* 1993) in length. Fusion to pVIII appears more restricted to peptides of relatively short (6–8 amino acids) length, (Kishchenko *et al.,* 1994). The displayed peptides have a free amino terminus thus allowing the peptide considerable flexibility much like peptides free in solution. Because these peptides are relatively short in length it is unlikely that they fold into stable secondary structures. In order to lower the entropy of such peptides for the sake of enhancing their affinity for binding to a target molecule, several constrained peptide libraries have been built. These include cyclization of the random peptide sequences by flanking cysteines residues which form disulfide bonds (O'Neil *et al.*, 1992; Luzzago *et al.*, 1993; McLafferty *et al.*, 1993) or by displaying the peptides in the loop of a stable scaffold protein (Martin *et al.*, 1994; McConnell *et al.*, 1994).

## Mapping Epitopes of Monoclonal and Polyclonal Antibodies

Random peptide libraries have been invaluable in mapping the specificity of the antibody binding sites. As proposed by Mario Geysen, random peptide libraries represent a source of sequences from which epitopes and mimotopes can be operationally defined (Geysen *et al.*, 1986). Phage-displayed random peptide libraries have been used for the same purpose, and examples are listed below. With these libraries scientists have identified immunodominant peptide sequences of antigens, generated peptide competitors of antigen–antibody interactions, and mapped accessible and/or functional sites of numerous antigens, some of which are listed below.

*Mapping Epitopes and Mimotopes*

*pIII fusions*
| | |
|---|---|
| mAb A2, M33–myohemerythrin | (Scott and Smith, 1990) |
| mAb–endorphin | (Cwirla *et al.*, 1990) |
| mA–HIV gp120 | (Christian *et al.*, 1992; Jellis *et al.*, 1993) |
| mAb B3–carbohydrate | (Hoess *et al.*, 1993) |
| mAb CB5B10–PAI-1 | (Hoess *et al.*, 1994) |
| mAb PAB240–p53 | (Stephen and Lane, 1992) |
| pAb goat–mouse Ig | (Kay *et al.*, 1993) |
| pAb goat–biotin | (Roberts *et al.*, 1993) |
| mAb 5.5–AChR | (Balass *et al.*, 1993) |
| mAb DG2, DE6–bFGF | (Yayon *et al.*, 1993) |
| mAb–keratin | (Bottger and Lane, 1994; Bottger *et al.*, 1995) |
| mAb A16–HSV gpD | (Schellendens *et al.*, 1994) |
| pAb – TNFα | (Sioud *et al.*, 1994) |
| mAb SOM-018–somatostatin | (Wright *et al.*, 1995) |

*pVIII fusions*
| | |
|---|---|
| mAb 1B7–pertussin toxin | (Felici *et al.*, 1993) |
| mAb H107–ferritin | (Luzzago *et al.*, 1993) |
| mAb MGr2–HER2/neu | (Orlandi *et al.*, 1994) |
| pAb–HBsAg | (Folgori *et al.*, 1994; Motti *et al.*, 1994) |

## Identifying Peptide Ligands

Random peptide libraries are a rich source of peptide ligands for a variety of receptors (Cortese *et al.*, 1995). In some cases the peptides resemble the primary structure of receptor ligands, and in other cases the peptides mimic the binding of nonpeptide ligands. A number of results published in the literature are listed below, where random peptide libraries have provided antagonists for protein–protein and protein–nonpeptide ligand interactions, *in vitro* and *in vivo*.

*Identifying Peptide Ligands*

| | |
|---|---|
| Streptavidin | (Devlin *et al.*, 1990;  Kay *et al.*, 1993; McLafferty *et al.*, 1993) |
| Concanavalin A | (Oldenburg *et al.*, 1992; Scott *et al.*, 1992) |
| HLA-DR's | (Hammer *et al.*, 1992, 1993) |
| Integrin ($\alpha_{II}\beta_3$, $\alpha_V\beta_3$, $\alpha_5\beta_1$) | (O'Neil *et al.*, 1992; Koivunen *et al.*, 1993; Healy *et al.*, 1995) |
| S fragment of RNase | (Smith *et al.*, 1993) |
| Calmodulin | (Dedman *et al.*, 1993; Brennan *et al.*, 1996) |
| BiP | (Blond-Elguindi *et al.*, 1993) |
| Urokinase cell surface receptor | (Goodson *et al.*, 1994) |
| Src SH3 domain | (Cheadle *et al.*, 1994; Rickles *et al.*, 1994; Sparks *et al.*, 1994) |
| Fyn, Lyn, PI3K, Abl SH3 domains | (Rickles *et al.*, 1994) |
| DnaK | (Gragerov *et al.*, 1994) |
| p53 | (Daniels and Lane, 1994) |
| Single-stranded DNA | (Krook *et al.*, 1994) |
| Thrombin receptor | (Doorbar and Winter, 1994) |
| Hepatitis B virus capsid | (Dyson and Murray, 1995) |

## Defining Post-translational Substrate Sequences

Another important use of random peptide libraries has been the development of "substrate phage" (Matthews and Wells, 1993). In this case the libraries are used as a means of defining substrate specificity rather than simply binding to a target molecule. A number of groups have been able to define optimal protease cleavage sites (see table below). In theory, many different post-translational modifications (i.e., phosphorylation, glycosylation, acetylation, methylation, ubiquitination) can be mapped the same way.

*Mapping Post-translational Modification Substrate Sites*

| | |
|---|---|
| Subtilisin cleavage sequence | (Matthews and Wells, 1993) |
| Kinase substrate sequence (recognized by mAb MPM2) | (Westendorf *et al.*, 1994) |
| Furin cleavage sequence | (Matthews *et al.*, 1994) |
| Matrilysin cleavage sequence | (Smith *et al.*, 1995) |

## PHAGE DISPLAY OF PROTEIN AND PROTEIN DOMAINS

Many different proteins and protein domains have been displayed on M13 virus particles for a variety of purposes. Table 1 lists many of the proteins and domains which have been successfully displayed. In many cases, the phage-displayed protein retains its normal binding or enzymatic activity even when fused to the N-terminus of mature pIII or pVIII. Several other proteins, including staphylococcal nuclease, CD4, and PDGF receptor, have been displayed (Chiswell and McCafferty, 1992).

**TABLE I**   Displayed Proteins and Protein Domains

| Displayed protein | Size | Reference |
|---|---|---|
| Alkaline phosphatase | Entire protein | (McCafferty et al., 1991) |
| Human growth hormone | 1–191 aa | (Lowman et al., 1991) |
| Antibody[a] | Fab, scFv | (McCafferty et al., 1990; Barbas et al., 1991; Roberts et al., 1992) |
| Bovine pancreas trypsin inhibitor | Entire protein | (Swimmer et al., 1992) |
| Ricin | 1–262, 46–143, 81–262 | (Lehar et al., 1994) |
| Conotoxin | Entire protein | (Olivera et al., 1992) |
| Plasminogen-activator inhibitor | Entire protein | (Pannekoek et al., 1993) |
| Trypsin | Entire protein | (Corey et al., 1993) |
| IgE receptor | a Subunit | (Robertson, 1993) |
| "Minibody" | 61 aa | (Martin et al., 1994) |
| cDNA | Segments | (Crameri and Suter, 1993) |
| Bacterial Rop | Four helices | (Santiago-Vispo et al., 1993) |
| Prostate specific antigen | Entire protein | (Eerola et al., 1994) |
| Protein A | Domain B | (Djojonegoro et al., 1994; Kushwaha et al., 1994) |
| Atrial natriuretic peptide | Entire protein | (Cunningham et al., 1994) |
| Zif268 zinc-finger protein | Three domains | (Choo and Klug, 1994; Jamieson et al., 1994; Rebar and Pabo, 1994; Wu et al., 1995) |
| β-Lactamase | Entire protein | (Soumillion et al., 1994) |
| APPI Kunitz domain | Domain | (Dennis and Lazarus, 1994) |
| Tendamistat | Entire protein | (McConnell and Hoess, 1995) |
| CD4 | 1–176 aa | (Abrol et al., 1994) |
| Cytokine IL-3 | Entire protein | (Gram et al., 1993) |
| Digoxin-binding protein | Entire protein | (Tang et al., 1995) |
| Ecotin | Entire protein | (Wang et al., 1995) |
| Ciliary neurotrophic factor | Entire protein | (Saggio et al., 1995) |
| Glutathione S-transferase | Entire protein | (Widersten and Mannervik, 1995) |

[a] Many different antibodies have been displayed on phage; see Web Page for additional references.

There have been three major uses of phage-displayed proteins and protein domains: directed evolution, selection of high-affinity antibodies, and cDNA expression screening. Large numbers of recombinants can be screened quickly to identify those proteins or domains which have altered or improved affinity for a target.

## Directed Evolution of Proteins

It has been possible to display a protein or protein domain on phage in such a manner that it has its normal binding activity for a target molecule. The coding region for a protein or protein domain can be mutagenized by cassette mutagenesis, error-prone PCR, or shuffling to generate a plurality of altered sequences. From populations of $>10^8$ recombinant phage, those members with desired binding properties can be isolated by affinity selection and their primary structure deduced by DNA sequencing. It has been possible to identify stronger binding ligands for a receptor (Lowman *et al.,* 1991; Lowman and Wells, 1993), novel enzyme inhibitors (Roberts *et al.,* 1992; Dennis and Lazarus, 1994), novel DNA binding proteins (Choo *et al.,* 1994), novel antagonists (Martin *et al.,* 1994), and potentially novel enzymes (Soumillion *et al.,* 1994).

## Isolation of High-Affinity Antibodies

One of the more powerful applications of phage-display has been in the arena of antibody engineering. It has been possible to express both Fab and single-chain Fv (scFv) antibody fragments on the surface of M13 viral particle with no apparent loss of the antibody's affinity and specificity (McCafferty *et al.,* 1990; Barbas *et al.,* 1991; Clackson *et al.,* 1991). The coding regions of the $V_L$ and $V_H$ chains can be obtained from naive mice (Gram *et al.,* 1992; Nissim *et al.,* 1994), immunized mice (Clackson *et al.,* 1991), or germline genes (Hoogenboom and Winter, 1992). Antibodies to many diverse antigens have been successfully isolated using phage display technology.

In many respects the natural immune system is mimicked by the types of manipulation possible with phage-displayed antibodies (Marks *et al.,* 1992; Winter *et al.,* 1994). Antigen-driven stimulation of antibody production can be achieved by selecting for high-affinity binders from a phage-display library of antibodies. The large number of chain permutations that occur during recombination of heavy and light chain genes in developing B cells can be mimicked by shuffling the cloned heavy and light chains as DNA (Marks *et al.,* 1992), and protein (Fiugini *et al.,* 1994) and through the use of prokaryotic site-specific recombination (Waterhouse *et al.,* 1993; Geoffroy *et al.,* 1994; Griffiths *et al.,* 1994). Finally, somatic mutation can be matched by the introduction of mutations in the complementarity-determining regions of phage-displayed antibodies (Barbas *et al.,* 1992a, 1994; Glaser *et al.,* 1992).

While these techniques are comparable to mouse monoclonal antibody technology, their application to the generation of human antibodies (Marks *et al.,* 1991; Hoogenboom and Winter, 1992) is helping to overcome current limitations in generating human monoclonal antibodies or humanizing mouse antibodies. It is possible to isolate human antibodies from patients exposed to certain viral pathogens such as Hepatitis B (Zebedee *et al.,* 1992), respiratory syncytial virus (Barbas *et al.,* 1992b), and HIV (Barbas *et al.,* 1994) for the purposes of understanding the immune response during infection and generating protective antibodies. Finally, this may be a valuable method of elucidating the specificity of autoimmune antibodies (Calcutt *et al.,* 1993; Portolano *et al.,* 1993).

## cDNA Expression Screening

An area of increasing interest in molecular biology is the vast communication network among proteins in the cell. This is best exemplified by signal transduction pathways in which a whole series of protein domains interact with one another. In trying to deduce these interactions the investigator is not interested in exploring all of sequence space as defined by random peptide libraries but merely the sequence space defined by the genome under study. The yeast two-hybrid system (Chien *et al.,* 1991) has proven to be a useful tool in defining protein–protein interactions. Recently, a number of phage display systems have also been described which will be able to perform a similar function. It is possible to express cDNA-encoded proteins on the surface of phage which can then be tested against a particular immobilized target *in vitro* using biopanning enrichment. These systems include a vector where cDNA segments are expressed at the C-terminus of a *fos*-dimerization domain-containing protein which then is assembled on phage capsids containing the *jun*-dimerization domain (Crameri and Suter, 1993). Another approach is to fuse cDNA's directly to the carboxy terminus of pVI (Jespers *et al.,* 1995). In addition, two systems have been developed using bacteriophage λ that would also be useful for the display and affinity selection of expressed cDNAs (Maruyama *et al.,* 1994; Sternberg and Hoess, 1995).

Direct selection of protein–protein interactions has also proven possible with phage-displayed libraries. For M13 phage particles to be infective both amino and carboxy domains of pIII must be present. Thus, it has been possible to express the target protein as a fusion protein with the carboxy-terminal half of pIII separately from cDNAs segments which have been expressed and fused to the amino terminal domain of pIII; when the target and particular cDNA-expressed protein domains interact, the two pIII domains are brought together and the resulting phage particles regain infectivity (Dueñas and Borrebaeck, 1994; Gramatikoff *et al.,* 1994). All of these systems have the potential to complement the yeast two-hybrid system (Phizicky and Fields, 1995). While phage systems will obviously lack post-translational modifications, they offer a number of advantages including the capability of constructing larger library sizes then can be currently done in yeast, as well as demonstrating a direct physical interaction when the *in vitro* biopanning procedure is done. The power of this approach was made evident in a recent publication, where a lymphocyte cDNA library cloned in pIII was used to demonstrate an interaction between β-actin and HIV reverse transcriptase (Hottiger *et al.,* 1995).

## CONCLUSION

Phage display is a methodology that is rapidly making inroads in the fields of immunology, protein biochemistry, protein engineering, and cell biology. It is a technique that most laboratories comfortable in molecular biology can adopt quickly and smoothly. Phage display has been used to create libraries of random peptides, proteins, and protein domains for the purposes of mapping epitopes and mimotopes, identifying antagonists and agonists of various target molecules, engineering human antibodies, optimizing antibody specificities, and creating novel binding activities.

## WORLD WIDE WEB PAGE

The world of phage display literature is rapidly expanding. To include this information for the reader after this book has been published, a World Wide Web page will be established (http://www.unc.edu/depts/biology/bkay/phagedisplay.html) that will attempt to keep a comprehensive list. This site will also contain corrections to various protocols covered in the book, as well as links to other relevant home pages.

## References

Abrol, S., Sampath, A., Arora, K., and Chaudhary, V. K. (1994). Construction and characterization of M13 bacteriophages displaying gp120 binding domains of human CD4. *Indian J. Biochem. Biophys.* **31,** 302–309.

Balass, M., Heldman, Y., Cabilly, S., Givol, D., Katchalski-Katzir, E., and Fuchs, S. (1993). Identification of a hexapeptide that mimics a conformation-dependent binding site of acetycholine receptor by use of a phage-epitope library. *Proc. Natl. Acad. Sci. U.S.A.* **90,** 10638–10642.

Barbas, C., Kang, A., Lerner, R., and Benkovic, S. (1991). Assembly of combinatorial antibody libraries on phage surfaces: The gene III site. *Proc. Natl. Acad. Sci. U.S.A.* **88,** 7978–7982.

Barbas, C., Bain, J., Hoesktra, D., and Lerner, R. (1992a). Semisynthetic combinatorial antibody libraries: A chemical solution to the diversity problem. *Proc. Natl. Acad. Sci. U.S.A.* **89,** 4457–4461.

Barbas, C. F., Crowe, J. E., Cababa, D., Jones, T. M., Zebeddee, S. L., Murphay, B. R., Chanock, R. M., and Burton, D. R. (1992b). Human monoclonal Fab fragments derived from a combinatorial library bind to respiratory syncytial virus F glycoprotein and neutralize infectivity. *Proc. Natl. Acad. Sci. U.S.A.* **89,** 10164–10168.

Barbas, C. F., Hu, D., Dunlop, N., Sawyer, L., Cababa, D., Hendry, R. M., Nara, P. L., and Burton, D. R. (1994). In vitro evolution of a neutralizing human antibody to human immunodeficiency virus type 1 to enhance affinity and broaden strain cross-reactivity. *Proc. Natl. Acad. Sci. U.S.A.* **91,** 3809–3813.

Bass, S. H., Greene, R., and Wells, J. A. (1990). Hormone phage: An enrichment method for variant proteins with altered binding properties. *Proteins* **8,** 309–314.

Blond-Elguindi, S., Cwirla, S., Dower, W., Lipshutz, R., Sprang, S., Sambrook, J., and Gething, M.-J. (1993). Affinity panning of a library of peptides displayed on bacteriophages reveals the binding specificity of BiP. *Cell (Cambridge, Mass.)* **75,** 717–728.

Bottger, V., and Lane, E. B. (1994). A monoclonal antibody epitope on keratin 8 identified using a phage peptide library. *J. Mol. Biol.* **235,** 61–67.

Bottger, V., Bottger, A., Lane, E. B., and Spruce, B. A. (1995). Comprehensive epitope analysis of monoclonal anti-proenkephalin antibodies using phage display libraries and synthetic peptides: Revelation of antibody fine specificities caused by somatic mutations in the variable region genes. *J. Mol. Biol.* **247,** 932–946.

Brennan, J. D., Clark, I. D., Szabo, A. G., Adey, N. B., Hanson, H. L., and Kay, B. K. (1996). Characterization of calmodulin-binding peptides using phage-display random peptide libraries. (submitted for puplication).

Calcutt, M. J., Kremer, M. T., Giblin, M. F., Quinn, T. P., and Deutscher, S. L. (1993). Isolation and characterization of nucleic acid-binding antibody fragments from autoimmune mice-derived bacteriophage display libraries. *Gene* **137,** 77–83.

Cheadle, C., Ivashchenko, Y., South, V., Searfoss, G. H., French, S., Howk, R., Ricca, G. A., and Jaye, M. (1994). Identification of a Src SH3 domain binding motif by screening a random phage display library. *J. Biol. Chem.* **269,** 24034–24039.

Chien, C. T., Bartel, P. L., Sternglanz, R., and Fields, S. (1991). The two-hybrid system: A method to identify and clone genes for proteins that interact with a protein of interest. *Proc. Natl. Acad. Sci. U.S.A.* **88,** 9578–9582.

Chiswell, D. J., and McCafferty, J. (1992). Phage antibodies; will new 'coliclonal' antibodies replace monoclonal antibodies? *Trends Biotechnol.* **10,** 80–84.

Choo, Y., and Klug, A. (1994). Selection of DNA binding sites for zinc fingers using rationally randmized DNA reveals coded interactions. *Proc. Natl. Acad. Sci. U.S.A.* **91,** 11168–11172.

Choo, Y., Sánchez-Garcia, I., and Klug, A. (1994). *In vivo* repression by a site-specific DNA-binding protein designed against an oncogenic sequence. *Nature (London)* **372,** 642–645.

Christian, R. B., Zuckermann, R. N., Kerr, J. M., Wang, L., and Malcolm, B. A. (1992). Simplified methods for construction, assessment, and rapid screening of peptide libraries in bacteriophage M13. *J. Mol. Biol.* **227,** 711–718.

Clackson, T., Hoogenboom, H. R., Griffiths, A. D., and Winter, G. (1991). Making antibody fragments using phage display libraries. *Nature (London)* **352,** 624–628.

Corey, D. R., Shiau, A. K., Yang, Q., Janowksi, B. A., and Craik, C. S. (1993). Trypsin display on the surface of bacteriophage. *Gene* **128,** 129–134.

Cortese, R., Monaci, P., Nicosia, A., Luzzago, A., Felici, F., Galfre, G., Pessi, A., Tramontano, A., and Sollazzo, M. (1995). Identification of biologically active peptides using random libraries displayed on phage. *Curr. Opin. Biotechnol.* **6,** 73–80.

Crameri, R., and Suter, M. (1993). Display of biologically active proteins on the surface of filamentous phages: A cDNA cloning system for selection of functional gene products linked to the genetic information responsible for their production. *Gene* **137,** 69–75.

Cunningham, B. C., Lowe, D. G., Li, B., Bennett, B. D., and Wells, J. A. (1994). Production of an atrial natriuretic peptide variant that is specific for type A receptor. *EMBO J.* **13,** 2508–2525.

Cwirla, S. E., Peters, E. A., Barrett, R. W., and Dower, W. J. (1990). Peptides of phage: A vast library of peptides for identifying ligands. *Proc. Natl. Acad. Sci. U.S.A.* **87,** 6378–6382.

Daniels, D. A., and Lane, D. P. (1994). The characterization of p53 binding phage isolated from phage peptide display libraries. *J. Mol. Biol.* **243,** 639–652.

Dedman, J. R., Kaetzel, M. A., Chan, H. C., Nelson, D. J., and Jamieson, J. G A. (1993). Selection of target biological modifiers from a bacteriophage library of random peptides: The identification of novel calmodulin regulatory peptides. *J. Biol. Chem.* **268,** 23025–23030.

de la Cruz, V., Laa, A., and McCutchan, T. (1988). Immunogenicity and epitope mapping of foreign sequences via a genetically engineered filamentous phage. *J. Biol. Chem.* **263,** 4318–4322.

Dennis, M. S., and Lazarus, R. A. (1994). Kunitz domain inhibitors of tissue-factor factor VIIa. I. Potent inhibitors selected from libraries by phage display. *J. Biol. Chem.* **269,** 22129–22136.

Devlin, J. J., Panganiban, L. C., and Devlin, P. E. (1990). Random peptide libraries: A source of specific protein binding molecules. *Science* **249,** 404–406.

Djojonegoro, B. M., Benedik, M. J., and Wilson, R. C. (1994). Bacteriophage surface display of an immunoglobulin-binding domain of Staphylococcus aureus protein A. *BioTechniques* **12,** 169–172.

Doorbar, J., and Winter, G. (1994). Isolation of a peptide antagonist to the thrombin receptor using phage display. *J. Mol. Biol.* **244,** 361–369.

Dueñas, M., and Borrebaeck, C. (1994). Clonal selection and amplification of phage displayed antibodies by linking antigen recognition and phage replication. *BioTechniques* **12,** 999–1002.

Dyson, M., and Murray, K. (1995). Selection of peptide inhibitors of interactions involved in complex protein assemblies: Association of the core and surface antigens of hepatitis B virus. *Proc. Natl. Acad. Sci. U.S.A.* **92,** 2194–2198.

Eerola, R., Saviranta, P., Lilja, H., Pettersson, K., Lövgren, T., and Karp, M. (1994). Expression of prostate specific antigen on the surface of filamentous phage. *Biochem. Biophys. Res. Commun.* **200,** 1346–1352.

Felici, F., Luzzago, A., Folgoir, A., and Cortese, R. (1993). Mimicking of discontinuous epitopes by phage displayed peptides. II. Selection of clones recognized by a protective monoclonal antibody against the *Bordetella pertussis* toxin from phage peptide libraries. *Gene* **128,** 21–27.

Fiugini, M., Marks, J., Winter, G., and Griffiths, A. (1994). *In vitro* assembly of repertoires of antibody chains on the surface of phage by renaturation. *J. Mol. Biol.* **239,** 68–78.

Folgori, A., Tafi, R., Meola, A., Felici, F., Galfre, G., Cortese, R., Monaci, P., and Nicosia, A. (1994). A general strategy to identify mimotopes of pathological antigns using only random peptide libraries and human sera. *EMBO J.* **13,** 2236–2243.

Geoffroy, F., Sodoyer, R., and Aujame, L. (1994). A new phage display system to construct multicombinatorial libraries of very large antibody repertoires. *Gene* **151,** 109–113.

Geysen, H. M., Rodda, S. J., and Mason, T. J. (1986). *A priori* delineation of a peptide which mimics a discontinuous antigenic determinant. *Mol. Immunol.* **23,** 709–715.

Glaser, S. M., Yelton, D. E., and Huse, W. D. (1992). Antibody engineering by codon-based mutagenesis in a filamentous phage vector system. *J. Immunol.* **149,** 3903–3913.

Goodson, R. J., Doyle, M. V., Kaufman, S. E., and Rosenberg, S. (1994). High-affinity urokinase receptor antagonists identified with bacteriophage peptide display. *Proc. Natl. Acad. Sci. U.S.A.* **91,** 7129–7133.

Gragerov, A., Zeng, L., Zhao, X., Burkholder, W., and Gottesman, M. E. (1994). Specificity of DnaK-peptide binding. *J. Mol. Biol.* **235,** 848–854.

Gram, H., Marconi, L. A., Barbas, C. F., Collet, T. A., Lerner, R. A., and Kang, A. S. (1992). *In vitro* selection and affinity maturation of antibodies from a naive combinatorial immunoglobulin library. *Proc. Natl. Acad. Sci. U.S.A.* **89,** 3576–3580.

Gram, H., Strittmatter, U., Lorenz, M., Gluck, D., and Zenke, G. (1993). Phage display as a rapid gene expression system: Production of bioactive cytokine-phage and generation of neutralizing monoclonal antibodies. *J. Immunol. Methods* **161,** 169–176.

Gramatikoff, K., Georgiev, O., and Shaffner, W. (1994). Direct interaction rescue, a novel filamentous phage technique to study protein-protein interactions. *Nucleic Acids Res.* **22,** 5761–5762.

Greenwood, J., Willis, A., and Perham, R. (1991). Multiple display of foreign peptides on a filamentous bacteriophage: Peptides from *Plasmodium falciparum* circumsporozoite protein as antigens. *J. Mol. Biol.* **220,** 821–827.

Griffiths, A. D., Williams, S. C., Hartley, O., Tomlinson, I. M., Waterhouse, P., Crosby, W. L., Kontermann, R. E., Jones, P. T., Low, N. M., Allison, T. J., Prospero, T. D., Hoggenboom, H. R., Nissim, A., Cox, J. P. L., Harrison, J. L., Zaccolo, M., Gheradi, E., and Winter, G. (1994). Isolation of high affinity human antibodies directly from large synthetic repertoires. *EMBO J.* **13,** 3245–3260.

Hammer, J., Takacs, B., and Sinigaglia, F. (1992). Identification of a motif for HLA-DR1 binding peptides using M13 display libraries. *J. Exp. Med.* **176,** 1007–1013.

Hammer, J., Valsasnini, P., Tolba, K., Bolin, D., Higelin, J., Takacs, B., and Sinigaglia, F. (1993). Promiscuous and allele-specific anchors in HLA-DR-binding peptides. *Cell (Cambridge, Mass.)* **74,** 197–203.

Healy, J., Murayama, O., Maeda, T., Yoshino, K., Sekiguchi, K., and Kikuchi, M. (1995). Peptide ligands for integrin alpha v beta 3 selected from random phage display libraries. *Biochemistry* **34,** 3948–3955.

Hoess, R. H., Brinkmann, U., Handel, T., and Pastan, I. (1993). Identification of a peptide which binds to the carbohydrate-specific monlconal antibody B3. *Gene* **128,** 43–49.

Hoess, R. H., Mack, A. J., Walton, H., and Reilly, T. M. (1994). Identification of a structural epitope by using a peptide library displayed on filamentous phage. *J. Immunol.* **22,** 724–729.

Hoogenboom, H. R., and Winter, G. (1992). By-passing immunization. Human antibodies from synthetic repertoires of germline VH gene segments rearranged in vitro. *J. Mol. Biol.* **227,** 381–388.

Hottiger, M., Gramatikoff, K., Georgiev, O., Chaponnier, C., Schaffner, W., and Hubscher, U. (1995). The large subunit of HIV-1 reverse transcriptase interacts with beta-actin. *Nucleic Acids Res.* **23,** 736–7341.

Jamieson, A. C., Kim, S. H., and Wells, J. A. (1994). *In vitro* selection of zinc fingers with altered DNA-binding specificity. *Biochemistry* **33,** 5689–5695.

Jellis, C. L., Cradick, T. J., Rennert, P., Salinas, P., Boyd, J., Amirault, T., and Gray, G. S. (1993). Defining critical residues in the epitope for a HIV-neutralizing monoclonal antibody using phage display and peptide array technologies. *Gene* **137,** 63–68.

Jespers, L., Messens, J., De Keyser, A., Eeckhout, D., Van Den Brande, I., Gansemans, Y., Lauwereys, M., Vlasuk GP, and Stanssens, P. E. (1995). Surface expression and ligand-based selection of cDNAs fused to filamentous phage gene VI. *Bio/Technology* **13,** 378–382.

Kay, B. K. (1995). Mapping protein-protein interactions with biologically expressed random peptide libraries. *Perspect. Drug Discovery Des.* **2,** 251–268.

Kay, B. K., Adey, N. B., He, Y.-S., Manfredi, J. P., Mataragnon, A. H., and Fowlkes, D. M. (1993). An M13 library displaying 38-amino-acid peptides as a source of novel sequences with affinity to selected targets. *Gene* **128,** 59–65.

Kishchenko, G., Batliwala, H., and Makowski, L. (1994). Structure of a foreign peptide displayed on the surface of bacteriophage M13. *J. Mol. Biol.* **241,** 208–213.

Koivunen, E., Gay, D. A., and Ruoslahti, E. (1993). Selection of peptides binding to the alpha 5 beta 1 integrin from phage display library. *J. Biol. Chem.* **268,** 20205–20210.

Krook, M., Mosbach, K., and Lindbladh, C. (1994). Selection of peptides with affinity for single-stranded DNA using a phage display library. *Biochem. Biophys. Res. Commun.* **204,** 849–854.

Kushwaha, A., Chowdhury, P., Arora, K., Abrol, S., and Chaudhary, V. (1994). Construction and characterization of M13 bacteriophages displaying functional IgG-binding domains of staphylococcal protein A. *Gene* **151,** 45–51.

Lehar, S., Pedersen, J., Kamath, R., Swimmer, C., Goldmacher, V., Lambert, J., Blattler, W., and Guild, B. (1994). Mutational and structural analysis of the lectin activity in binding domain 2 of ricin B chain. *Protein Eng.* **7,** 1261–1266.

Lowman, H. B., and Wells, J. A. (1993). Affinity maturation of human growth hormone by monovalent phage display. *J. Mol. Biol.* **234,** 564–578.

Lowman, H. B., Bass, S. H., Simpson, N., and Wells, J. A. (1991). Selecting high-affinity binding proteins by monovalent phage display. *Biochemistry* **30,** 10832–10838.

Luzzago, A., Felici, F., Tramontano, A., Pessi, A., and Cortese, R. (1993). Mimicking of discontinuous epitopes by phage-displayed peptides. I. Epitope mapping of human H ferritin using a phage library of constrained peptides. *Gene* **128,** 51–57.

Marks, J. D., Hoogenboom, H. R., Bonnert, T. P., McCafferty, J., Griffiths, A. D., and Winter, G. (1991). By-passing immunization. Human antibodies from V-gene libraries displayed on phage. *J. Mol. Biol.* **222,** 581–597.

Marks, J. D., Hoogenboom, H. R., Griffiths, A. D., and Winter, G. (1992). Molecular evolution of proteins on filamentous phage. *J. Biol. Chem.* **267,** 16007–16010.

Martin, F., Toniatti, C., Salvati, A. L., Venturini, S., Ciliberto, G., Cortese, R., and Sollazzo, M. (1994). The affinity-selection of a minibody polypeptide inhibitor of human interleukin-6. *EMBO J.* **13,** 5303–5309.

Maruyama, I. N., Maruyama, H., and Brenner, S. (1994). λfoo: A λ phage vector for the expression of foreign proteins. *Proc. Natl. Acad. Sci. U.S.A.* **91,** 8273–8277.

Matthews, D. J., and Wells, J. A. (1993). Substrate phage: Selection of protease substrates by monovalent phage display. *Science* **260,** 1113–1117.

Matthews, D. J., Goodman, L. J., Gorman, C. M., and Wells, J. A. (1994). A survey of furin substrate specificity using substrate phage display. *Protein Sci.* **3,** 1197–1205.

McCafferty, J., Griffiths, A. D., Winter, G., and Chiswell, D. J. (1990). Phage antibodies: Filamentous phage displaying antibody variable domains. *Nature (London)* **348,** 552–554.

McCafferty, J., Jackson, R. H., and Chiswell, D. J. (1991). Phage-enzymes: Expression and affinity chromatography of functional alkaline phosphatase on the surface of bacteriophage. *Protein Eng.* **4,** 955–961.

McConnell, S. J., and Hoess, R. H. (1995). Tendamistat as a scaffold for conformationally constrained phage peptide libraries. *J. Mol. Biol.* **250,** 460–470.

McConnell, S. J., Kendell, M. L., Reilly, T. M., and Hoess, R. H. (1994). Constrained libraries as a tool for finding mimotopes. *Gene* **151,** 115–118.

McLafferty, M. A., Kent, R. B., Ladner, R. C., and Markland, W. (1993). M13 bacteriophage displaying disulfide-constrained microproteins. *Gene* **128,** 29–36.

Minnenkova, O. O., Ilyichev, G. P., Kishchenko, G. P., and Petrenko, V. A. (1993). Design of specific immunogens using filamentous phage as the carrier. *Gene* **128,** 85–88.

Motti, C., Nuzzo, M., Meola, A., Galfre, G., Felici, F., Cortese, R., Nicosia, A., and Monai, P. (1994). Recognition by human sera and immunogenicity of HBsAg mimotopes select from an M13 phage display library. *Gene* **146,** 191–198.

Needels, M. C., Jones, D. G., Tate, E. H., Heinkel, G. L., Kochersperger, L. M., Dower, W. J., Barrett, R. W., and Gallop, M. A. (1993). Generation and screening of an oligonucleotide-encoded synthetic peptide library. *Proc. Natl. Acad. Sci. U.S.A.* **90,** 10700–10704.

Nissim, A., Hoogenboom, H. R., Tomlinson, I. M., Flynn, G., Midgley, C., Lane, D., and Winter, G. (1994). Antibody fragments from a 'single pot' phage display library as immunochemical reagents. *EMBO J.* **13,** 692–698.

Ohlmeyer, M., Swanson, R., Dillard, L., Reader, J., Asouline, G., Kobayashi, R., Wigler, M., and Still, W. (1993). Complex synthetic chemical libraries indexed with molecular tags. *Proc. Natl. Acad. Sci. U.S.A.* **90,** 10922–10926.

Oldenburg, K., Loganathan, D., Goldstein, I., Schultz, P., and Gallop, M. (1992). Peptide ligands for a sugar-binding protein isolated from a random peptide library. *Proc. Natl. Acad. Sci. U.S.A.* **89,** 5393–5397.

Olivera, B. M., Cruz, L. J., Myers, R. A., Hillyard, D. R., Rivier, J., and Scott, J. K. (1992). *Conus* peptides and biotechnology. *In* "Molecular Basis of Drug & Pesticide Action" (I. Duce, ed.), **pp. ?** Elsevier, New York.

O'Neil, K. T., Hoess, R. H., Jackson, S. A., Ramachandran, N. S., Mousa, S. A., and DeGrado, W. F. (1992). Identification of novel peptide antagonists for GPIIb/IIIa from a conformationally constrained phage peptide library. *Proteins* **14,** 509–515.

Orlandi, R., Ménard, S., Colnaghi, M., Boyer, C. M., and Felici, F. (1994). Antigenic and immunogenic mimicry of the HER2/neu oncoprotein by phage-displayed peptides. *Eur. J. Immunol.* **24,** 2868–2873.

Pannekoek, H., van Meijer, M., Schleef, R. R., Loskutoff, D. J., and Barbas, C. F., III (1993). Functional display of human plasminogen-activator inhibitor 1 (PAI-1) on phages: Novel perspectives for structure-function analysis by error-prone DNA synthesis. *Gene* **128,** 135–140.

Parmley, S. F., and Smith, G. P. (1988). Antibody-selectable filamentous fd phage vectors: Affinity purification of target genes. *Gene* **73,** 305–318.

Phizicky, E., and Fields, S. (1995). Protein-protein interactions: Methods for detection and analysis. *Microbiol. Rev.* **59,** 94–123.

Portolano, S., McLachlan, S. M., and Rapoport, B. (1993). High affinity, thyroid-specific human autoantibodies displayed on the surface of filamentous phage use V genes similar to other autoantibodies. *J. Immunol.* **151,** 2839–2851.

Rebar, E., and Pabo, C. O. (1994). Zinc finger phage: Affinity selection of fingers with new DNA-binding specificity. *Science* **263,** 671–673.

Rickles, R. J., Botfield, M. C., Weng, Z., Taylor, J. A., Green, O. M., Brugge, J. S., and Zoller, M. J. (1994). Identification of Src, Fyn, Lyn, PI3K, and Abl SH3 domain ligands using phage display libraries. *EMBO J.* **13,** 5598–5604.

Roberts, B., Markland, W., Ley, A., Kent, R., White, D., Guterman, S., and Ladner, R. (1992). Directed evolution of a protein: Selection of potent neutrophil elastase inhibitors displayed on M13 fusion phage. *Proc. Natl. Acad. Sci. U.S.A.* **89,** 2429–2433.

Roberts, D., Guegler, K., and Winter, J. (1993). Antibody as a surrogate receptor in the screening of a phage display library. *Gene* **128,** 67–69.

Robertson, M. W. (1993). Phage and *Escherichia coli* expression of the human high affinity immunglobulin E receptor alpha-subunit ectodomain. Domain localization of the IgE-binding site. *J. Biol. Chem.* **268,** 12736–123743.

Saggio, I., Gloaguen, I., and Laufer, R. (1995). Functional phage display of ciliary neurotrophic factor. *Gene* **152,** 35–39.

Santiago-Vispo, N., Felici, F., Castagnoli, L., and Cesareni, G. (1993). Hybrid Rop-pIII proteins for the display of constrained peptides on filamentous phage capsids. *Ann. Biol. Clin. (Paris)* **51,** 917–922.

Schellendens, G. A., Lasonder, E., Feijlbrief, M., Koedijk, D. G., Drijfhout, J. W., Scheffer, A. J., Welling-Wester, S., and Welling, G. W. (1994). Identification of the core residues of the epitope of a monoclonal antibody raised against glycoprotein D of the herpes simplex virus type 1 by screening of a random peptide library. *Eur. J. Immunol.* **24,** 3188–3193.

Scott, J. K., Loganathan, D., Easley, R., Gong, X., and Goldstein, I. (1992). A family of concanavalin A-binding peptides from a hexapeptide epitope library. *Proc. Natl. Acad. Sci. U.S.A.* **89,** 5398–5402.

Scott, J. K., and Smith, G. P. (1990). Searching for peptide ligands with an epitope library. *Science* **249,** 386–390.

Sioud, M., Dybwad, A., Jespersen, L., Suleyman, S., Natvig, J. B., and Forre, O. (1994). Characterization of naturally occurring autoantibodies against tumour necrosis factor-alpha (TNF-alpha): *In vitro* function and precise epitope mapping by phage epitope library. *Clin. Exp. Immunol.* **98,** 520–525.

Smith, G. P. (1985). Filamentous fusion phage: Novel expression vectors that display cloned antigens on the surface of the virion. *Science* **228,** 1315–1317.

Smith, G. P., Schultz, D. A., and Ladbury, J. E. (1993). A ribonuclease S-peptide antagonist discovered with a bacteriophage display library. *Gene* **128,** 37–42.

Smith, M., Shi, L., and Navre, M. (1995). Rapid identification of highly active and selective substrates for stromelysin and matrilysin using bacteriophage peptide display libraries. *J. Biol. Chem.* **270,** 6440–6449.

Soumillion, P., Jespers, L., Bouchet, M., Marchand-Brynaert, J., Winter, G., and Fastrez, J. (1994). Selection of beta-lactamase on filamentous bacteriophage by catalytic activity. *J. Mol. Biol.* **237,** 415–422.

Sparks, A. B., Quilliam, L. A., Thorn, J. M., Der, C. J., and Kay, B. K. (1994). Identification and characterization of Src SH3 ligands from phage-displayed random peptide libraries. *J. Biol. Chem.* **269,** 23853–23856.

Stephen, C. W., and Lane, D. P. (1992). Mutant conformation of p53: Precise epitope mapping using a filamentous phage epitope library. *J. Mol. Biol.* **225,** 577–583.

Sternberg, N., and Hoess, R. (1995). Display of peptides and proteins on the surface of bacteriophage lambda. *Proc. Natl. Acad. Sci. U.S.A.* **92,** 1609–1613.

Swimmer, C., Lehar, S. M., McCafferty, J., Chiswell, D. J., Blattler, W. A., and Guild, B. C. (1992). Phage display of ricin B and its single binding domains: System for screening galactose-binding mutants. *Proc. Natl. Acad. Sci. U.S.A.* **89,** 3756–3760.

Tang, P., Foltz, L., Mahoney, W., and Schueler, P. (1995). A high affinity digoxin-binding protein displayed on M13 is functionally identical to the native protein. *J. Biol. Chem.* **270,** 7829–7835.

Tsunetsugu-Yokota, Y., Tatsumi, M., Robert, V., Devaus, C., Sprite, B., Chermann, J.-C., and Hirsch, I. (1991). Expression of an immunogenic region of HIV by a filamentous bacteriophage vector. *Gene* **99,** 323–326.

Veronese, F. D. M., Willis, A. E., Boyer-Thompson, C., Appella, E., and Perham, R. N. (1994). Structural mimicry and enhanced immunogenecity of peptide epitopes displayed on filamentous bacteriophage. *J. Mol. Biol.* **243,** 167–172.

Wang, C., Yang, Q., and Craik, C. (1995). Isolation of a high affinity inhibitor of urokinase-type plasminogen activator by phage display of ecotin. *J. Biol. Chem.* **270,** 12250–12256.

Wang, L.-F., Du Plessis, D. H., White, J. R., Hyatt, A. D., and Eaton, B. T. (1995). Use of a gene-targeted phage display random epitope library to map an antigenic determinant on the bluetongue virus outer capsid protein VP5. *J. Immunol. Methods* **178,** 1–12.

Waterhouse, P., Griffiths, A. D., Johnson, K. S., and Winter, G. (1993). Combinatorial infection and *in vivo* recombination: A strategy for making large phage antibody repertoires. *Nucleic Acids Res.* **21,** 2265–2266.

Westendorf, J. M., Rao, P. N., and Gerace, L. (1994). Cloning of cDNAs for M-phase phosphoproteins recognized by the MPM2 monoclonal antibody and determination of the phosphorylated epitope. *Proc. Natl. Acad. Sci. U.S.A.* **91,** 714–718.

Widersten, M., and Mannervik, B. (1995). Glutathione transferases with novel active sites isolated by phage display from a library of random mutants. *J. Mol. Biol.* **250,** 115–122.

Willis, A. E., Perham, R. N., and Wraith, D. (1993). Immunological properties of foreign peptides in multiple display on a filamentous bacteriophage. *Gene* **128,** 79–83.

Winter, G., Griffiths, A. D., Hawkins, R. E., and Hoogenboom, H. R. (1994). Making antibodies by phage display technology. *Annu. Rev. Immunol.* **12,** 433–455.

Wright, R. M., Gram, H., Vattay, A., Byrne, S., Lake, P., and Dottavio, D. (1995). Binding epitope of somatostatin defined by phage-displayed peptide libraries. *Bio/Technology* **13,** 165–169.

Wu, H., Yang, W.-P., and Barbas, C. F. (1995). Building zinc fingers by selection: Toward a therapeutic application. *Proc. Natl. Acad. Sci. U.S.A.* **92,** 344–348.

Yayon, A., Aviezer, D., Safran, M., Gross, J. L., Heldman, Y., Cabilly, S., Givol, D., and Katchalski-Katzir, E. (1993). Isolation of peptides that inhibit binding of basic fibroblast growth factor to its receptor from a random phage-epitope library. *Proc. Natl. Acad. Sci. U.S.A.* **90,** 10643–10647.

Zebedee, S. L., Barbas, C. F., Hom, Y. L., Caothien, R. H., Graff, R., DeGraw, J., Pyati, J., LaPolla, R., Burton, D. R., and Lerner, R. A. (1992). Human combinatorial antibody libraries to hepatitis B surface antigen. *Proc. Natl. Acad. Sci. U.S.A.* **89,** 3175–3179.

# 3

# Vectors for
# Phage Display

*Norris Armstrong, Nils B. Adey,*
*Stephen J. McConnell, and Brian K. Kay*

A range of vectors are available for exogenous expression on the surface of bacteriophage M13 virus particles. The display sites most commonly used are within genes III or VIII, although there have been attempts at cloning in genes VII and IX (Makowski, 1993). Viral vectors that accept and display fusions for genes III and VIII have been termed type 3 and 8 (Smith, 1993), and are shown in Fig. 1. Many short peptides and a variety of proteins have been displayed at the N-terminus of mature pIII. pVIII, on the other hand, appears to tolerate only short inserts of five (Il'ichev *et al.,* 1989) or six amino acid additional residues (Greenwood *et al.,* 1991); this may be due to the close packed nature of the viral surface (Kishchenko *et al.,* 1994). Additional information regarding viral morphogenesis, pIII, and pVIII can be found in Chapter 1.

Many investigators have observed that some sequences are not well displayed on the surface of M13 phage, due to defects in viral particle assembly, stability, and infectivity. While most of this biological intolerance is unclear, two compensating vector systems have been designed. In one system, phagemid vectors carry a copy of either gene III or VIII; these vectors have been termed type 3+3 or 8+8 (Smith, 1993; Fig. 1). When bacterial cells harboring these phagemids are infected with M13 helper phage, which carry the full complement of capsid-encoding genes but are defective in replication, the secreted phage particles carry the phagemid genome and a mixture of wild-type and fusion pIIIs or pVIIIs (Fig. 2). In another system, the

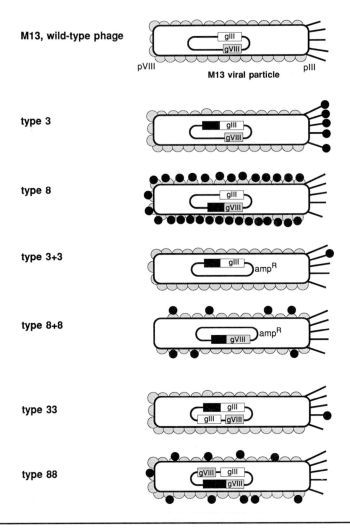

**M13, wild-type phage**

pVIII                    pIII
**M13 viral particle**

**type 3**

**type 8**

**type 3+3**                amp^R

**type 8+8**                amp^R

**type 33**

**type 88**

**FIGURE I**     Comparison of different M13 phage-display vectors and viral particles. The black boxes and spheres correspond to the foreign genetic elements and their encoded peptides, respectively. The displayed peptides, proteins, or protein domains are fused to either proteins III or VIII. Each vector system is then classified as type 3 or 8 depending on whether the fusion is with proteins III or VIII, respectively. Type 3 and 8 viral vectors have single copies of the fusion gene, whereas types 33 and 88 have two copies. Types 3+3 and 8+8 are phagemid based; virus particles are only formed when cells carrying the phagemid genomes are infected with M13 helper virus particles.

vectors are phage but carry two copies of gene III or VIII; bacterial cells infected with these phage incorporate both wild-type and fusion copies of pIII or pVIII into the same viral particles. George Smith has named this second class of vectors as "33" or "88," to designate phage genomes that carry both wild-type and recombinant copies of genes III or VIII, respectively (Smith, 1993).

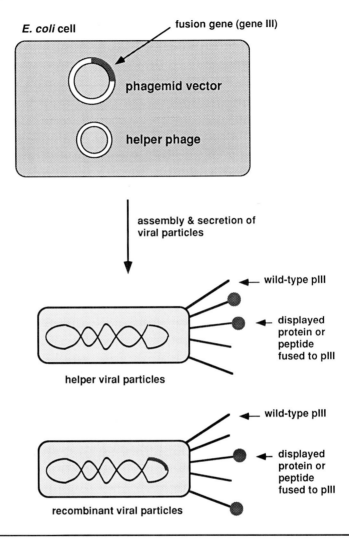

**FIGURE 2**   Diagram of bacterial cells infected with type 3+3 (or 8+8) viral genomes and secretion of viral particles. Bacteria carrying the phagemid genome are infected with M13 helper viruses which supply the missing M13 viral proteins in *trans*. The phagemid genome is packaged and secreted more efficiently than the helper genome which is defective in replication.

One major feature of the 3+3 vector system is that the displayed element is often monovalent. The concentration of displayed proteins or peptides is diluted by the wild-type copies contributed by the helper phage. Thus both helper and phagemid viral particles will have 0–5 copies of the pIII fusion protein per particle. This can be advantageous when the pIII fusions yield phage which are not stable or infectious. The reduced valency is also valuable in eliminating avidity, thus permitting discrimination between low- and high-affinity display elements (see below).

In the 8+8 vector system, the valency of the displayed element is between 0 and 2400, with an average of several hundred reported per viral particle. The viral particles have a mixture of both fusion and wild-type pVIII. The increased valency of pVIII display, as compared to pIII, has advantages in experiments in which low-affinity receptors are the target. In spite of this increase in valency, people also use the pVIII phagemid when they want a greater distance between display units. Since only some fraction of the 2400 pVIIIs will be fusion protein, the probability of two fusions being next to each other is thought to be less than in the pIII display system.

A variety of commonly used display vectors are tabulated below, with their name, site of expression, restriction site used, marker carried on the vector, and reference.

| Vector | Gene | Rest. Site(s) | Marker | Reference |
|--------|------|---------------|--------|-----------|
| fUSE5 | III | *Bgl*I–S–*Bgl*I | tet$^R$ | (Scott and Smith, 1990) |
| fAFF1 | III | *Bst*XI–S–*Bst*XI | tet$^R$ | (Cwirla *et al.*, 1990) |
| fd-CAT1 | III | *Pst*I–S–*Xho*I | tet$^R$ | (McCafferty *et al.*, 1990) |
| m663 | III | *Xho*I–S–*Xba*I | *lacZ*$^+$ | (Fowlkes *et al.*, 1992) |
| fdtetDOG | III | *Apa*LI–S–*Not*I | tetR | (Hoogenboom *et al.*, 1991) |
| 33 | III | | | (Corey *et al.*, 1993) |
| 33 | III | *Sfi*I–S–*Not*I | | (McConnell *et al.*, 1994) |
| 88 | VIII | | | G. Smith, Univ. Missouri |
| 88 | VIII | | | (Haaparanta and Huse, 1995) |
| Phagemid | III | | amp$^R$ | (Bass *et al.*, 1990) |
| pHEN1 | III | *Sfi*I–S–*Not*I | amp$^R$ | (Hoogenboom *et al.*, 1991) |
| pComb3 | III | | | (Gram *et al.*, 1992) |
| pComb8 | VIII | | | (Gram *et al.*, 1992) |
| pCANTAB 5E | III | *Sfi*I–S–*Not*I | amp$^R$ | Pharmacia |
| p8V5 | VIII | *Bst*XI–S–*Bst*XI | ampR | Affymax |
| λSurfZap | III | *Not*I–S–*Spe*I | amp$^R$ | (Hogrefe *et al.*, 1993) |

NOTE: S, stuffer fragment with internal restriction sites in some cases; λSurfZap is commercially available from Stratagene (Cat. No. 240211). pCANTAB 5E is part of a recombinant phage antibody system sold by Pharmacia (Cat. No. 27-9401); the GenBank accession number for pCANTAB 5E is U14321.

Nucleotide and protein coding sequences are shown in Figs. 3–8 for M13 and several vectors listed in the table above. Only the regions covering the site of insertion for display are shown.

## PHAGEMID

The putative positions of the signal peptidase cleavage site for several constructs are noted, although in most cases they have not been confirmed.

There has been one report that protein domains are more accessible when linked to domain 2 of pIII. Human growth hormone was more accessible to two different monoclonal antibodies when attached to a truncated form of pIII (aa 198–406) (Lowman *et al.*, 1991). Most antibody display experiments have utilized this truncated form of pIII for efficient display. However, replacement of domain I (aa 1–198) of pIII with an exogenous sequence leads to noninfectious phage; these par-

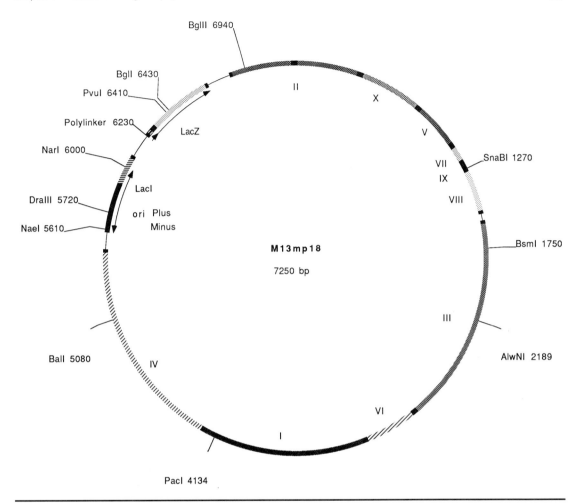

**FIGURE 3**   Map of M13mp18 genome. Unique restriction sites are shown on the circular map of M13mp18 (Accession No. X02513). The 10 viral genes are shown as boxes on the map; they are transcribed in a clockwise fashion. The polylinker within the *lac* Z gene contains the following restriction sites: *Eco*RI, *Sac*I, *Kpn*I, *Sma*I, *Xma*I, *Bam*HI, *Sal*I, *Hinc*II, *Acc*I, *Sph*I, and *Hin*dIII. The origin of replication can be in either orientation to direct the replication and packaging of either the plus or minus strand. The figure was generated with the program MacPlasmap, written by Jingdong Liu (Salt Lake City, UT).

ticles must also carry wild-type pIII, expressed from a second gene III in the vector (i.e., type 33) or a helper phage genome.

In several vectors, the displayed element is separated from the remainder of pIII or pVIII, by short linkers. While there have been no formal experiments on the best linker sequences to use, many vectors use some variation of the sequence GGGGS. There are two glycine-rich regions in pIII (i.e., 68–86 and 218–256), consisting of many GGGS and EGGGS repeats. It is also possible to include a proteolytic cleavage site

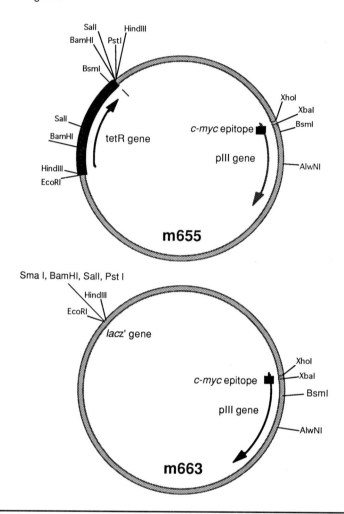

**FIGURE 4**     Map of two M13 type 3 viral vectors. These vectors were derived from M13mp8 according to Fowlkes *et al.* (1992). The vectors carry cloning sites *Xho*I and *Xba*I, in gene III for expression of foreign sequences; these sites flank the c-myc epitope recognized by mAb 9E10. In addition, one vector (m663) carries a segment of the alpha fragment of β-galactosidase; bacterial cells expressing the omega fragment of β-galactosidase convert the clear XGal substrate into an insoluble blue precipitate and the plaques appear blue. The other vector (m655) carries the tetracycline resistance gene of pBR322 at the polylinker site.

between the displayed peptide/protein and the capsid protein. A Factor Xa cleavage site, IEGR, has been engineered in the M13mp18Xa vector (Fig. 8; of Cytogen Corporation, Princeton, N. J.) which permits viral particles bound to any target to be recovered by treatment with Factor Xa. The presence of a string of five glycines downstream of the Factor Xa cleavage site may enhance its cleavage efficiency (Rodriguez and Carrasco, 1994).

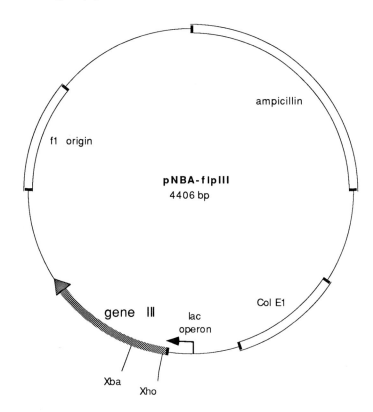

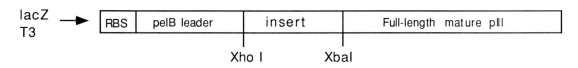

**FIGURE 5**   Map of a phagemid (type 3+3) vector. This vector was constructed by N. B. Adey, by cloning gene III from m663 into the *lac* Z gene of pBluescript. It expresses the full-length form of mature protein III with the pelB leader sequence. The gene is under the control of *lac* O and has a ribosome binding sequence (RBS).

To reduce the level of nonrecombinants in phage display libraries, the reading frames of genes III or VIII have been intentionally destroyed in the stuffer sequence of some vectors. This has been accomplished by frame-shifting (adding or deleting a nucleotide) the gene when it was engineered with restriction sites. Thus, the reading frames are restored in the vector when the stuffer fragments are replaced with DNA encoding peptides or protein domains which have the appropriate reading

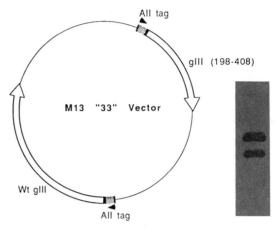

angiotensin II tag

| S | H | S | A | D | N | R | V | Y | I | H | P | F | G | G | D |
|---|---|---|---|---|---|---|---|---|---|---|---|---|---|---|---|

..TCT CAC TCC GCT GAA AAC CGT GTT TAC ATC CAC CCG TTC GGT GGG GAT

aa 198

| R | P | G | G | P | A | L | R | G | R | M | G |
|---|---|---|---|---|---|---|---|---|---|---|---|

CGG CCA GGT GGG CCA GCA TTA AGC GGC CGC ATG TGG..

  *Sfi* I                              *Not* I

---

**FIGURE 6**     Map of a type 33 vector. This vector was constructed by Stephen McConnell and Ronald Hoess (DuPont-Merck, Willmington, DE) and described in a publication (McConnell *et al.*, 1994). It was derived from M13 and carries two copies of gene III [i.e., full-length and truncated (aa 198–408)]. Each copy of pIII is epitope-tagged with a short segment of angiotensin II protein which is recognized by a mouse monoclonal antibody. Inset shows the results of a Western blot performed with this antibody protein; viral particles express both the full-length and the truncated forms of pIII at a ratio of 5:1, respectively.

frames. As pIII and pVIII are required components of the virus particle, this strategy permits positive selection of recombinant versus parental genomes. One caveat of this selection scheme, however, is that reversion of the defective genes occurs frequently, leading to a loss of discrimination between recombinant and parental genomes.

Another means of selecting for recombinant *versus* parental phage is to include a stop codon in the stuffer fragments of genes III or VIII. The stop codon, TAG, can be suppressed efficiently in bacteria containing *sup*E or *sup*F; when these vectors are propagated in such bacterial strains either a glutamine (Q) or tyrosine (Y) is inserted at the TAG codon, respectively. However, when TAG containing genes III or VIII are in bacterial strains that lack either *sup*E or *sup*F, no full-length pIII or pVIII accumulate and no virus particles are generated. This is a very powerful method of selecting for recombinant phage. In our own hands, almost no (i.e., <10$^{-6}$) parental phage exist in a library when ligation mixes (double-cut vector and open reading frame fragments) are introduced into suppressor-less F' bacteria.

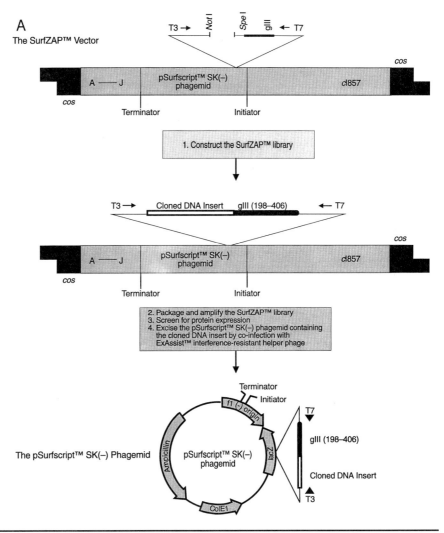

**FIGURE 7**   Maps and sequences of Stratagene's SurfZAP vector. Diagram of how inserts are cloned into λ and excised as phagemid genomes. Maps were kindly provided by Dr. Holly Hofgrefe (Stratagene, La Jolla, CA).

*continues*

One vector that deserves additional comment is λSurfZap. This vector takes advantage of the efficient cloning of DNA segments by *in vitro* packaging. In this system, inserts are cloned into a truncated gene III (aa 198–406) that resides in the excision portion of the λ vector (Fig. 7A, B). Once a library of λ recombinants has been generated, the phagemid genomes are excised by coinfecting cells with M13 helper phage; at the same time the secreted phage particles are displaying a mixture of wild-type and fusion pIIIs. Thus, λSurfZap yields 3+3 viral particles.

**B**

## Left Arm of the SurfZAP™ Vector

The 5′ end of your insert must include a complete *Not* I site (include at least 15 extra nucleotides 5′ of the *Not* I site for help in enzyme recognition). After the *Not* I site, the remainder of the pelB leader sequence must be included.

Vector    5′ AATTAACCCTCACTAAAGGGAACAAAAGCTGGAGCTTGAATTCTTAACTACTCGCCAAGGAGACAGTCATAATGAAATACCTATTGCCTACGGCGGCCGC

       T3 Promoter                 EcoR I               RBS            pelB Leader Sequence     Not I

                                           M   K   Y   L   L   P   T   A   A

### Example Insert

                       Remaining pelB Leader Sequence

5′ CTCGCTCGCCCATATGCGGCCGCAGGTCTCCTCCTCTTAGCAGCACAACCAGCAATGGCCXXXXXX - 3′

                  Not I    A   G   L   L   L   A   A   Q   P   A   M   A

## Right Arm of the SurfZAP™ Vector

The 3′ end of your insert (XXX) should be in-frame and should include the downstream *Spe* I restriction site. Be sure to include four to six additional nucleotides (e.g., GATGCT) on the 3′ end to aid in efficient endonuclease digestion of the insert template.

Vector      Flexible linker      gIII (198–406)

        Spe I    G   G   G   G   S

5′ - ACTAGTGGAGGTGGAGGTAGCCCATTCGTTTGTGAATATCAGGGCCAATCGTCTGACCTGCCTCAACCTCCTGTCAATGCTGGCGGCGGCTCTGGTG

                             198

GTGGTTCTGGTGGCGGCTCTGAGGGTGGCGGTTCTGAGGGTGGCGGCTCTGAGGGAGGCGGTTCCGGTGGCTCTGGTTCCGG

TGATTTTGATTATGAAAAGATGCAAACGCTAATAAGGGGCTATGACCGAAAATGCCGATGAAAACGCGCTACAGTCTGACGCTAAAGGCAAACTTGAT

TCTGTCGCTACTGATTACGGTGCTGCTATCGATGGTTTCATTGGTGACGTTTCCGGCCTTGCTAATGGTAATGGTGCTACTGGTGATTTTGCTGGCTCTA

ATTCCCAAAATGCTCAAGTCGGTGACGGTGATAATTCACCTTTAATGAATAATTTCCGTCAAATAAATTATTTCTATTGTGACAAAATAAACTTATTCCGT

TTTTGTCTTTGGCGCTGGTAAACCATATGAATTTTCTATTGATTGTGACAAAATAAACTAAACTTATTCCGTGGTGTCTTTGCGTTTCTTTTATATGTTGCCACC

TTTATGTATGTATTTTTCACGTTTGCTAACATACTGCGTAATAAGGAGTCTTAATCATGCCAGTTCAAAAGGGTATTCCATTATTCTAGAGTTAAGCGGC

                         406 stop                        3′ untranslated        Xba I

CGTCGAGGGGGGCCCGGTACCCAATTCGCCCTATAGTGAGTCGTATTA

                         T7 Promoter

**FIGURE 7**     *Continued*

## Cloning Sites in M13 genes

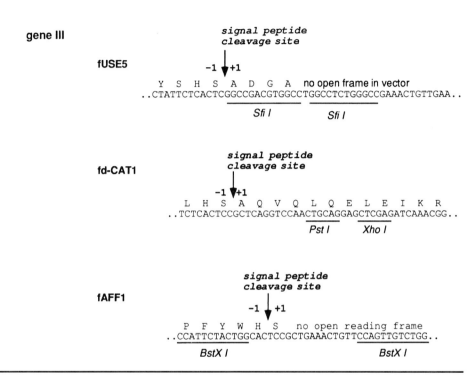

**FIGURE 8**   Nucleotide and coding sequences surrounding cloning sites in genes III or VIII of various vectors. The names of the vectors are listed along with the nucleotide and coding sequences of each vector.

*continues*

While most phage display vectors have been used to express single peptides or protein subunits, it is possible to express two heterodimeric protein subunits. In pComb3, there are two *lac* promoters followed by pelB leader sequences that can drive expression of different protein subunits, one of which is fused to pIII and the other is secreted. In the periplasmic space of cells carrying pComb3, the pIII fusion and the secreted protein have the opportunity to associate, prior to secretion of the virus particle. This vector, and other similar ones, has been used to display assembled immunoglobulin heavy and light chains (Barbas *et al.,* 1991; Kang *et al.,* 1991; Winter and Milstein, 1991).

## MARKERS

In choosing a vector for display, genetic markers can be a consideration. These include antibiotic resistance, epitope expression, *lac*Z expression, and stop codons. Each of these various attributes is described below.

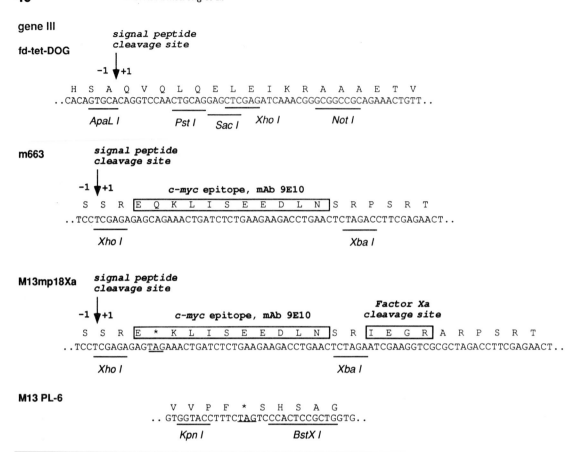

**FIGURE 8**          *Continued*

## Drug Resistance

In a number of vectors, there are drug resistance genes which have been introduced by bacterial transposons. Under tetracycline selection, the fd-tet vector can be grown in bacterial cells as a plasmid, independent of phage production. This permits the propagation of engineered M13 genomes which normally would not yield viable phage (i.e., frameshifted in gene III) as plasmids inside bacteria. The fd-tet vector has served as the starting point for the construction of the fUSE vectors of G. Smith. These vectors not only carry the tetracycline resistance gene, but also have a very low (i.e., ~1) intracellular RF copy number. Tetracycline selection minimizes the loss from a recombinant library of those phage that replicate slowly due to their inserts. The fd-tet vector can be made kanamycin resistant by replacing a *Nsi*I fragment with a *Pst*I fragment from the vector pUC-4K (Pharmacia, Cat. No. 27-4958) which encodes kanamycin resistance (Doorbar and Winter, 1994).

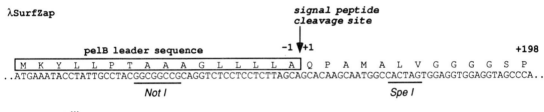

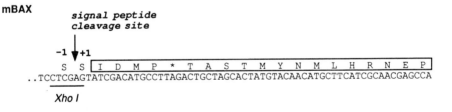

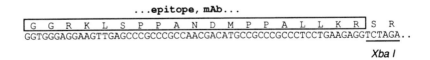

**FIGURE 8**        *Continued*

## Epitope

Several vectors have been engineered to express short peptide "tags" that can be recognized by particular monoclonal antibodies. One such sequence is the *c-myc*

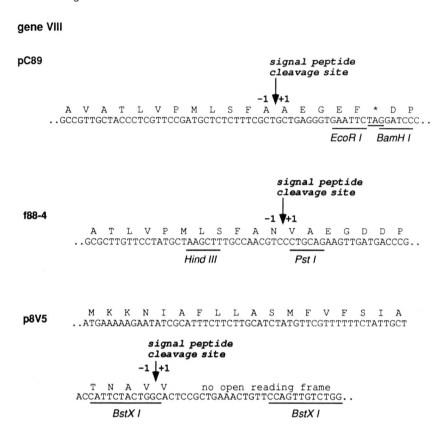

**gene VIII**

**pC89**

*signal peptide cleavage site*

-1 ▼+1

A  V  A  T  L  V  P  M  L  S  F  A  A  E  G  E  F  *  D  P
..GCCGTTGCTACCCTCGTTCCGATGCTCTCTTTCGCTGCTGAGGGTGAATTCTAGGATCCC..

*EcoR I*    *BamH I*

*signal peptide cleavage site*

**f88-4**                    -1 ▼+1

A  T  L  V  P  M  L  S  F  A  N  V  A  E  G  D  D  P
..GCGCTTGTTCCTATGCTAAGCTTTGCCAACGTCCCTGCAGAAGTTGATGACCCG..

*Hind III*              *Pst I*

M  K  K  N  I  A  F  L  L  A  S  M  F  V  F  S  I  A
**p8V5**    ..ATGAAAAAGAATATCGCATTTCTTCTTGCATCTATGTTCGTTTTTTCTATTGCT

*signal peptide cleavage site*

-1 ▼+1

T  N  A  V  V        no open reading frame
ACCATTCTACTGGCACTCCGCTGAAACTGTTCCAGTTGTCTGG..

*BstX I*                        *BstX I*

**FIGURE 8**    *Continued*

epitope, EQKLISEEDLN, which is recognized by mAb 9E10. In some vectors (m663, M13mp18Xa) the *c-myc* epitope is in the stuffer fragment where it is useful in immunologically discriminating between parental and recombinant phage particles, for the purposes of estimating the percentage of recombinants in a library. In the pHEN vector (Hoogenboom *et al.*, 1991), it is upstream of the cloning site in gene III; this arrangement is useful in detecting chimeric pIII molecules on Western blots and verifying that full-length chimeric molecules exist on phage.

## LacZ

The vectors m663 (Fowlkes *et al.*, 1992) and M13mp18Xa (Cytogen) carry *lacZ* genes. When these vectors are in the appropriate bacterial host *(lacIqZΔM15)* that has been exposed to the inducer isopropyl-β-D-thiogalactopyranoside (IPTG), functional β-galactosidase activity is present. Thus, this gene can be detected easily by the occurrence of blue plaques in bacterial lawns containing the indicator bromo-4-chloro-3-indoyl-β-D-galactoside (XGal). Vectors with the *lacZ* gene permit simple discrimination between plaques formed by different phage genomes (i.e., original and evolved phage; see Chapter 14).

## Stop Codons

Suppressed stop codons have also been placed between displayed protein domains and domain II of pIII. When such recombinants are propagated in a suppressor carrying bacterial strain, virus particles are produced incorporating chimeric protein domain pIII into the capsids. After screening phage populations for virus particles with certain properties (based on the particular protein domain displayed), individual isolates can be introduced into bacteria strains that lack suppressors. In these hosts, there is efficient translational termination at the stop codon in the chimeric gene III, and consequently the protein domain is secreted by itself. This has been a very effective means of generating soluble forms of the protein, such as antibodies (Hoogenboom *et al.,* 1991) or growth hormone (Lowman and Wells, 1993), for testing purposes.

## OTHER VIRAL DISPLAY VECTORS

Recently, another viral display system has been described based on bacteriophage λ. The entire β-galactosidase protein has been expressed at the C-terminus of protein V with full enzymatic activity and without loss of phage infectivity (Maruyama *et al.,* 1994). There have also been reports that exogenous segments can be displayed as fusions with proteins J (Waclaw Szybalski, University of Wisconsin, Madison, WI) and D (Sternberg and Hoess, 1995) of λ. Bacteriophage λ may be a useful vector for the display of cDNA segments, and because large recombinant libraries can be generated in this vector with commercially available *in vitro* packaging systems.

Other viruses remain to be exploited, such as RNA phage Qβ (Kozlovska *et al.,* 1993), baculovirus, and retroviruses. There have been two reports of expressing erythropoietin (Kasahara *et al.,* 1994) or RGD peptides (Valsesia-Wittmann *et al.,* 1994) on the surface of retroviruses for the purposes of improving or redirecting viral uptake by animal cells.

## PROTEOLYSIS

An important issue to keep in mind is that the displayed peptide or protein domain is susceptible to proteolysis. While M13 phage are resistant to many proteases (i.e., trypsin, chymotrypsin) (Schwind *et al.,* 1992), the displayed peptides and protein domains are not. Stocks of recombinant phage can lose their displayed element in hours to days when stored at 4°C in culture supernatant. Proteolysis can be minimized by storing phage supernatants with 20% glycerol at -80°C, polyethylene glycol precipitation, or CsCl banding. For short-term experiments monitoring binding properties of particular phage isolates, it is important to work with fresh (i.e., 6–18 hr) culture supernatants whenever possible. It is likely that the addition of protease inhibitors would lengthen the storage life of the displayed protein or peptide of phage kept in the refrigerator. However, in the long run, phage stocks should be maintained as phage in 20% glycerol at -80°C or as DNA in the -20°C freezer.

## VALENCY

The display of peptides or proteins in M13 can have a valency of one to thousands of copies. This is dependent on the display site and type of vector. There are a number of important considerations in choosing display systems. The type of fusion, expression levels, and experiment will dictate which display system is appropriate.

| Multivalent display | Monovalent display |
|---|---|
| pIII or pVIII phage | Type 3+3 and 8+8 phagemids, type 33 and 88 phage |
| Displayed element might affect phage infectivity | The negative effect of the displayed element is minimized |
| Isolation of modest affinity binders | Discrimination between modest and high-affinity binders |
| Virus particles display the sequence in 3-5 (i.e., pIII fusion) 2400 copies (i.e., pVIII fusion) | Many particles are displaying 0–1 copies (i.e., pIII fusions) or 0–2400 copies (i.e., pVIII) of the fusion protein |

The phagemid and 33 vector systems have proven to be useful in discriminating between peptides or proteins of moderate versus high binding affinity. McConnell *et al.* (1994) have demonstrated that a collection of phage, displaying peptides that bind to a monoclonal antibody, can bind equivalently, even though corresponding synthetic peptides differ as much as 20-fold in relative affinity by radioimmunoassay. However, when the same peptide is displayed by a 33 vector, the relative binding of the phage correlates well with the synthetic peptide results. When McConnell measured the valency of the fusions displayed on the "33" viral particles, there was an average of 0.74 chimeric proteins per phage particle. These experiments suggest that monovalent phage display can be used to discriminate between modest and high affinity binding peptides.

In summary, monovalent display can be a useful tool for the display of peptides and proteins that may not be well tolerated by M13 in 5 or 2800 copies. Monovalent display also allows the discrimination of displayed molecules that bind targets with moderate versus high affinity. The use of polyvalent display vectors should not be precluded, however, since in the initial stages of selection of binding peptides it is often more important to accumulate a broad spectrum of peptides as potential leads than to identify a single high-affinity lead. The peptides identified can then be ordered in terms of their affinities using monovalent display or other methods.

## ISSUE OF BIOLOGICAL SELECTION

A number of investigators have characterized their random peptide libraries by observing the frequency of amino acids at each random position and the frequency of all possible dipeptide frequencies. It appears that most, if not all, possible dipeptide combinations can be found in phage display libraries (Cwirla *et al.,* 1990; DeGraaf *et al.,* 1993).

During our analysis of random peptides expressed by M13 within pIII, we have observed that there is a strong negative selection against odd numbers of cysteine residues (Kay *et al.,* 1993). A similar conclusion has been drawn elsewhere (Matthews and Wells, 1993). We hypothesize that phage that express unpaired or lone cysteines may be detrimental to phage propagation because the cysteines have the potential to form disulfide cross-bridges between cysteine residues on different peptides, proteins, and phage. Whenever possible, scientists should display proteins and protein domains that have an even number of cysteines. Recently, the eight cysteines of pIII have been demonstrated to be all in disulfide bridged; the cystine sites are Cys7–Cys36, Cys46–Cys53, Cys188–Cys201, and Cys354–Cys371 (Kremser and Rasched, 1994). Protein VIII on the other hand has no cysteines. At the moment, there are no data on the effect of unpaired cysteines in sequences displayed by pVIII.

When phage display random peptides with even numbers of cysteines, the distance between the cysteines tends to be short. In a population of 228 phage displaying random peptides 38 residues in length there were 75 peptides with two cysteines, with the distance between cysteines of 0 to 21 residues. For the majority of inserts, however, the number of intervening residues was 4–6. This arrangement is nonrandom, because if the pairs of cysteines were positioned at random in a peptide of 38 residues, the mean interval would be 12 residues. The $\chi^2$ value for the null hypothesis is 71.3, which is highly significant ($p<0.0001$; C. E. Smith, personal communication). The short interval among the phage-displayed peptides is most likely due to an increasing negative selection of phage for intervening distances greater or less than 5 residues; this may be due to inefficient disulfide bond formation.

We should point out, however, that we have isolated some phage that display peptides that carry single cysteine residues. These phage are stable and grow quite well; curiously, the unpaired cysteines appear to be tolerated in these peptides. Presumably, these lone cysteines are buried in an inaccessible region of random peptides which fold up on themselves.

In the course of our experiments, we have also observed biological selection against certain recombinants that is not based on cysteine frequency. Andrew Sparks, in one of our laboratories (B.K.K.), has attempted to clone oligonucleotides encoding the peptide sequence RRLIEDAEYAARG into gene III of M13. The recombinants formed *tiny* plaques (only visible in the presence of XGal and IPTG because the vector carrying *lac*Z formed blue plaques). When the phage in the small plaques were propagated in bacterial cultures for 6 hr, they yielded both large and small plaques. When both types of plaques were repropagated, all the large plaque-forming phage were stable and had deletions of the insert, whereas the tiny plaques continuously generated large plaques. The DNA in small plaques was verified to carry the desired insert by PCR; however, we were never able to isolate an adequate amount of DNA to confirm this by sequencing. Thus, certain peptide (and protein) sequences can be problematic in phage display experiments.

Whenever one finds it difficult to generate recombinant phage that carry the insert of interest, there are a number of possible problems. Some of these defects will be a reflection of the sequence character of the displayed protein or peptide. Charged peptides may not pass through the inner bacterial membrane efficiently, for example, during assembly of pIII into viral particles. However, this may be compensated for through the use of bacterial hosts that carry mutations in the secretory system (Peters *et al.,* 1994) or are deficient in certain proteases (McLafferty *et al.,* 1993). Several different hosts can be tested to identify the ideal genetic background which permits proper expression of the displayed element of interest. Another avenue to overcome cloning limitations observed during display of large foreign peptides or protein domains is to use phagemid, 33, or 88 vectors. There are many options available to investigators.

## References

Barbas, C., Kang, A., Lerner, R., and Benkovic, S. (1991). Assembly of combinatorial antibody libraries on phage surfaces: The gene III site. *Proc. Natl. Acad. Sci. U.S.A.* **88,** 7978–7982.

Bass, S. H., Greene, R., and Wells, J. A. (1990). Hormone phage: An enrichment method for variant proteins with altered binding properties. *Proteins* **8,** 309–314.

Corey, D. R., Shiau, A. K., Yang, Q., Janowski, B. A., and Craik, C. S. (1993). Trypsin display on the surface of bacteriophage. *Gene* **128,** 129–134.

Cwirla, S. E., Peters, E. A., Barrett, R. W., and Dower, W. J. (1990). Peptides of phage: A vast library of peptides for identifying ligands. *Proc. Natl. Acad. Sci. U.S.A.* **87,** 6378–6382.

DeGraaf, M. E., Miceli, R. M., Mott, J. E., and Fischer, H. D. (1993). Biochemical diversity in a phage display library of random decapeptides. *Gene* **128,** 13–17.

Doorbar, J., and Winter, G. (1994). Isolation of a peptide antagonist to the thrombin receptor using phage display. *J. Mol. Biol.* **244,** 361–369.

Fowlkes, D., Adams, M., Fowler, V., and Kay, B. (1992). Multipurpose vectors for peptide expression on the M13 viral surface. *BioTechniques* **13,** 422–427.

Gram, H., Marconi, L. A., Barbas, C. F., Collet, T. A., Lerner, R. A., and Kang, A. S. (1992). *In vitro* selection and affinity maturation of antibodies from a naive combinatorial immunoglobulin library. *Proc. Natl. Acad. Sci. U.S.A.* **89,** 3576–3580.

Greenwood, J., Willis, A., and Perham, R. (1991). Multiple display of foreign peptides on a filamentous bacteriophage: Peptides from *Plasmodium falciparum* circumsporozoite protein as antigens. *J. Mol. Biol.* **220,** 821–827.

Haaparanta, T., and Huse, W. (1995). A combinatorial method for constructing libraries of long peptides displayed by filamentous phage. *Mol. Div.* **1,** 39–52.

Hogrefe, H. H., Amberg, J. R., Hay, B. N., Sorge, J. A., and Shopes, B. (1993). Cloning in a bacteriophage lambda vector for the display of binding proteins on filamentous phage. *Gene* **137,** 85–91.

Hoogenboom, H., Griffiths, A., Johnson, K., Chisswell, D., Hudson, P., and Winter, G. (1991). Multisubunit proteins on the surfaces of filamentous phage: Methodologies for displaying antibody (Fab) heavy and light chains. *Nucleic Acids Res.* **19,** 4133–4137.

Il'ichev, A. A., Minenkova, O. O., Tat'kov, S. I., Karpyshev, N. N., Eroshkin, A. M., Petrenko, V. A., and Sandakhchiev, L. S. (1989). Production of a viable variant of the M13 phage with a foreign peptide inserted into the basic coat protein. *Dokl. Akad. Nauk. SSSR* **307,** 481–483.

Kang, A. S., Barbas, C. F., Janda, K. D., Benkovic, S. J., and Lerner, R. A. (1991). Linkage of recognition and replication functions by assembling combinatorial antibody Fab libraries along phage surfaces. *Proc. Natl. Acad. Sci. U.S.A.* **88,** 4363–4366.

Kasahara, N., Dozy, A. M., and Kan, Y. W. (1994). Tissue-specific targeting of retroviral vectors through ligand-receptor interactions. *Science* **266,** 1373–1376.

Kay, B. K., Adey, N. B., He, Y.-S., Manfredi, J. P., Mataragnon, A. H., and Fowlkes, D. M. (1993). An M13 library displaying 38-amino-acid peptides as a source of novel sequences with affinity to se-lected targets. *Gene* **128,** 59–65.

Kishchenko, G., Batliwala, H., and Makowski, L. (1994). Structure of a foreign peptide displayed on the surface of bacteriophage M13. *J. Mol. Biol.* **241,** 208–213.

Kozlovska, T. M., Cielens, I., Dreilinna, D., Dislers, A., Baumanis, V., Ose, V., and Pumpens, P. (1993). Recombinant RNA phage Qb capsid particles synthesized and self-assembled in *Escherichia coli.* *Gene* **137,** 133–137.

Kremser, A., and Rasched, I. (1994). The adsorption of filamentous phage fd: Assignment of its disulfide bridges and identification of the domain incorporated in the coat. *Biochemistry* **33,** 13954–13958.

Lowman, H. B., and Wells, J. A. (1993). Affinity maturation of human growth hormone by monovalent phage display. *J. Mol. Biol.* **234,** 564–578.

Lowman, H. B., Bass, S. H., Simpson, N., and Wells, J. A. (1991). Selecting high-affinity binding pro-teins by monovalent phage display. *Biochemistry* **30,** 10832–10838.

Maruyama, I. N., Maruyama, H., and Brenner, S. (1994). λfoo: A λ phage vector for the expression of foreign proteins. *Proc. Natl. Acad. Sci. U.S.A.* **91,** 8273–8277.

Matthews, D. J., and Wells, J. A. (1993). Substrate phage: Selection of protease substrates by monovalent phage display. *Science* **260,** 1113–1117.

McCafferty, J., Griffiths, A. D., Winter, G., and Chiswell, D. J. (1990). Phage antibodies: Filamentous phage displaying antibody variable domains. *Nature (London)* **348,** 552–554.

McConnell, S. J., Kendell, M. L., Reilly, T. M., and Hoess, R. H. (1994). Constrained libraries as a tool for finding mimotopes. *Gene* **151,** 115–118.

McLafferty, M. A., Kent, R. B., Ladner, R. C., and Markland, W. (1993). M13 bacteriophage displaying disulfide-constrained microproteins. *Gene* **128,** 29–36.

Peters, E. A., Schatz, P. J., Johnson, S. S., and Dower, W. J. (1994). Membrane insertion defects caused by positive charges in the early mature region of protein pIII of filamentous phage fd can be cor-rected by prlA suppressors. *J. Bacteriol* **176,** 4296–4305.

Rodriguez, P. I., and Carrasco, L. (1994). Improved factor Xa cleavage of fusion proteins containing maltose binding proteins. *BioTechniques* **18,** 238.

Schwind, P., Kramer, H., Kremser, A., Ramsberger, U., and Rasched, I. (1992). Subtilisin removes the surface layer of the phage fd coat. *Eur. J. Biochem.* **210,** 431–436.

Scott, J. K., and Smith, G. P. (1990). Searching for peptide ligands with an epitope library. *Science* **249,** 386–390.

Smith, G. P. (1993). Surface display and peptide libraries. *Gene* **128,** 1–2.

Sternberg, N., and Hoess, R. (1995). Display of peptides and proteins on the surface of bacteriophage lambda. *Proc. Natl. Acad. Sci. U.S.A.* **92,** 1609–1613.

Valsesia-Wittmann, S., Drynda, A., Deleage, G., Aumailley, M., Heard, J. M., Danos, O., Verdier, G., and Cosset, F. L. (1994). Modifications in the binding domain of avian retrovirus envelope protein to redirect the host range of retroviral vectors. *J. Virol.* **68,** 4609–4619.

Winter, G., and Milstein, C. (1991). Man-made antibodies. *Nature (London)* **349,** 293–299.

# 4

# Microbiological Methods

## James E. Rider, Andrew B. Sparks, Nils B. Adey, and Brian K. Kay

Most aspects of phage display require the use of basic microbiological methods. This section covers the basics in handling bacteria and phage.

---

**PROTOCOL 1**   **Spreading techniques**

### Procedure A: For bacteria

1. Heat-sterilize a platinum inoculating loop and cool by waving the loop in the air or plunging it into sterile culture medium. Transfer a bacterial colony from an agar petri plate or from a frozen or stab culture to 1 ml of 2xYT bacteriological medium. Vortex or tap the tube vigorously to disperse the transferred bacterial cells.
2. Dilute the cells into another culture tube containing 2x YT bacteriological medium. Typically 10 µl of liquid is transferred with a sterile yellow pipet tip to a tube with 1 ml of culture medium.
3. Repeat step 2 serially two or three times.
4. Spot a small aliquot (10–50 µl) of each dilution in the center of a 2x YT petri plate containing agar along with ~200 µl of 2x YT liquid culture medium on its surface. Spread the cells over the entire surface of the medium by moving a sterile, bent glass rod back and forth gently over the agar surface and, at the same

time, rotating the plate by hand or using a rotating wheel. The glass spreader can be sterilized by dipping it into a beaker containing 95% ethanol and then holding it in the flame of a Bunsen burner to ignite the ethanol. The spreader should then be cooled, first in air and then by touching the surface of a plate of sterile agar medium.

**CAUTION:** Do not place the beaker of ethanol near the lighted Bunsen burner. Do not return the hot glass rod directly to the beaker of ethanol.

5. Incubate the petri plates upside-down at 37°C overnight to permit formation of colonies (colony forming units, CFU).

### Procedure B: For phagemids

1. Place 200 µl of F′ bacterial cells and diluted virus into a sterile test tube.
2. Incubate at 37°C in growth broth for 15–20 min to permit viral infection and Amp r gene expression.
3. Transfer the contents of the tube into the center of a petri plate containing hardened agar medium and antibiotic (i.e., 100 µg/µl ampicillin, 10 µg/µl tetracycline). Spread the cells over the entire surface of the medium by moving a sterile, bent glass rod back and forth gently over the agar surface and, at the same time, rotating the plate by hand or using a rotating wheel. The glass spreader can be sterilized by dipping it into a beaker containing 95% ethanol and then holding it in the flame of a Bunsen burner to ignite the ethanol. The spreader should then be cooled, first in air and then by touching the surface of a plate of sterile agar medium.

**CAUTION:** Do not place the beaker of ethanol near the lighted Bunsen burner. Do not return the hot glass rod directly to the beaker of ethanol.

4. Incubate the petri plates upside-down at 37°C overnight to permit formation of colonies.

### Procedure C: For phagemids

1. Distribute 10–20 glass beads into several Eppendorf tubes. Sterilize by autoclaving. The beads are 3 mm in diameter and can be purchased from Fisher Scientific (Catalog No. 11-312A).
2. To a sterile tube, add 200 µl of F′ cells and an aliquot of phagemid viral particles. Incubate at 37°C for 5–10 min.
3. To the middle of a prewarmed 2x YT petri plate containing antibiotics (i.e., 100 µg/µl ampicillin, 10 µg/µl tetracycline) add the contents of the tube prepared in step 2 along with 10–20 sterile glass beads.
4. Manually move the plate in a circular motion to distribute the liquid. The beads help to disperse the liquid evenly over the surface of the plate without the need of a sterile, bent glass rod or a rotating wheel.
5. Invert the plates after spreading the liquid.
6. Invert the plate and incubate upside-down overnight at 37°C. The beads will fall down and not interfere with formation of the bacterial colonies. When handling the plates, decant the beads out into an appropriate waste container.

## PROTOCOL 2    Titering bacteriophage M13

It is often necessary to quantitate the number of phage particles in experiments. A convenient method of accomplishing this is to infect *Escherichia coli* bacterial cells and then count the number of plaques in the bacterial lawn on a petri plate the following day. This is a very sensitive method of evaluating the number of M13 phage particles as the efficiency of plaque formation is >50% of the number of actual phage particles. Titering PFU or CFU is an effective means of evaluating the input and output of phage particles in affinity selection experiments.

In working with M13 phage, one can quantitate the number of phage particles in several manners. Typically wild-type M13 generates cultures that contain >$10^{12}$ particles per milliliter, whereas the phage display vectors yield $10^{10}$ to $10^{12}$ particles per milliliter. Methods for estimating particle number are described below. Another method which may be of interest to readers is based on infected bacterial cell growth curves (Zhong and Smith, 1994).

In some cases, *lacZ* carrying viral or phagemid vectors can be followed by monitoring the expression of ß-galactosidase activity. To each petri plate, 20 µl IPTG (20 mg/ml) and 20 µl XGal (20 mg/ml) are either added to the top agar for plaque formation or spread evenly on the surface of a petri plate (along with 200 µl 2x YT) for colony formation.

### Procedure A: Phage titering by plating

1. Prewarm petri plates in a 37°C incubator. Remove any excess moisture by flicking the plate, wiping the top with a clean Kimwipe, or leaving the plate open in a laminar flow hood for a few minutes. (To minimize the amount of moisture in the plates, they should be at least several days old.) The number of plates should be equal to the number of dilutions that will be tested, although a few extra plates can be added to the incubator in order to anticipate any *ad hoc* increases in the size of the experiment.

2. Transfer the bottle of molten top agar to a 55°C bath. Double-check that everything is dissolved—swirl the bottle and examine by eye. If not, microwave the bottle and cool to 55°C; otherwise, chunks of agar in the lawn will prevent formation of visible plaques.

3. Distribute 200 µl of F' *E. coli* bacterial cells (i.e., JM105, XL1-Blue, DH5αF', JS5; overnight culture or grown to $OD_{550nm}$ = 0.6) to 6-ml sterile plastic-capped tubes (Falcon 2059; 12 x 75 mm). Anticipate the number of dilutions that will be tested; this number should equal the number of plates in the incubator.

4. Set up dilutions in sterile PBS or TBS. Estimate the concentration of phage in the sample (i.e., $10^{10}$ to $10^{14}$) and make the appropriate dilutions. Typically one takes 10 µl and dilutes it into 1 ml of PBS ($10^2$x dilution) in an Eppendorf tube. This is repeated several times in serial fashion.

5. Take 100 µl from selected dilutions and add to sterile plastic 6-ml tubes that were prepared in step 3.

6. Add 25 µl of XGal and 25 µl of IPTG to the tubes if you wish to monitor the *lacZ* gene carried by the introduced vector.

7.  Add 3 ml of top agar, prewarmed to 55°C, to each tube. Mix gently with a vortex or by tapping with a finger.

8.  Pour the contents of the tube over the top of the 2x YT Petri plates quickly. Tilt the plate back and forth to cover the entire surface without leaving any blank spaces or bubbles.

9.  Leave the plate on the lab bench for a minimum of 5 min to allow the top agar to harden.

10. Carefully pick up one plate and determine that the top agar has hardened. If so, stack the plates in an inverted fashion in a 37°C incubator. Leave overnight in the incubator. Plaques will be small, but apparent, within 6 hr of incubation. If, by 6 hr, it is evident that the dilutions were too great or insufficient to give an adequate number of plaques on the plates, repeat (and adjust) the dilutions accordingly.

11. The number of plaques can be counted manually (Fig. 1). It is often useful to mark the plaques as they are counted through the bottom of the Petri plate with a felt pen. One handy device [Monostat colony counter; 81-520-000] is both a counter and a pen.

12. Determine the titer of the starting sample by multiplying the number of plaques by the extent of dilution and the volume plated. If XGal and IPTG are used, count the number of individual blue and white plaques.

**Procedure B: Phage titering by pronging**

1.  Using an 8-channel pipetter, add 180 µl PBS to each well in the microtiter plate. (One can also use a 12-channel pipettor, except that the direction of the dilutions would be orthogonal.)

2.  Using an 8-channel pipetter and a non-ELISA microtiter plate, perform serial dilutions as shown in Fig. 2.

3.  Sterilize the replica-pronger by dipping it 95% ethanol and flaming it with a Bunsen burner. Be careful not to drip the ethanol onto your hand or the jar of ethanol. Tilt the pronger while it is ignited.

4.  Stamp the pronger onto the inside lid of a petri plate to cool off the metal prongs.

5.  Carefully place the pronger into one side of the dish. Each prong should be partially submerged in the wells.

6.  Pick up the pronger and transfer it to the surface of a bacterial lawn in a petri plate. [The petri plates can be prepared in advance. Add 200 µl of bacteria (F′) to 3 ml of 2x YT agar and pour over the surface of a 2x YT plate. Allow the liquid to harden. Use immediately or store sealed in the refrigerator until needed. We have successfully used plates that were over a month old.]

7.  Allow the transferred droplets to dry. The lid of the petri plate can be placed askew, so that the plate is partially exposed to air. This takes 5–15 min.

8.  Incubate the petri plate at 37°C overnight.

9.  Count the number of phage in the array formed in the bacterial lawn. Adjust the number of plaques by the dilution to calculate the final titer.

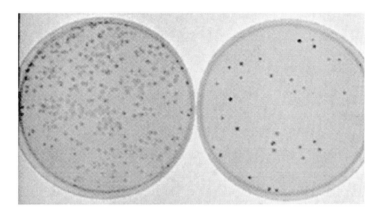

**FIGURE 1**   Phage supernatant was serially diluted to produce a countable plaque density; 100 μl of two different dilutions was added to sterile 6-ml plastic capped tubes containing 200 μl DH5αF′ *E. coli* overnight. Twenty-five microliters IPTG and XGal were added to the tubes, along with 3 ml of melted 2x YT top agar. The tubes were quickly vortexed and poured onto a 2x YT agar plate. Plates were cooled for 5 min at room temperature, inverted, and incubated at 37°C overnight. Plaques were counted in order to determine phage titers.

### Procedure C: Phage titering by ELISA

1. Add 1 μl rabbit anti-M13 antibody (1 mg/ml) + 100 μl sodium bicarbonate (0.1 M NaHCO$_3$, pH 8.5) per well of a 96-well ELISA plate. To coat an entire microtiter dish, add 20 μl of antibody to 10 ml of sodium bicarbonate (pH 8.5) and distribute 100 μl per well. (The nonconjugated rabbit anti-M13 antibody can be purchased from Sigma Chemical Co., Catalog No. B 7786).
2. Incubate plate at 37°C for 1 hr. The plates can be prepared in advance and stored at 4°.
3. Add 380 μl Block solution (0.1 M NaHCO$_3$, pH 8.5, 1% BSA) per well to block all nonspecific binding sites in the wells.
4. Incubate plate at 37°C for ~30 min.
5. Wash plate with standard wash buffer (PBS–0.1% Tween 20) several times. Flick the contents of the wells into the sink and remove all residual liquid by smacking the plate onto a stack of paper towels after each wash. Be careful not to spray phage around the laboratory, thereby contaminating it.
6. Add 100 μl wash buffer to each well and then add 1 μl culture supernatant or phage containing solution. Incubate at room temperature for ~30 min.
7. Discard the solution into the sink and wash the plate five times with wash buffer. Flick the contents of the wells into the sink and remove all residual liquid by smacking the plate onto a stack of paper towels after each wash.
8. Add 100 μl anti-M13-HRP conjugate solution (Pharmacia, Catalog No. 27-9402-01; diluted 1:5000 in wash solution) to each well and incubate at room temperature for 1 hr. (Note: do not have any sodium azide present as it inactivates HRP by chelating the enzyme's zinc cofactor.)

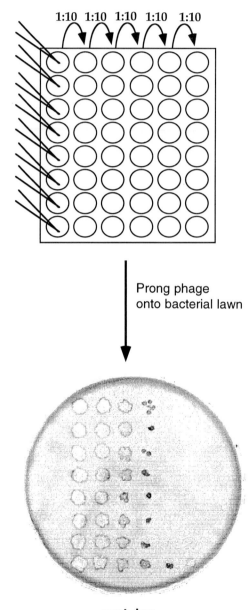

1:10  1:10  1:10  1:10  1:10

Prong phage
onto bacterial lawn

**next day**

**FIGURE 2**      Phage supernatants were serially diluted 1:10 in 1X PBS five times in a non-ELISA microtiter plate. A replica-pronger was sterilized by flaming with 95% ethanol. The pronger was allowed to cool and then dipped in the wells of the microtiter plate. The pronger was then gently applied to a pre-prepared DH5F′ bacterial lawn containing IPTG and XGal. After allowing the transfer droplet (~2 µl) to dry, the plate was inverted and incubated overnight at 37°C. (To see photo of pronging method, see Ch. 13, Fig. 3.)

9. Wash wells with standard wash five times. Flick the contents of the wells into the sink and remove all residual liquid by smacking the plate onto a stack of paper towels after each wash.

10. Add 100 µl of substrate solution [2′, 2′-azino-bis(3-ethylbenzthiazoline-6-sulfonic acid) (ABTS), 0.05% hydrogen peroxide, 50 mM sodium citrate, pH 5.0]. Do not spill $H_2O_2$ onto your skin as it is a caustic.

11. Incubate the microtiter plate for approximately 15 min at room temperature, add stop solution (0.2 N $H_2SO_4$), and then read on a microtiter plate reader at 405 nm.

**NOTES:** To establish a standard curve, include dilutions of a control phage of a known titer. This will allow a rough estimation of how many PFUs produce a given ELISA signal (Fig. 3).

### Procedure D: Colony-forming units

For many phage-display experiments, phagemid vectors are used in place of phage. One important difference between the two vector systems is that quantitation of the number of phage particles is done by monitoring their ability to form colonies in the presence of an antibiotic. The phagemids generally encode an ampicillin-resistance gene; thus, bacterial cells which take up the phagemid genome

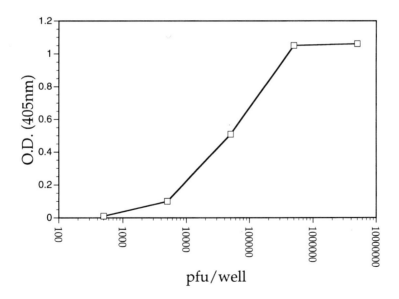

**FIGURE 3**     Optical density curve of different concentrations of a phage culture of known titer detected by phage ELISA. Once a standard OD curve is established with a phage culture of known titer, additional phage cultures can be compared to establish a rough estimation of the titer. See Protocol 2, procedure C.

become ampicillin resistant. This same technique can be used with phage vectors that carry drug resistance markers (*e.g.,* fUSE5, fAFF1, fd-tet) as bacteria infected with M13 phage can form drug-resistant colonies.

1. Aliquot 200 µl of an overnight culture of F′ bacteria into 6-ml sterile plastic-capped tubes (Falcon 2063; 12 x 75 mm).
2. Dilute the phage in PBS. Estimate the concentration of phage in the sample (i.e., $10^{10}$ to $10^{14}$) and make the appropriate dilutions. Typically one takes 10 µl and dilutes it into 1 ml of PBS ($10^2$x dilution) in an Eppendorf tube. This is repeated several times in serial fashion.
3. Add 100 µl of each dilution, 100 µl of bacterial cells (F′, overnight culture) and 800 µl of antibiotic-free 2x YT broth. Incubate for 15–20 min at 37°C.
4. Pipet 100 µl of the 1-ml infection onto the surface of petri plates containing antibiotics.
5. Spread the 100 µl with a glass rod or with sterile glass beads (above).
6. After the liquid has dried, invert the plates, and transfer into a 37°C incubator.
7. The next day, count the number of colonies and deduce CFU concentrations.
8. Note that your calculations will bring you back to PFU or CFU/ml.

---

## PROTOCOL 3    Preparing phage particles

### Procedure A: PEG precipitation

1. Spin down the bacterial culture containing the phage particles. To remove cells, fill a 1.5-ml Eppendorf tube with liquid from the bacterial culture. Spin at >6000 rpm in a Eppendorf microfuge for 5 min.
2. Carefully transfer 1.2 ml of the supernatant to a clean tube.
3. Add 300 µl 30% PEG–1.5 *M* NaCl to the tube. Seal the tube and invert several times to mix everything. Use 8000 MW polyethylene glycol.
4. Chill the tube for 1 hr on ice or overnight at 4°C.
5. Centrifuge the tube at 15,000 rpm in an Eppendorf microfuge for 15 min to pellet the precipitated phage particles.
6. Carefully pour off and discard the supernatant. Invert the tube over some paper towels to drain off the residual liquid. Any residual liquid can be collected by spinning the tube in a microfuge for a few seconds and removing with a glass capillary tube.
7. Resuspend the pellet, often dispersed along the sides of the tube as well as the bottom, in 100 µl 1x TBS. Transfer the phage suspension to a clean tube. If desired, the phage can be reprecipitated a second time.
8. To pasteurize the sample, heat for 5 min at 65°C. The sample can be stored indefinitely at 4°C; however, proteolysis may occur along the surface. Alternatively, add glycerol to 15% and store the phage at -70°C.

This protocol can be scaled up to process larger (i.e., 10–1000 ml) of culture supernatant. Note that various extracellular products coprecipitate with the phage.

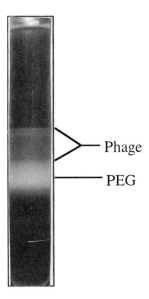

**FIGURE 4**        Photograph of a CsCl equilibrium density gradient of M13 bacteriophage particles. Courtesy of Dr. Steve McConnell (Cytogen Corporation). See protocol in Chapter 8.

### Procedure B: CsCl gradient centrifugation

In some cases, it is advantageous to purify phage from the polyethylene glycol using CsCl equilibrium gradients. Trace amounts of PEG have been thought to interfere with the binding of phage to various targets. Phage can be purified from PEG precipitates by CsCl gradient centrifugation. Readers are directed to the work of Pratap et al. for a detailed protocol for this technique (Smith and Scott, 1993). A photograph of a CsCl gradient is shown in Fig. 4 (courtesy of Dr. Steve McConnell, Cytogen Corporation, Princeton, NJ).

**PROTOCOL 4        Preparation of single-stranded DNA from viral particles**

1. Prepare a solution of phage particles (Protocol 3, Procedures A or B).
2. To 100 μl of phage resuspended in 1x TBS in a 1.5-ml Eppendorf tube, add 100 μl of a phenol:chloroform mixture (1:1). Vortex the sample vigorously to generate an emulsion. Wear gloves and eye and skin protection when working with phenol, chloroform, or phenol:chloroform.
3. Centrifuge the tube at 15,000 rpm for 5 min in an Eppendorf centrifuge to separate the phases.

4. Carefully transfer the upper, aqueous phase to a clean Eppendorf tube. Leave behind any white material at the interface of the lower organic layer.

5. To the transferred liquid, add 10 µl of 3 *M* sodium acetate (pH 5.0) and 200 µl of 95 or 100% ice-cold ethanol. Cap and invert the tube several times.

6. Chill the tube to precipitate the DNA (i.e., on dry ice for 30 min, overnight at -20°C).

7. Collect the precipitate by spinning the tube 15,000 rpm for 15 min in an Eppendorf centrifuge.

8. Carefully decant the supernatant away. A small opaque pellet may be visible at the bottom of the tube.

9. Add ~300 µl of ice-cold 70% ethanol to the tube to help wash away residual salt. Spin the tube at 15,000 rpm for 5 min in an Eppendorf centrifuge.

10. Carefully decant the supernatant away. Invert the tube to drain away all the ethanol.

11. Once the pellet is dry, add 25 µl 1x TE to dissolve the DNA.

12. Analyze 2 µl of the DNA solution by electrophoresis on a 1% agarose gel and store the remainder at -20°C. The mobility of the single-stranded, circular, viral DNA in a gel is comparable to a 2.5-kb linear DNA molecule.

---

## PROTOCOL 5    Preparation of vector RF DNA

### Procedure

1. Grow a 2-ml, overnight culture of a strain of *E. coli* carrying an F factor (i.e., DH5αF') at 37°C overnight in 2x YT. If you are propagating a form of filamentous bacteriophage that carries a drug resistance marker (i.e., fd-tet, M13amp$^R$), then the appropriate antibiotic should be included in all culture media.

2. Plate M13 phage with bacterial cells to obtain single plaques. Use a sterile toothpick or inoculating needle to transfer phage from a single plaque into a sterile test tube containing 1 ml of 2x YT with 10 µl of cells from an overnight culture. Grow the infected culture overnight at 37°C with vigorous aeration.

3. Centrifuge a small aliquot of the culture to pellet the cells. Transfer 10 µl of the clarified supernatant to a sterile tube containing 1 ml of an overnight culture. After incubating the tube for 10 min at room temperature, divide it in half and inoculate two flasks (≥2-liter size), each containing 500 ml 2x YT. Incubate the culture at 37°C (220 rpm) until the cells reach stationary phase (overnight) to ensure a high RF (replicating form) DNA copy number.

4. Pellet cells by centrifugation at 4000g (~6000 rpm using a GSA rotor in Sorvall centrifuge) for 10 min at 4°C.

5. Prepare RF DNA from the cell pellet using the Qiagen Maxi Prep Kit (Catalog No. 12162), Promega Wizard kit (Catalog No. A7270), CsCl banding (ref), or any other convenient method (Ausubel *et al.*, 1987).

6. Extract the resultant DNA twice with a 1:1 mixture of phenol:chloroform and once with chloroform alone.

7. Precipitate the DNA by addition of 0.1 vol 3 *M* sodium acetate (pH 5.2) plus 2 vol of cold 100% ethanol. Recover the DNA by centrifugation at 14,000 rpm in a microfuge for 30 min at 4°C. Decant the supernatant and wash the DNA pellet with 70% ethanol. Recentrifuge and remove all traces of the ethanol. Air dry the pellet and dissolve in 500 μl TE.

8. Determine the DNA concentration by measuring the optical absorbance of the solution at 260 nm (i.e., 1 $OD_{260nm}$ unit = 50 μg DNA/ml). Note that the yields of the fd-tet vector are much lower than most type 3 vectors.

**NOTE:** One problem preparing DNA from phage vectors that lack selectable markers is potential contamination by wild-type M13 phage. Wild-type phage grow significantly better than most vectors. Therefore, it is advisable to use disposable, sterile cultureware whenever possible. Avoid the use of pipetmen when growing M13 cultures as they can be a major source of wild-type contamination; phage contamination can be limited with the use of aerosol resistant pipet tips. Alternatively, use a drug-resistant form of M13, such as the tetracycline resistance-carrying vectors fUSE5, (Scott and Smith, 1990), fd-CAT1 (McCafferty *et al.*, 1990), fdtetDOG (Hoogenboom *et al.*, 1991), or the ampicillin resistant vector, M13amp$^R$ (Devlin *et al.*, 1990).

---

### References

Ausubel, F. M., Brent, R., Kingston, R. E., Moore, D. D., Seidman, J. G., Smith, J. A., and Struhl, K., Eds. (1987). *"Current Protocols in Molecular Biology."* Wiley, New York.

Devlin, J. J., Panganiban, L. C., and Devlin, P. E. (1990). Random peptide libraries: A source of specific protein binding molecules. *Science* **249**, 404–406.

Hoogenboom, H., Griffiths, A., Johnson, K., Chiswell, D., Hudson, P., and Winter, G. (1991). Multisubunit proteins on the surfaces of filamentous phage: Methodologies for displaying antibody (Fab) heavy and light chains. *Nucleic Acids Res.* **19**, 4133–4137.

McCafferty, J., Griffiths, A. D., Winter, G., and Chiswell, D. J. (1990). Phage antibodies: Filamentous phage displaying antibody variable domains. *Nature* **348**, 552–554.

Scott, J. K., and Smith, G. P. (1990). Searching for peptide ligands with an epitope library. *Science* **249**, 386–390.

Smith, G. P., and Scott, J. K. (1993). Libraries of peptides and proteins displayed on filamentous phage. *Methods Enzymol.* **217**, 228–257.

Zhong, G., and Smith, G. P. (1994). Kinetic microplate assay for titering microbial cells. *Biotechnology* **16**, 838–839.

# 5

# Construction of Random Peptide Libraries in Bacteriophage M13

## Nils B. Adey, Andrew B. Sparks, Jim Beasley, and Brian K. Kay

## INTRODUCTION

Bacteriophage M13 has been adapted for the expression of diverse populations of peptides in a manner that affords the rapid purification of active peptides by affinity selection (Scott and Smith, 1990; Cwirla *et al.,* 1990; Devlin *et al.,* 1990). We describe herein the construction of libraries of peptides expressed as N-terminal fusions to the M13 minor coat protein pIII. The random peptides are encoded by a DNA insert assembled from synthetic degenerate oligonucleotides and cloned into gIII. As outlined in Fig. 1, we discuss the assembly of double-stranded DNA inserts from degenerate oligonucleotides (Protocol 1), the preparation of vector DNA to accept said inserts (Protocol 2), the ligation of these DNAs (Protocol 3), their introduction into *Escherichia coli* by electroporation (Protocol 4), and the amplification, recovery, and storage of the resulting phage library (Protocol 5). Using these techniques, it is possible to construct libraries composed of billions of different peptide sequences in as little as 2 weeks.

Peptides have been expressed as fusions to a number of M13 capsid proteins, including pIII (Scott and Smith, 1990; Cwirla *et al.,* 1990; Devlin *et al.,* 1990) and the major capsid protein pVIII (Felici *et al.,* 1991; Markland *et al.,* 1991). Furthermore, phagemid vectors have been employed for the purpose of displaying peptides or entire proteins on the surface of phage M13 (Bass *et al.,* 1990; McCafferty *et*

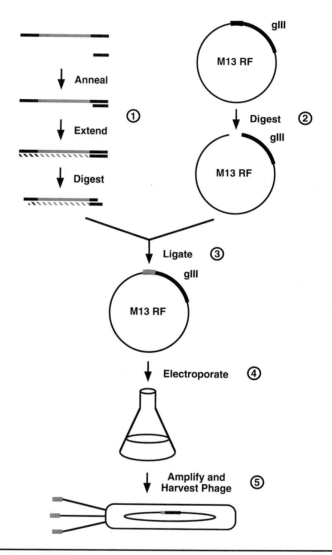

**FIGURE 1**    Schematic of random peptide library construction. Circled numbers refer to Protocols within which each procedure is described in detail. Degenerate DNA or amino acid sequences are represented by shaded lines. Regions filled in by DNA polymerase are indicated by slashed lines. gIII, gene for minor coat protein pIII; M13 RF, M13 RF vector DNA molecule.

*al.,* 1990). Minor modifications of the procedures discussed below afford the use of any of these permutations of phage display technology. Additional discussion of issues surrounding phage display may be found elsewhere in this book.

**PROTOCOL 1    Assembly of double-stranded DNA insert from degenerate oligonucleotides**

Several different codon schemes may be used to encode unspecified amino acid sequences. For example, NNN (where N is an equimolar representation of all four bases) produces all 64 possible codons, and therefore all 20 amino acids. Unfortunately, NNN also encodes 3 stop codons, which leads to nonproductive clones. Two popular codon schemes that address this problem are NN(G/T) and NN(G/C); these schemes use 32 codons to encode all 20 amino acids and 1 stop codon (TAG). Alternatively, degenerate sequences may be synthesized from mixtures of trinucleotide codons representing all 20 amino acids and no stop codons (Sondek and Shortle, 1992; Virnekäs et al., 1994). In practice, the NN(G/T) codon scheme is popular because it only yields an unacceptably high frequency of stop codons when used to encode very large (>50 amino acids) peptides and it does not require sophisticated oligonucleotide synthesis technologies.

A variety of strategies have been used to produce clonable degenerate DNA fragments appropriate for encoding random peptides. Several of these approaches are summarized in Fig. 2. Each method relies upon synthetic oligonucleotides to generate the degenerate regions within the DNA insert. This Protocol represents an implementation of the strategy described by Sparks et al., (1996) but can be readily adapted to any of the approaches listed in Fig. 2.

Synthesis of long (>75 nt) oligonucleotides often results in a crude product containing a large fraction of contaminating $n$-1 and smaller products. We therefore recommend purification of oligonucleotides by denaturing polyacrylamide gel electrophoresis (PAGE) or high-performance liquid chromatography (HPLC) prior to their use in this Protocol. Many oligonucleotide synthesis facilities offer HPLC purification for a nominal charge.

**Procedure**

1. Anneal the oligonucleotides by combining the following in a microcentrifuge tube:

   | | | |
   |---|---|---|
   | X µl | Upper oligo | (400 pmol) |
   | X µl | Lower oligo | (400 pmol) |
   | 40 µl | 5X Sequenase buffer | (1X) |
   | X µl | ddH$_2$O | (200 µl total volume) |

   Incubate the mixture at 75°C for 15 min. Allow the reaction to cool slowly to <35°C (~30 min). The exact times and temperatures may vary depending on the T$_m$ of the oligonucleotides (Ausubel et al., 1987). Collect any condensation by briefly pulsing the tube in a microcentrifuge (5 sec at 14,000 rpm).

2. Convert the annealed oligonucleotides to fully double-stranded DNA with Sequenase T7 DNA polymerase by adding the following:

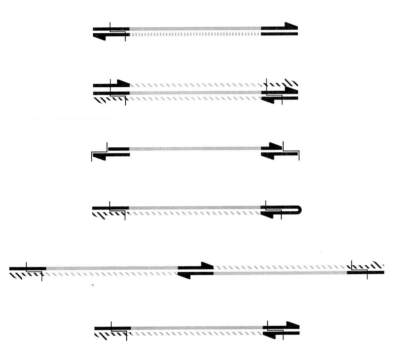

Devlin *et al*, 1990
1. Anneal oligos
2. 1X restriction digest
3. Ligate into vector

Scott and Smith, 1990
1. Anneal oligos
2. PCR amplify
3. 2X restriction digest
4. Ligate into vector

Cwirla *et al*, 1990
1. Anneal three oligos
2. Ligate into vector

Christian *et al*, 1993
1. Self anneal oligo to form hairpin
2. DNA polymerase extend oligo
3. 2X restriction digest
4. Ligate into vector

Kay *et al*, 1993
1. Anneal oligos
2. DNA polymerase extend oligos
3. 2X restriction digest
4. Ligate into vector

Sparks *et al*, 1996
1. Anneal oligos
2. DNA polymerase extend oligos
3. 2X restriction digest
4. Ligate into vector

**FIGURE 2**　Schematic of strategies for assembling double-stranded inserts from degenerate oligonucleotides. Fixed and degenerate sequences are represented by black and shaded lines, respectively. Regions filled in by DNA polymerase are indicated by slashed lines, whereas sequences corresponding to poly-inosine are indicated by dashed lines. Arrows indicate 3′ ends of DNA fragments, and restriction enzyme recognition sequences are indicated by ⌐. References and steps for each process are listed.

| 2.0 µl | 20 mM dNTPs | (0.2 mM) |
|---|---|---|
| 2.0 µl | 10 µg/µl acetylated BSA | (0.1 µg/µl) |
| 2.0 µl | 100 mM DTT | (1 mM) |
| 4.0 µl | 13 U/µl Sequenase | (50 U) |

Incubate the reaction at 37°C for 30 min.

**NOTE:** We have found that the Klenow fragment of *E. coli* DNA polymerase and *Taq* DNA polymerase also work well in converting the oligonucleotides into double-stranded DNA. Appropriate reaction conditions should be used for each enzyme.

3. Heat inactivate the polymerase by incubating the tube at 65°C for 1 hr. Complete inactivation of polymerase is necessary to avoid filling in cohesive ends generated by restriction digestion. We avoid phenol–chloroform extraction and ethanol precipitation steps at this stage to minimize loss of material.

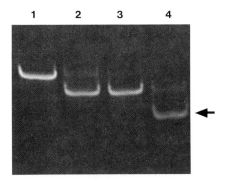

**FIGURE 3**     Polyacrylamide gel of products of assemby of degenerate oligonucleotides into double-stranded DNA insert. Oligonucleotides (Sparks *et al.,* 1996) were annealed; extended with T7 DNA polymerase (lane 1); cleaved with *Xho*I (lane 2), *Xba*I (lane 3), or *Xho*I + *Xba*I (lane 4); and electrophoresed in a 15% nondenaturing polyacrylamide gel. The DNA was visualized by ethidium bromide staining.

4. Prepare the restriction digestion reaction by adding the following:

    | 40 µl | 10X restriction enzyme buffer | (1X) |
    | 4 µl | 10 µg/ml acetylated BSA | (0.1 µg/µl) |
    | 4 µl | 100 mM DTT | (1 mM) |
    | X µl | ddH$_2$O | (400 µl final volume) |

    Remove a 20 µl aliquot as a no-enzyme control. Divide the remaining solution into two tubes and add 100 U *Xho*I to one tube and 100 U *Xba*I to the second tube. Remove all but 20 µl from each of the single-digest tubes and combine in a new tube. Incubate all tubes at 37°C for 3 hr.

    **NOTE:** Ensure that the amount of enzyme does not exceed 10% of the total reaction volume. Excessive concentrations of glycerol, which is used to stabilize the restriction enzyme during storage, may result in star activity.

5. We use nondenaturing PAGE to separate the double-digested central fragment of DNA from contaminating uncut, single-cut, and terminal DNA fragments. To this end, prepare a nondenaturing 15% polyacrylamide gel (Ausubel *et al.,* 1987). Typically, we use preparative gels of 15 cm X 15 cm X 1 mm with 12 1-cm wells.

6. Add 1/10 vol loading dye to the digested DNA sample. Load approximately 50 µl of the digested DNA sample in each of eight wells. Save 50 µl as a control for Step 10 (below). Run undigested and single enzyme-digested DNA samples as controls (Fig. 3). Electrophorese at 10 W until the digested fragment is separated from uncut DNA (~1.5 hr).

7. Stain the gel with 0.2 µg/ml ethidium bromide and photograph the gel using a long-wavelength UV transilluminator. Excise the band of interest with a razor blade. Be careful to avoid partially digested DNA and to minimize exposure of the DNA to UV light. Minimize exposure of eyes and skin to UV light by wearing protective eyewear, etc.

8. Recover the DNA insert by crushing the gel slice into a fine paste. Suspend the gel in 3.0 ml 0.5 *M* ammonium acetate and tumble the suspension overnight at 37°C. Pellet the gel fragments by centrifugation and recover the liquid. Filter the liquid through silanized glass wool to remove residual polyacrylamide fragments. Reduce the vol to 0.5 ml with repeated 1-butanol extractions. The DNA remains in the aqueous (lower) phase.

NOTE: Alternatively, one may use electroelution to recover the DNA from the gel fragment (Ausubel *et al.,* 1987).

9. Extract the sample with an equal volume of 50:50 phenol: chloroform. Precipitate the DNA by addition of 0.1 vol 3 *M* sodium acetate (pH 5.2) plus 2.5 vol of cold 100% ethanol. Recover the DNA by centrifugation at 14,000 rpm in a microcentrifuge for 30 min at 4°C. Decant the supernatant and wash the DNA pellet with 0.5 ml cold 80% ethanol. Recentrifuge and remove the ethanol, being careful not to aspirate the DNA pellet. Air dry the pellet and dissolve it in 200 μl TE. Aliquot the assembled insert into several tubes and store at -70°C. Avoid subjecting the DNA insert to repeated freeze/thaw cycles, as this may lead to reduced cloning efficiency.

10. Assess the integrity and concentration of the DNA insert by running 10% of the sample on a nondenaturing 15% polyacrylamide gel. Run the double-digested DNA sample from Step 6 as a control. Visualize the DNA by ethidium bromide staining as above. The concentration of the DNA insert may be estimated by comparing its ethidium bromide staining with that of standards of known quantity (e.g., molecular weight markers or assembled oligonucleotides).

NOTE: The use of 400 pmol of each oligo in this procedure should result in the recovery of at least 100 pmol of purified double-digested insert DNA. This corresponds to approximately 5 μg of an insert 75 bp in length. The final concentration of an insert this size should be on the order of 25 ng/μl.

---

**PROTOCOL 2　Preparation of linearized vector DNA**

Because the number of permutations of a peptide of given length scales exponentially with respect to the size of the peptide, a principal objective in random peptide library construction has been the maximization of library complexity. This has been accomplished primarily by maximizing the total number of individual recombinants during library construction. Thus, a large amount of double-stranded [replicative form (RF)] M13 vector DNA capable of efficiently accepting double-stranded degenerate inserts is required.

In addition to preparing an adequate quantity of vector DNA, steps should be taken to limit the number of parental clones among the recombinants in the library. Several strategies have been employed to minimize the recovery of parental clones by vector reclosure, including the use of two different restriction enzymes that produce noncompatible cohesive ends (described herein), the treatment of digested vector with alkaline phosphatase, and the gel purification of digested vector from undigested vector and parental insert fragments. Additionally, the fact that pIII is required for phage infection has given rise to several different vectors that are only capable of producing viable phage when they have acquired an appropriately de-signed recombinant insert (Smith and Scott, 1993). For example, we have used a vector (mBAX) possessing a TAG stop codon within its parental insert (Sparks *et al.*, 1996). This vector may be propagated in strains carrying suppresser tRNAs (e.g., DH5αF') but not in strains lacking suppressor tRNAs (e.g., JS5). Libraries con-structed with mBAX and inserts lacking stop codons may be amplified in nonsup-pressor hosts, imposing a strong selection against parental clones. We have observed better than $10^6$-fold selection against parental clones in JS5 relative to DH5αF'.

Large-scale preparation of M13 RF DNA is described in Chapter 4. In this pro-tocol, RF DNA of an M13 vector (mBAX) which gIII has been engineered to accept degenerate oligonucleotide inserts with *Xho*I- and *Xba*I-compatible cohesive ends is cleaved with *Xho*I and *Xba*I restriction enzymes.

### Procedure

1. Assemble the restriction digestion reaction by combining the following in a mi-crocentrifuge tube:

    | | | |
    |---|---|---|
    | $X$ μl | RF DNA | (250 μg) |
    | 100 μl | 10X restriction enzyme buffer | (1X) |
    | 10 μl | 10 μg/μl acetylated BSA | (0.1 μg/μl) |
    | 10 μl | 100 mM DTT | (1 mM) |
    | $X$ μl | ddH$_2$O | (to 1000 μl total volume) |

    Remove a 20-μl aliquot as a no-enzyme control. Divide the remaining solution into two tubes and add 300 U *Xho*I to one tube and 300 U *Xba*I to the second tube. Remove all but 20 μl from each of the single-digest tubes and combine in a new tube. Incubate all tubes at 37°C for 3 hr.

    **NOTE:** Ensure that the amount of enzyme does not exceed 10% of the total reac-tion volume. Excessive concentrations of glycerol, which is used to stabilize the restriction enzyme during storage, may result in star activity.

2. Extract the digested DNA twice with an equal volume of 50:50 phenol:chloro-form to remove residual enzyme activity. Precipitate the DNA by addition of 0.1 vol 3 M sodium acetate (pH 5.2) plus 2.5 vol of cold 100% ethanol. Recover the DNA by centrifugation at 14,000 rpm in a microcentrifuge for 30 min at 4°C. Decant the supernatant and wash the DNA pellet with 0.5 ml cold 80% ethanol. Recentrifuge and remove the ethanol, being careful not to aspirate the DNA pellet. Air dry the pellet and dissolve it in 500 μl TE.

3. Confirm complete digestion by resolving the digested DNA on a 0.8% agarose gel. Add 1 μl loading dye to 10 μl of the double-digested DNA sample. Run a

molecular weight marker and the undigested and single enzyme-digested DNA samples as controls. Visualize the DNA by ethidium bromide staining as above. Determine the DNA concentration by measuring its optical absorbance at 260 nm.

---

**PROTOCOL 3    Ligation of the vector and insert DNA**

In this Protocol, double-digested M13 vector DNA and double-stranded degenerated insert DNA are ligated to form the DNA population to be transformed into *E. coli,* ultimately giving rise to the phage library. Although a 1:1 vector:insert (V:I) ratio often produces good ligation efficiencies, it is prudent to perform a series of test ligations to determine the optimal V:I ratio for the particular vector and insert preps being used. Furthermore, test ligations provide useful information regarding the scale of ligation required to achieve a desired library complexity.

**Procedure**

1  To determine the appropriate ratio of insert to vector DNA, perform a series of test ligations. Assemble five separate ligation reactions composed of the following components:

| | | |
|---|---|---|
| X μl | linearized vector DNA | (1 μg) |
| 2 μl | 10X T4 DNA ligase buffer | (1X) |
| X μl | ddH$_2$O | (18 μl total volume) |

2. Perform three twofold dilutions of a small aliquot of the insert DNA. Add 2 μl 1X insert DNA to the first ligation reaction, 2 μl 0.5X insert DNA to the second reaction, 2 μl 0.25X insert DNA to the third reaction, 2 μl 0.125X insert DNA to the fourth reaction, and 2 μl H$_2$O (no insert DNA control) to the fifth reaction. Add 2 Weiss U T4 DNA ligase to each tube and incubate the tubes at 15°C overnight. In addition, test ligase from a variety of suppliers since quality can vary.

**NOTE:** Assuming an insert concentration of 0.5 pmol/μl (see Protocol 1, Step 10 Note), the V:I ratio for the ligation reaction with the highest concentration of insert should be 1:5. Therefore, the different reactions should provide a range of V:I ratios of from 1:5 to 2:1. This range is typically sufficient for defining high-efficiency ligation conditions.

3. Precipitate the DNA by addition of 0.1 vol 3 *M* sodium acetate (pH 5.2) plus 2.5 vol of cold 100% ethanol. Recover the DNA by centrifugation at 14,000 rpm in a microcentrifuge for 30 min at 4°C. Decant the supernatant and wash the DNA pellet with 0.5 ml cold 80% ethanol. Recentrifuge and remove the ethanol, being careful not to aspirate the DNA pellet. Air dry the pellet and dissolve it in 10 μl TE.

4. Transform electrocompetent F′ *E. coli* (Protocol 5) with the contents of each tube and plate for isolated plaques (see Chapter 4). Determine the V:I ratio that

produced the largest number of recombinants. Also confirm that the background of parental clones (as indicated by the no insert control) is acceptably low, i.e., at least 2 logs lower than the insert-vector ligation.

5. Assemble a large-scale ligation of vector and insert at the optimal ratio as determined in Step 2:

| | | |
|---|---|---|
| $X$ µl | linearized vector DNA | (100 µg = 20 pmol) |
| $X$ µl | insert DNA | (to produce optimal V:I ratio) |
| 400 µl | 10X T4 DNA ligase buffer | (1X) |
| 25 µl | 5 Weiss U/µl T4 DNA ligase | (125 Weiss U total) |
| $X$ µl | ddH$_2$O | (4000 µl total volume) |

Incubate the reaction at 15°C overnight.

6. Extract the reaction twice with an equal vol 50:50 phenol:chloroform to remove residual ligase. Precipitate the DNA by addition of 0.1 vol 3 M sodium acetate (pH 5.2) plus 2.5 vol of cold 100% ethanol. Recover the DNA by centrifugation at 14,000 rpm in a microcentrifuge for 30 min at 4°C. Decant the supernatant and wash the DNA pellet with 0.5 ml cold 80% ethanol. Recentrifuge and remove the ethanol, being careful not to aspirate the DNA pellet. Air dry the pellet and resuspend it in 200 µl TE. Store the large scale ligation at -70°C.

**NOTE:** Informative ligation controls include a "no insert" reaction and a "no ligase" reaction.  A high background in "no insert" implies single or uncut vector.  A high background in "unligated" must be from uncut vector.

---

**PROTOCOL 4    Electroporation of ligated DNA into bacteria**

Electroporation produces high efficiencies of transformation by subjecting a cell/DNA mixture to a brief but intense electrical field of exponential decay. Because of the strong electrical fields used in this procedure, the cell/DNA mixture must be free of salt to avoid arcing during electroporation. Thus, a culture of E. coli is made electrocompetent by a series of washes with salt-free H$_2$O or buffer.

Using JS5 E. coli and uncut M13 RF DNA, we routinely achieve efficiencies of >10$^8$ transformants/µg DNA. Electroporation of ligated M13 vector DNA typically yields efficiencies ~10-fold lower than this. Thus, the 100-µg large-scale ligation performed in Protocol 4 should result in >10$^9$ unique transformants. Because transformation efficiency decays considerably with electroporation of >2 µg ligated M13 vector DNA, electroporation of the large-scale ligation entails 50 separate electroporations of 2 µg ligated DNA into 100-µl electrocompetent cells. The following protocol, adapted from Dower et al., (1988), yields >10 ml electrocompetent cells, enough for at least 100 electroporations.

## Procedure

### Preparation of Competent Cells

1. Innoculate 4 liters 2xYT with 4 ml of a fresh overnight culture of the appropriate *E. coli* strain (e.g., DH5αF' for TAG suppression or JS5 for TAG nonsuppression). Grow the culture at 37°C with shaking to an $OD_{600}$ of 0.5.

2. Chill the cells on ice. It is important to keep the cells at 4°C for the duration of the procedure. Transfer the culture to eight sterile 500-ml centrifuge bottles. Pellet the cells by centrifugation at 2500g (4000 rpm in a GSA rotor in a Sorvall RC-5B centrifuge) for 10 min at 4°C. Decant the supernatant.

3. Resuspend the cells in 4 liters ice-cold sterile double-distilled or deionized $H_2O$, e.g., by rocking the bottles gently back and forth. Pellet the cells by centrifugation at 4000g (5000 rpm in a GSA rotor in a Sorvall RC-5B centrifuge) for 10 min at 4°C. Aspirate the supernatant, taking care not to disturb the cell pellet.

4. Gently resuspend the cells in 2 liters ice-cold sterile double-distilled or deionized $H_2O$. Pellet the cells by centrifugation at 6000g (6000 rpm in a GSA rotor in a Sorvall RC-5B centrifuge) for 10 min at 4°C. Aspirate the supernatant.

5. Gently resuspend the cells in 80 ml ice-cold sterile 10% glycerol. Pellet the cells by centrifugation at 6000g (6000 rpm in a GSA rotor in a Sorvall RC-5B centrifuge) for 10 min at 4°C. Carefully aspirate the supernatant.

6. Gently resuspend the cells in 10 ml ice-cold sterile 10% glycerol. Flash-freeze small (500 μl) aliquots in a dry ice/ethanol bath and store at -70°C.

**NOTE:** Frozen electrocompetent cells are also available from a variety of commercial sources.

### Electroporation Using the Bio-Rad Gene Pulser

7. For each electroporation, aliquot 100 μl of cells into a microcentrifuge tube. Add 2 μg ligated DNA in less than 10 μl ddH$_2$O. Alternatively, the large-scale ligation DNA may be mixed with electrocompetent cells *en masse* at a ratio of 2 μg ligated DNA/100 μl electrocompetent cells. Keep the cells on ice until electroporation. Place the electroporation cuvettes (0.2 cm pathlength) on ice for at least 10 min. Set the following parameters on the Gene Pulser: V = 2.0, kV, C = 25 μF, R = 400 W. This combination, when used with 0.2-cm cuvettes, will yield a pulse of 10 kV cm$^{-1}$, and theoretical time constant of 10 msec.

8. Carefully transfer the cell–DNA mixture into a cold cuvette, place the cuvette into the cuvette chamber, and electroporate. Immediately add 1 ml 2xYT to the cells and transfer the cells to a flask containing 1 liter sterile 2x YT at 37°C. Continue this process for the remaining 49 electroporations. Save the resulting culture for Step 1 of Protocol 5.

9. To determine the complexity of the library (the total number of individual recombinants), plate six 10-fold serial dilutions of one of the electroporations for isolated plaques (see Chapter 4). Also plate cells electroporated with uncut M13 vector DNA and no DNA as positive and negative controls, respectively.

## PROTOCOL 5   Amplification and harvest of library phage

### Procedure

1. Amplify the library by growing the culture from Step 8 of Protocol 4 with aeration at 37°C for 8–10 hr. Harvest the library phage as soon after the culture has reached stationary phase as possible, as expressed peptides may be susceptible to proteolysis.

2. Transfer the culture to four sterile 250-ml centrifuge bottles. Pellet the bacterial cells by centrifugation at 6000*g* (6000 rpm in a GSA rotor in a Sorval RC-5B centrifuge) for 10 min at 4°C. Carefully decant the supernatant into new sterile 250-ml centrifuge bottles.

   **NOTE:** The library may also be amplified by plating the electroporated cells on petri plates and allowing for phage propagation on solid media. Amplification on solid media may limit biases resulting from clones with different rates of infection (although see McConnel *et al.,* 1995). We routinely plate the large-scale electroporation on a total of one hundred 100-mm plates. Accordingly, split each electroporation into two sterile 6-ml Falcon 2063 tubes. Add 3 ml 42°C top agar (2x YT, 0.8% agar) to each tube and immediately pour the contents of the tube onto a prewarmed 2x YT petri plate. Incubate the plates inverted at 37°C for 8 hr. Elute the phage particles from the top agar by placing 5 ml sterile PBS onto each plate and incubating the plates at 4°C with gentle rocking for 2–4 hr. Decant the supernatant into sterile 250-ml centrifuge bottles, and pellet any bacterial cells by centrifugation at 6000*g* (6000 rpm in a GSA rotor in a Sorval RC-5B centrifuge) for 10 min at 4°C. Carefully decant the supernatant into new sterile 250-ml centrifuge bottles.

3. Precipitate the phage by adding 0.20 vol 30% PEG 8000 / 1.6 *M* NaCl. Mix well and incubate at 4°C for 1 hr. Pellet the precipitated phage by centrifugation at 10,000*g* (~8000 rpm using a GSA rotor in a Sorvall RC-5B centrifuge) for 20 min at 4°C. Carefully decant the supernatant, recentrifuge the tube for 5 min, and pipette off any residual liquid.

4. Gently resuspend the pellets in a total of 20 ml sterile PBS + 20% glycerol by pipetting up and down with a large bore pipette. Centrifuge the resuspended phage at 6000*g* (6000 rpm in a GSA rotor in a Sorvall RC-5B centrifuge) for 10 min at 4°C to eliminate any insoluble material. Dispense the phage into 100 to 500-μl aliquots in sterile microfuge tubes. Flash freeze the aliquots using a dry ice:ethanol bath and store at -70°C. Thaw one aliquot and determine the phage titer by plating. The titer should be on the order of $10^{13}$ pfu/ml.

### References

Ausubel, F., Brent, R., Kingston, R., Moore, D., Seidman, J., Smith, J., and Struhl, K. (1987). "Current Protocols in Molecular Biology." Wiley, New York.

Bass, S. H., Greene, R., and Wells, J. A. (1990). Hormone phage: An enrichment method for variant proteins with altered binding properties. *Proteins* **8,** 309–314.

Christian, R. B., Zuckermann, R. N., Kerr, J. M., Wang, L., and Malcolm, B. A. (1992). Simplified methods for construction, assessment, and rapid screening of peptide libraries in bacteriophage. *J. Mol. Biol.* **227,** 711–718.

Cwirla, S. E., Peters, E. A., Barrett, R. W., and Dower, W. J. (1990). Peptides of phage: a vast library of peptides for identifying ligands. *Proc. Natl. Acad. Sci. U.S.A.* **87,** 6378–6382.

Devlin, J. J., Panganiban, L. C., and Devlin, P. E. (1990). Random peptide libraries: A source of specific protein binding molecules. *Science* **249,** 404–406.

Dower, W.J., Miller, J.F., and Ragsdale, C.W. (1988). High efficiency transformation of *E. coli* by high voltage electroporation. *Nucleic Acids Res.* **16,** 6127–6145.

Felici, F., Castagnoli, L., MU.S.Acchil, A., Jappelli, R., and Cesareni, G. (1991). Selection of antibody ligands from a large library of oligopeptides expressed on a multivalent exposition vector. *J. Mol. Biol.* **222,** 301–310.

Kay, B. K., Adey, N. B., He, Y.-S., Manfredi, J. P., Mataragnon, A. H., and Fowlkes, D. M. (1993). An M13 library displaying 38-amino-acid peptides as a source of novel sequences with affinity to selected targets. *Gene* **128,** 59–65.

Markland, W., Roberts, B. L., Saxena, M. J., Guterman, S. K., and Ladner, R. C. (1991). Design, construction and function of a multicopy display vector using fusions to the major coat protein of bacteriophage M13. *Gene* **109,** 13–19.

McCafferty, J., Griffiths, A. D., Winter, G., and Chiswell, D. J. (1990). Phage antibodies: Filamentous phage displaying antibody variable domains. *Nature (London)* **348,** 552–554.

McConnell, S., Uvegas, A., and Spinella, D. (1995). Comparison of plate versus liquid amplification of M13 phage display libraries. *BioTechniques* **18,** 803–804.

Scott, J. K., and Smith, G. P. (1990). Searching for peptide ligands with an epitope library. *Science* **249,** 386–390.

Smith, G. P., and Scott, J. K. (1993). Libraries of peptides and proteins displayed on filamentous phage. *In* "*Methods in Enzymology*" **217,** pp. 228–257. (R. Wu, ed.), Academic Press, San Diego, CA.

Sondek, J., and Shortle, D. (1992). A general strategy for random insertions and substitution mutagenesis: Substoichiometric coupling of trinucleotide phosphoramidites. *Proc. Natl. Acad. Sci. U.S.A.* **89,** 3584–3585.

Sparks, A. B., Rider, J. E., Hoffman, N. G., Fowlkes, D. M., Quilliam, L. A., and Kay, B. K. (1996). Distinct ligand preferences of SH3 domains from Src, Yes, Abl, Cortactin, p53bp2, PLCγ, Crk, and Grb2. *Proc. Natl. Acad. Sci. U.S.A.* **93,** 1540–1544.

Virnekäs, B., Ge, L., Plückthun, A., Schneider, K. C., Wellnhofer, G., and Moroney, S. E. (1994). Trinucleotide phosphoramidites: Ideal reagents for the synthesis of mixed oligonucleotides for random mutagenesis. *Nucleic Acids Res.* **22,** 5600–5607.

**6**

# Construction and Screening of Antibody Display Libraries

## *John McCafferty and Kevin S. Johnson*

## GENERAL INTRODUCTION AND BACKGROUND

Antibodies were the first functional proteins to be displayed on the surface of phage (McCafferty *et al.,* 1990) following the initial demonstrations of peptide display (Parmley and Smith, 1988). Antibody display was initially demonstrated using antibody variable domain fragments. Antibodies have a tetrameric structure consisting of two identical heavy chains and two identical light chains. The region which defines the binding specificity of an antibody is referred to as the variable domain and is at the N terminal end of each of each chain. The association of a heavy- and light-chain variable domain (VH and VL, respectively) gives rise to a heterodimeric molecule, termed an Fv fragment, which retains the binding specificity of the parental antibody (Fig. 1).

Antibodies can be expressed in *Escherichia coli* in the form of an Fv fragment by expressing separate VH and VL domains, under the influence of a signal peptide. This directs the chains into the periplasmic space where they associate to form an Fv with the binding characteristics of the parental antibody. One problem with this approach is the tendency for the Fv fragment to dissociate in some clones. One solution to this problem is to join the genes encoding the heavy- and light-chain variable regions with DNA encoding a flexible linker peptide. This gives rise to a single protein with covalently linked heavy- and light-chain variable regions termed a single-chain

### A. Protein structure

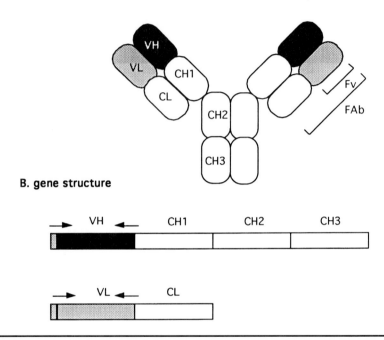

### B. gene structure

**FIGURE I**     Protein and gene structure of an IgG molecule. (A) The globes represent the various domains in the heavy- and light-chain antibody molecule which is a tetrameric structure made up of two heavy chains and two light chains. The binding specificity of the antibody is determined by the VH and VL segments. (B) The structure of processed mRNA is shown. The structural gene is preceded by a signal sequence. The arrows represent the position of the primers used in the primary PCR amplification.

Fv molecule (scFv). As an alternative, antibodies can be expressed in the form of an FAb fragment in which the variable heavy chain, along with its first constant domain, associates with a whole light chain (Fig. 1). The association between the constant domains means that this heterodimer is also more stable than an Fv fragment. All three forms (Fv, scFv, and FAb) have been expressed on the surface of phage.

Alignment of known variable domain gene sequences demonstrates that there are regions of conservation within the variable domain, particularly at the 5' and 3' termini. This has led to the determination of species-specific consensus sequences which have been used in the design of PCR primers. In this way whole repertoires of antibody genes have been prepared by PCR. This includes repertoires derived from immunized, nonimmunized, and synthetic sources (Clackson *et al.,* 1991; Marks *et al.,* 1991; Hoogenboom and Winter, 1992; Griffiths *et al.,* 1994; Barbas *et al.,* 1992).

Antibodies are by no means a special case and a number of other functional proteins have been displayed on the surface of phage, including alkaline phosphatase, ricin B chain, CD4, protein A, growth hormone, and BTPI, zinc finger domains, β-lactamase (Bass *et al.,* 1990; McCafferty *et al.,* 1992; Roberts *et al.,* 1992; Swimmer *et al.*, 1992; Abrol *et al.*, 1994; Kushwaha *et al.*, 1994; Jamieson *et al.*, 1994; Rebar and Pabo, 1994; Soumillion *et al.*, 1994) Many of the methods described here will be directly applicable to work with other proteins.

The original phage vector for antibody display fd-CAT1 (McCafferty *et al.*, 1990) was derived from fd-tet (Parmley and Smith, 1988) and enabled the insertion of antibody genes in frame with gene 3 and downstream of the gene 3 signal sequence which normally directs the gene 3 protein (g3p) to the periplasm. These vectors carry all the genetic information encoding the phage life cycle. An alternative system involves cloning into phagemid vectors which have a copy of gene 3 and a phage packaging signal sequence. Thus antibody fragments can be displayed as a fusion with the gene 3 protein and the genetic information is packaged thanks to the packaging signal. These phagemid vectors require infection of the host cell with a helper virus to generate phage particles.

For simplicity this chapter will concentrate on the vectors used at Cambridge Antibody Technology, but again the methods apply to whatever vectors are used. The phagemid vectors pCANTAB5-E/pCANTAB6 [Recombinant Phage Antibody System (Pharmacia); McCafferty *et al.*, 1994] allow cloning of antibody genes using rare-cutting *Sfi*I and *Not*I restriction enzymes (figure in Appendix). They incorporate an amber codon between the C-terminus of the cloned antibody and the start of gene 3 allowing the antibody to be made as a soluble fragment in appropriate (non-suppressing) *E.coli* strains such as HB2151 (see Appendix for genotype). They also include a peptide tag, allowing detection of the soluble antibody fragment, and in the case of pCANTAB6 a hexahistidine tag is included, enabling rapid purification and concentration of the antibody fragment by immobilized metal affinity chromatography.

Many of the methods we employ utilize commercially available kits for convenience and our own experience is based on using these. If required, alternative methods (e.g., RNA preparation, ligations) may be found in laboratory manuals such as Sambrook *et al.*, (1989) or Current Protocols in Molecular Biology (Ausubel *et al.*, 1995).

Figure 2 shows the overall relationship of the steps involved in preparing and panning antibody display library and attempts to give approximate time scales. The actual time taken will depend on the skill and experience of the molecular biologist as well as the degree of pressure being applied from above!

# PRODUCTION OF ANTIBODY GENES AND PREPARATION OF LIBRARIES

This section deals with preparing immunoglobulin RNA, converting it into a cDNA template for PCR amplification, and cloning the resulting product.

## Preparation of PCR Template

The source of mRNA may be B-lymphocytes from the spleens of hyperimmunized mice. In this case it should be confirmed that an appropriate immune response

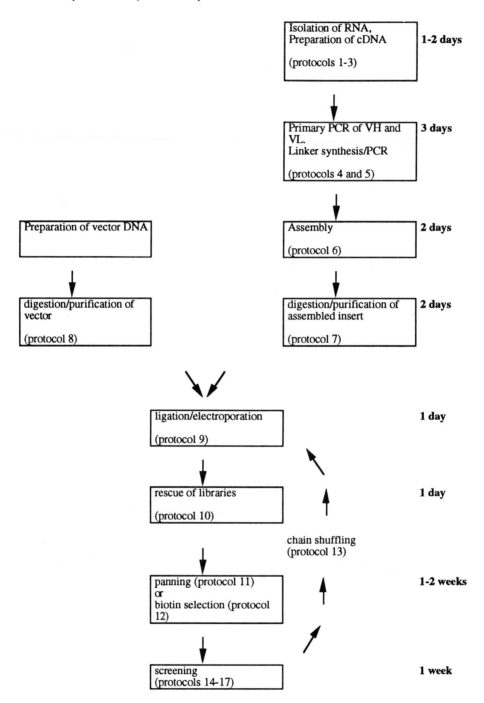

**FIGURE 2**     Preparation and selection of phage antibody libraries.

has been elicited by testing in ELISA for the presence of serum antibodies which bind the antigen used for immunization. Human B-lymphocytes may be obtained from peripheral blood or even from other lymphoid tissues such as tonsils/spleen if available. Methods for total RNA or mRNA preparation are given in Sambrook *et al.*, (1989).

---

**PROTOCOL 1**   **Preparation of lymphocytes as single-cell suspensions from mouse spleen cells**

The aim of this step is to finely disaggregate the spleen into individual cells so that they are readily disrupted by the denaturants used in the RNA extraction procedure.

1. Remove spleen from the mouse and place in a sterile petri dish containing 15 ml PBS. Tease tissue apart using either sterile forceps and a disposable syringe needle or disrupt between frosted ends of microscope slides to produce a single-cell suspension of lymphocytes and erythrocytes.
2. Transfer to a tube using a 22-gauge needle and syringe and let the clumps settle to the bottom (5 min). Remove the supernatant to a fresh, sterile container, carefully avoiding the tissue clumps that have settled at the bottom of the tubes.
3. Centrifuge in a bench top centrifuge for 7 min at 1100g at room temperature and discard supernatant.
4. It is not necessary to purify B cells before purifying mRNA but it is advantageous to reduce the bulk of of red blood cells by hypotonic shock. Resuspend cells in 900 µl of $H_2O$ and immediately add 100 µl of 10x PBS followed by 9 ml of 1X PBS. Transfer to a fresh tube and repellet.
5. Place the cell pellet on ice and use for RNA preparation immediately or freeze at -70°C.

---

**PROTOCOL 2**   **Preparation of human peripheral blood lymphocytes**

1. Dilute 20 ml of heparinized blood with an equal volume of PBS.
2. Divide the diluted blood into two 20-ml fractions and overlay each, onto a 15-ml "cushion" of Ficoll Paque (Pharmacia 10-A-001-07).
3. To separate the lymphocytes from the red blood cells, the tubes are spun for 30 min at around 400g at room temperature (e.g., at 1800 rpm in an IEC CENTRA 3E table centrifuge).
4. The peripheral blood lymphocytes are then collected from the interface by aspiration with a Pasteur pipette.
5. The cells are diluted with an equal volume PBS and spun again at 400g for 5 min.
6. The supernatant is aspirated and the cell pellet is resuspended in 1 ml PBS and distributed into two Eppendorf tubes.

**TABLE 1** Specific Primers for cDNA Synthesis of Human Antibody Variable Regions (V Genes)[a]

| | |
|---|---|
| *Human heavy chain constant region primers* | |
| HuIgG1-4CH1FOR | 5'-GTC CAC CTT GGT GTT GCT GGG CTT-3' |
| HuIgMFOR | 5'-TGG AAG AGG CAC GTT CTT TTC TTT-3' |
| | |
| *Human k constant region primer* | |
| HuGkFOR | 5'-AGA CTC TCC CCT GTT GAA GCT CTT-3' |
| | |
| *Human lambda constant region primer* | |
| HuCλFOR | 5'-TGA AGA TTC TGT AGG GGC CAC TGT CTT-3' |

[a]All primers are anti-sense.

7. The cells are sedimented by a 2-min spin in a Micro-Centaur Centrifuge at the low-speed setting (~2000g). Following aspiration of supernatant, cells are used for RNA preparation immediately or frozen at -70°C.

**NOTE:** (a) From a mouse spleen the yield is $5 \times 10^7$–$2 \times 10^8$ white blood cells (WBC) of which 35–40% are B cells. In human blood there are approximately $4$–$11 \times 10^6$ WBC/ml of which approximately 30% are lymphocytes. (b) Quality of the RNA is very important, as we find that purity and integrity of the RNA to a large extent dictate the efficiency and ease with which subsequent steps are performed. (c) A range of commercially available RNA purification kits are also available for RNA preparation, e.g., RNAzol B (Biotecx Labs Inc.) or QuickPrep mRNA Purification Kit (Cat. No. 27-9254-01, Pharmacia). Following RNA isolation, cDNA is prepared and used as a template for PCR amplification.

Once RNA is prepared cDNA must be synthesized using reverse transcriptase to act as template in a PCR reaction. Reverse transcriptase uses a short RNA:DNA hybrid region as template and this can be generated on the RNA using random DNA hexamers. Some people prefer to use chain-specific primers which anneal with the various constant region. The primers required for each chain are shown in Tables 1 and 2.

---

**PROTOCOL 3**     **Preparation of mouse or human antibody cDNA**

1. Set up the following reverse transcription mix:

| | |
|---|---|
| [a]H$_2$O (DEPC-treated) | 20 μl |
| 1.25 mM each dNTP | 10 μl |
| 10x first strand buffer | 10 μl |
| 0.1 M dithiothreitol | 10 μl |
| Random hexamers, 80 ng/μl (Pharmacia) | 10 μl |
| RNasin (40 U/μl) (Promega Corp.) | 4 μl |

**TABLE 2**   Specific Primers for cDNA Synthesis of Mouse Antibody Variable Regions (V Genes)

*Mouse heavy chain constant region primers*

| | | |
|---|---|---|
| For IgG: | MOCG12FOR | 5′-CTC AAT TTT CTT GTC CAC CTT GGT GC-3′ |
| | MOCG3FOR | 5′-CTC GAT TCT CTT GAT CAA CTC AGT CT-3′ |
| For IgM: | MOCMFOR | 5′-TGG AAT GGG CAC ATG CAG ATC TCT-3′ |

*Mouse k constant region primer*

| | | |
|---|---|---|
| | CKFOR | 5′-CTC ATT CCT GTT GAA GCT CTT GAC-3′ |

If chain-specific primers are used, 20 pmol of each should be added per reaction in place of the random hexamers. These primers may be used as a mix; 10x first strand buffer is 1.4$M$ KCl, 0.5 $M$ Tris-HCl, pH 8.1, at 42°C, 80 m$M$ MgCl$_2$.

2. Dilute 10 μg total RNA to 40-μl final volume with DEPC-treated water. If mRNA has been prepared, then an amount of mRNA equivalent to 10 μg total RNA is used. Heat at 65°C for 3 min to remove secondary structure in the RNA and then place on ice for 1 min.

3. Add the reverse transcription mix to the RNA and add reverse transcriptase (e.g., 100 Units Pharmacia Reverse Transcriptase 27-0925-01). Incubate at 42°C for 1 hr.

4. Boil the reaction mix for 3 min, cool on ice for 1 min, and then spin in a microfuge at 13,000 rpm for 5 min to pellet debris. Transfer the supernatant to a new tube; cDNA may now be stored at -20°C until use.

**NOTE:** (a) DEPC-treated H$_2$O is prepared by adding diethylpyrocarbonate to H$_2$O (0.1% v/v), and incubating at 37°C for 2 hr. DEPC-treated H$_2$O should be auto-claved before use to inactivate the diethylpyrocarbonate. (b) DEPC is a carcinogen and a number of labs successfully prepare RNA and cDNA without using this.

## CONSTRUCTION OF ANTIBODY V GENE REPERTOIRE

Once cDNA is generated it is used as a template for PCR. PCR amplification of antibody variable regions is carried out using primers complementary with the ends of the heavy- and light-chain variable regions. The products of heavy- and light-chain PCR can be linked in turn by PCR to form a single-chain Fv fragment. Marks *et al.*, (1991) describe a series of primers for the amplification of the variable region of human light and heavy chains. Orlandi *et al.*, (1989) and Clackson *et al.*, (1991) describe primers for the PCR amplification of mouse variable regions. (These are reproduced in Tables 3 and 4). It was found that light chains from some monoclonal antibodies were difficult to amplify with these primers and they have since been redesigned as part of the Recombinant Phage Antibody System (Pharmacia; unpublished observations).

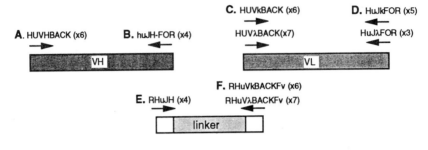

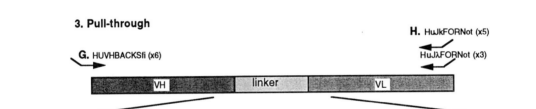

**FIGURE 3**    PCR amplification of antibody variable regions and assembly into scFv genes. The locations of antibody PCR primers on the single-chain Fv gene are shown. The sequences of all primers are given in Tables 3 and 4 where the same labeling of each group (A–H) is used. The names of the human antibody primers are shown on the figure. For PCR amplification of mouse V genes the following primers correspond: A, VH1BACK; B, VH1FOR-2; C, VK2BACK; D, VK4FOR (x4); E, LINKBACK; F, LINKFOR; G, VH1BACKSfi; H, VK4FORNot (x4). The DNA sequence encoding the flexible linker is shown at the bottom of the figure. For guidance the sequences flanking the linker sequence is shown. The huJH1-2FOR primer (antisense) is shown in lower case along with the corresponding sense strand. At the other side of the linker the HUVK1aBACK sequence is shown. Residues which are shared in common with all the other huJHFOR and HUVKBACK primers are shown underlined. The protein sequence is shown above the DNA sequence with the linker peptide shown in lower case.

The position and direction of the various groups of primers is shown in Fig. 3 and each group is represented by the letters A–H. The same letter code is used in Tables 3 and 4 where the actual sequence is given. Mouse heavy-chain are amplified in a single PCR reaction using VH1Back and VHFOR-2 (primer groups A and B).

**TABLE 3**   Primers for PCR Amplification of Human Antibody Variable Regions (V genes)

**1. V gene primary PCR**

*A. Human VH back primers (sense)*

| | |
|---|---|
| HuVH1aBACK | 5'-CAG GTG CAG CTG GTG CAG TCT GG-3' |
| HuVH2aBACK | 5'-CAG GTC AAC TTA AGG GAG TCT GG-3' |
| HuVH3aBACK | 5'-GAG GTG CAG CTG GTG GAG TCT GG-3' |
| HuVH4aBACK | 5'-CAG GTG CAG CTG CAG GAG TCG GG-3' |
| HuVH5aBACK | 5'-GAG GTG CAG CTG TTG CAG TCT GC-3' |
| HuVH6aBACK | 5'-CAG GTA CAG CTG CAG CAG TCA GG-3' |

*B. Human JH forward primers (anti-sense)*

| | |
|---|---|
| HuJH1-2FOR | 5'-TGA GGA GAC GGT GAC CAG GGT GCC-3' |
| HuJH3FOR | 5'-TGA AGA GAC GGT GAC CAT TGT CCC-3' |
| HuJH4-5FOR | 5'-TGA GGA GAC GGT GAC CAG GGT TCC-3' |
| HuJH6FOR | 5'-TGA GGA GAC GGT GAC CGT GGT CCC-3' |

*C. Human V kappa back primers (sense)*

| | |
|---|---|
| HuVk1aBACK | 5'-GAC ATC CAG ATG ACC CAG TCT CC-3' |
| HuVk2aBACK | 5'-GAT GTT GTG ATG ACT CAG TCT CC-3' |
| HuVk3aBACK | 5'-GAA ATT GTG TTG ACG CAG TCT CC-3' |
| HuVk4aBACK | 5'-GAC ATC GTG ATG ACC CAG TCT CC-3' |
| HuVk5aBACK | 5'-GAA ACG ACA CTC ACG CAG TCT CC-3' |
| HuVk6aBACK | 5'-GAA ATT GTG CTG ACT CAG TCT CC-3' |

*C. Human V lambda back primers (sense)*

| | |
|---|---|
| HuVλ1BACK | 5'-CAG TCT GTG TTG ACG CAG CCG CC-3' |
| HuVλ2BACK | 5'-CAG TCT GCC CTG ACT CAG CCT GC-3' |
| HuVλ3aBACK | 5'-TCC TAT GTG CTG ACT CAG CCA CC-3' |
| HuVλ3bBACK | 5'-TCT TCT GAG CTG ACT CAG GAC CC-3' |
| HuVλ4BACK | 5'-CAC GTT ATA CTG ACT CAA CCG CC-3' |
| HuVλ5BACK | 5'-CAG GCT GTG CTC ACT CAG CCG TC-3' |
| HuVλ6BACK | 5'-AAT TTT ATG CTG ACT CAG CCC CA-3' |

*D. Human J kappa forward primers (anti-sense)*

| | |
|---|---|
| HuJk1FOR | 5'-ACG TTT GAT TTC CAC CTT GGT CCC-3' |
| HuJk2FOR | 5'-ACG TTT GAT CTC CAG CTT GGT CCC-3' |
| HuJk3FOR | 5'-ACG TTT GAT ATC CAC TTT GGT CCC-3' |
| HuJk4FOR | 5'-ACG TTT GAT CTC CAC CTT GGT CCC-3' |
| HuJk5FOR | 5'-ACG TTT AAT CTC CAG TCG TGT CCC-3' |

*D. Human J lambda forward primers (anti-sense)*

| | |
|---|---|
| Hu Jλ1FOR | 5'-ACC TAG GAC GGT GAC CTT GGT CCC-3' |
| Hu Jλ2-3FOR | 5'-ACC TAG GAC GGT CAG CTT GGT CCC-3' |
| Hu Jλ4-5FOR | 5'-ACC TAA AAC GGT GAG CTG GGT CCC-3' |

*continues*

**TABLE 3** *Continued*

### 2. Linker fragment PCR

*E. Reverse JH for scFv linker (sense)*

|  | ——FR4 heavy————————||—linker |
| RHuJH1-2 | 5'-GC ACC CTG GTC ACC GTC TCC TCA GGT GG-3' |
| RHuJH3 | 5'-GG ACA ATG GTC ACC GTC TCT TCA GGT GG-3' |
| RHuJH4-5 | 5'-GA ACC CTG GTC ACC GTC TCC TCA GGT GG-3' |
| RHuJH6 | 5'-GG ACC ACG GTC ACC GTC TCC TCA GGT GG-3' |

*F. Reverse Vk for scFv linker (anti-sense)*

|  | ——FR1 light————————————||——linker— |
| RHuVk1aBACKFv | 5'-GG AGA CTG GGT CAT CTG GAT GTC CGA TCC GCC-3' |
| RHuVk2aBACKFv | 5'-GG AGA CTG AGT CAT CAC AAC ATC CGA TCC GCC-3' |
| RHuVk3aBACKFv | 5'-GG AGA CTG CGT CAA CAC AAT TTC CGA TCC GCC-3' |
| RHuVk4aBACKFv | 5'-GG AGA CTG GGT CAT CAC GAT GTC CGA TCC GCC-3' |
| RHuVk5aBACKFv | 5'-GG AGA CTG CGT GAG TGT CGT TTC CGA TCC GCC-3' |
| RHuVk6aBACKFv | 5'-GG AGA CTG AGT CAG CAC AAT TTC CGA TCC GCC-3' |

*F. Reverse Vλ for scFv linker (anti-sense)*

|  | ———FR1 light————————————||————linker———— |
| RHuVλBACK1Fv | 5'-GG CGG CTG CGT CAA CAC AGA CTG CGA TCC GCC ACC GCC AGA G-3' |
| RHuVλBACK2Fv | 5'-GC AGG CTG AGT CAG AGC AGA CTG CGA TCC GCC ACC GCC AGA G-3' |
| RHuVλBACK3aFv | 5'-GG TGG CTG AGT CAG CAC ATA GGA CGA TCC GCC ACC GCC AGA G-3' |
| RHuVλBACK3bFv | 5'-GG GTC CTG AGT CAG CTC AGA AGA CGA TCC GCC ACC GCC AGA G-3' |
| RHuVλBACK4Fv | 5'-GG CGG TTG AGT CAG TAT AAC GTG CGA TCC GCC ACC GCC AGA G-3' |
| RHuVλBACK5Fv | 5'-GA CGG CTG AGT CAG CAC AGA CTG CGA TCC GCC ACC GCC AGA G-3' |
| RHuVλBACK6Fv | 5'-TG GGG CTG AGT CAG CAT AAA ATT CGA TCC GCC ACC GCC AGA G-3' |

### 3. Pull-through primers for introduction of restriction sites[a]

*G. Human VH back (Sfi) primers (sense)*

HuVH1aBACKSfi                                                             |——FR1 heavy——————
5'-GTC CTC GCA ACT GC<u>G GCC</u> CAG CC<u>G GCC</u> ATG GCC CAG GTG CAG CTG GTG CAG TCT GG-3'
HuVH2aBACKSfi
5'-GTC CTC GCA ACT GC<u>G GCC</u> CAG CC<u>G GCC</u> ATG GCC CAG GTC AAC TTA AGG GAG TCT GG-3'
HuVH3aBACKSfi
5'-GTC CTC GCA ACT GC<u>G GCC</u> CAG CC<u>G GCC</u> ATG GCC GAG GTG CAG CTG GTG GAG TCT GG-3'
HuVH4aBACKSfi
5'-GTC CTC GCA ACT GC<u>G GCC</u> CAG CC<u>G GCC</u> ATG GCC CAG GTG CAG CTG CAG GAG TCG GG-3'
HuVH5aBACKSfi
5'-GTC CTC GCA ACT GC<u>G GCC</u> CAG CC<u>G GCC</u> ATG GCC CAG GTG CAG CTG TTG CAG TCT GC-3'
HuVH6aBACKSfi
5'-GTC CTC GCA ACT GC<u>G GCC</u> CAG CC<u>G GCC</u> ATG GCC CAG GTA CAG CTG CAG CAG TCA GG-3'

*H. Human J kappa forward (Not) primers (anti-sense)*

HuJk1FORNot                                                       |——FR4 light——————
5'-GAG TCA TTC TCG ACT T<u>GC GGC CGC</u> ACG TTT GAT TTC CAC CTT GGT CCC-3'
HuJk2FORNot
5'-GAG TCA TTC TCG ACT T<u>GC GGC CGC</u> ACG TTT GAT CTC CAG CTT GGT CCC-3'

*continues*

**TABLE 3** *Continued*

*H. Human J kappa forward (Not) primers (anti-sense) (Continued)*
HuJk3FORNot                                              |————FR4 light————————————
5′-GAG TCA TTC TCG ACT T<u>GC GGC CGC</u> ACG TTT GAT ATC CAC TTT GGT CCC-3′
HuJk4FORNot
5′-GAG TCA TTC TCG ACT T<u>GC GGC CGC</u> ACG TTT GAT CTC CAC CTT GGT CCC-3′
HuJk5FORNot
5′-GAG TCA TTC TCG ACT T<u>GC GGC CGC</u> ACG TTT AAT CTC CAG TCG TGT CCC-3′


*H. Human J lambda forward (Not) primers (anti-sense)*
Hu Jl1FORNOT                                             |————FR4 light————————————
5′-GAG TCA TTC TCG ACT T<u>GC GGC CGC</u> ACC TAG GAC GGT GAC CTT GGT CCC-3′
Hu Jl2-3FORNOT
5′-GAG TCA TTC TCG ACT T<u>GC GGC CGC</u> ACC TAG GAC GGT CAG CTT GGT CCC-3′
Hu Jl4-5FORNOT
5′-GAG TCA TTC TCG ACT T<u>GC GGC CGC</u> ACC TAA AAC GGT GAG CTG GGT CCC-3′

[a]Recognition site for restriction enzyme is underlined. (Primers are taken from Marks *et al.*, 1991).


**TABLE 4**    Primers for PCR Amplification of Mouse Antibody Variable Regions (V genes)

**1. V gene primary PCR**

*A. Mouse VH back primers (sense)*

VH1BACK     5′ AG GTS MAR <u>CTG CAG</u> SAG TCW GG 3′[a]


*B. Mouse JH forward primers (anti-sense)*

VH1FOR-2     5′ TGA GGA GAC <u>GGT GAC C</u>GT GGT CCC TTG GCC CC 3′


*C. Mouse V kappa back primers (sense)*

VK2BACK     5′ GAC ATT <u>GAG CTC</u> ACC CAG TCT CCA 3′


*D. Mouse J kappa forward primers (anti-sense)*[b]

MJK1FONX     5′ CCG TTT GAT TTC CAG CTT GGT GCC 3′
MJK2FONX     5′ CCG TTT TAT TTC CAG CTT GGT CCC 3′
MJK4FONX     5′ CCG TTT TAT TTC CAA CTT TGT CCC 3′
MJK5FONX     5′ CCG TTT CAG CTC CAG CTT GGT CCC 3′


**2. Linker fragment PCR**

*E. Reverse JH for scFv linker (sense)*

————————FR4 heavy————————
LINK BACK     5′ GGG ACC AC<u>G GTC ACC</u> GTC TCC TCA 3′


*F. Reverse Vk for scFv linker (anti-sense)*

————————FR1 light————————
LINKFOR     5′ TGG AGA CTG GGT <u>GAG CTC</u> AAT GTC 3′

*continues*

**TABLE 4** *Continued*

**3. Pull-through primers for introduction of restriction sites**

    *G. Mouse VH back (Sfi) primers[c]*

    VH1BACKSfi

|————————————FR1 heavy———————————|

5′ GTC CTC GCA ACT GC<u>G GCC</u> CAG CC<u>G GCC</u> ATG GCC CAG GTS MAR CTG CAG SAG TCW GG

    *H. Mouse J kappa forward (Not) primers[d]*

|————————————FR4 light————————————|

| | |
|---|---|
| JK1NOT10 | 5′ GAG TCA TTC T<u>GC GGC CGC</u> CCG TTT GAT TTC CAG CTT GGT GCC 3′ |
| JK2NOT10 | 5′ GAG TCA TTC T<u>GC GGC CGC</u> CCG TTT TAT TTC CAG CTT GGT CCC 3′ |
| JK4NOT10 | 5′ GAG TCA TTC T<u>GC GGC CGC</u> CCG TTT TAT TTC CAA CTT TGT CCC 3′ |
| JK5NOT10 | 5′ GAG TCA TTC T<u>GC GGC CGC</u> CCG TTT CAG CTC CAG CTT GGT CCC 3′ |

*Note:* Primers are taken from Clackson *et al.,* (1991). N.B. VHBACK and VHFOR and VK2BACK primers introduced additional *Pst*1, *Bst*EII, and *Sac*1 restriction sites, respectively (underlined).

[a]S=G,C; M=A,C; R=A,G; W=A,T.
[b]Together referred to as VK4FOR mix.
[c]Restriction site underlined.
[d]Together referred to as VK4FORNot mix.

For kappa light chains a mix of four different VKFOR primers with VK2BACK is used (primer groups C and D). We have not used primers to the mouse lambda light chains since only 5% of mouse antibodies use lambda light chains. Lack of sequence data makes it difficult to design primers for this group.

For human heavy chains six separate PCR reactions are set up using a different huVHBACK primer in each and with a mix of huJHFOR (primer groups A and B). For the kappa and lambda light chain the same approach is used with separate reactions defined by individual VBACK primers and a mix of huJFOR (i.e., individual HuJK-BACK/HuJKFOR mix and individual HuJλBACK/HuJλFOR mix, primer groups C and D). After PCR the various reactions from within a group can be mixed to give three pools (heavy chains, rappa chains, lambda chains).

---

**PROTOCOL 4**    **Preparation and purification of primary VH and VL PCR products**

1. Make up the following mix in 0.5-ml Eppendorf tubes.

| | |
|---|---|
| Individual BACK primer (10 pmol/µl) | 2.5 µl |
| FOR primer mix (10 pmol/µl) | 2.5 µl |
| Water | 30 µl |
| 10 × PCR buffer | 5 µl |
| Mix of dNTP at 10 mM each | 5 µl |
| BSA (10 mg/ml) | 0.5 µl |
| cDNA from protocol 2 | 5 µl |

(In negative control tube add 5 µl of water instead of cDNA).

2. Overlay with two drops of paraffin oil. Heat to 94°C for 5 min in the PCR block. At the end of this incubation, add 2.5 Units Taq polymerase (Cetus) under the oil. This "hot start" is to avoid the nonspecific extension that can occur as the reaction mix is heated through nonstringent temperatures.

3. Start temperature cycling, 30 times: 94°C for 1 min, 55–60°C for 1 min., 72°C for 2 min. This is followed at the end of cycling by an incubation of 10 min at 72°C. The cycling profiles of different thermal cyclers will vary and optimization of exact times and temperature may be required, particularly for the hybridization (55–60°C) step.

4. After the PCR, run a 5-µl aliquot of each reaction on a 2% agarose gel stained with ethidium bromide (see Appendix). The VH genes (~340 bp) and Vk/Vλ genes (~325 bp) should be clearly visible.

5. Primary amplification of human VH genes is carried out in six different reactions and these should be pooled at this stage to give a single human VH mix (see Fig. 3). Similarly, the six human VK and seven human Vλ primary PCRs can be separately pooled to give a human VK and a human Vλ mix ready for a VH/VK assembly and a VH/Vλ assembly. For mouse VHs and VLs a single reaction is used for each (see Fig. 3).

6. Before assembly, the amplified products must be separated from PCR primers; this is best achieved by gel electrophoresis. Add gel-loading dye, and fractionate the sample on a 2% LGT (low gelling temperature) agarose gel. The agarose gel-box, gel-tray, comb, and spacers should be depurinated with 0.25 M HCl (2 hr to overnight) and rinsed with water prior to use to reduce contamination with cloned DNA from previous analyses.

7. After electrophoresis transfer the gel to a long-wave transilluminator. Cover the screen first with Saran Wrap (to avoid DNA contamination). Excise both VH and VL bands using a fresh razor blade, and transfer the gel slices to 1.5-ml Eppendorf tubes.

8. Purify the VH and VL DNAs using a Geneclean II kit (Bio-101 *Inc.*) or Wizard Clean-Up kit (Promega) or other suitable clean-up method for gel slices.

9. Check the recovery of the DNA by analysis on an ethidium bromide-stained agarose gel and estimate the concentration against linearized DNA concentration standards. The sample may be stored at -20°C until required.

**NOTE:** (a) Most manufacturers of Taq DNA polymerase provide a suitable 10x PCR reaction buffer. We tend to use the high-fidelity buffer described by Eckert and Kunkel, (1990) (1x is 20 mM Tris-HCl, pH 7.3, at 70°C), 50 mM KCl, 4 mM magnesium chloride, 0.01% gelatin). With this buffer it is recommended that dNTP are used at a final concentration of 1 mM each. (b) It should be noted that the human primers HuVH2aBACK and HuVH6aBACK are designed for low-number and single-copy genes, respectively, and may give less product in the primary PCR. In addition, the products derived from the HuVK5aBACK and HuVK6aBACK primers may be fainter than the others.

The separate heavy- and light-chain fragments are converted to a single-chain Fv gene by inserting a DNA sequence encoding a flexible linker peptide. Figure 3 gives the sequence of a commonly used linker fragment encoding a (gly$_4$ser)$_3$ linker first described by Huston et al., (1988). Other linkers are described by Whitlow et al., (1993). The assembly of the VH, VL, and linker fragments is carried out by PCR and is driven by homologies at the ends of the various fragments (see Fig. 3). The 3' end of the heavy chain is complementary to the 5' end of the linker and the 3' end of this linker DNA is complementary to the 5' end of the light chain.

If an existing scFv clone is available the linker fragment can be prepared by PCR using this clone as template (Protocol 5). For mouse SCFv the primers for generating the linker fragment are LINKBACK and LINKFOR (see Fig. 3 and Table 4). With human SCFvs there are four RHuJH to be used (as a mix) in conjunction with seven RHuVλBACK and six RHuVkBACK primers to generate linker fragments (Fig. 3, and Table 3, primer sets E and F). The human primers were designed for use with the (gly$_4$ser)$_3$ linker and the primers straddle the junction between the variable regions and this linker (as shown in Table 3). The mouse primers shown for generating linker (LINKBACK and LINKFOR) are located wholly in the variable region and so are independent of the linker used.

If no existing scFv clone is available, the linker DNA fragment can be generated by synthesizing an oligonucleotide which encodes the linker peptide and is flanked by regions of homology with the VH and VL fragments, as shown at the bottom of Fig. 3. (The Recombinant Phage Antibody System (Pharmacia) includes a synthetic linker DNA fragment.)

---

**PROTOCOL 5**   **Preparation and purification of linker fragments**

1. For batch preparation of the human scFv linker fragments, set up the required number of PCR reactions as follows. Separate amplifications are performed for each RHuVkBACKFv or RHuVλBACKFv primer together with a mix of RHuJH primers (see Tables 3 and 4).

| | |
|---|---|
| RHuJH primer mix | 2.5 µl |
| Individual RHuVkBACKFv or RHuVλBACKFv primer | 2.5 µl |
| Water | 34.5 µl |
| 10x PCR buffer | 5 µl |
| Mix of dNTP at 10 mM each | 5 µl |
| BSA (10 mg/ml) | 0.5 µl |
| Template DNA, 1 ng | 1.0 µl |

   (See note (a) in Protocol 4 on choice of PCR buffer). For mouse scFvs replace RHuJH and RHuVkBACKFv/RHuVλBACKFv with LINK BACK and LINK FOR, respectively.

2. Overlay with two drops of paraffin oil. Heat to 94°C for 5 min in the PCR block. At the end of this incubation, add 0.5 µl Taq polymerase (5 units/µl; Cetus) under the oil. Then carry out 25 cycles at: 94°C for 1 min, 45°C for 1 min, 72°C for 1 min. This is followed by an elongation step at 72°C for 10 min.

3. Load 5 µl of each mix on a 2% agarose gel (see Appendix) to check if the PCR was successful. Since the DNA products might comigrate with the bromophenol blue dye, use 40% sucrose (sterile) instead of gel-loading dye or use an alternative dye such as Orange G or xylene cyanol.
4. The products are pooled according to their light chain family (lambda or kappa). These mixes can be run on 2–3% LGP agarose excised with a razor blade and purified using a Mermaid DNA purification kit (Bio 101) or other suitable method designed for purification of smaller fragments.

The generation of scFv genes requires the assembly of VH, linker, and VL fragments by PCR. The assembly step is often the most problematic step of the whole procedure and is helped by clean product from the primary PCRs. Assembly is a two-stage procedure: first a "mock" PCR without any primer is carried out where the short regions of complementarity built into the ends of the linker drive hybridization of the various fragments. In the second step, molecules which have been successfully linked are selectively amplified by the flanking primers. This step also introduces restriction enzyme sites at the termini for cloning (Fig. 3).

---

**PROTOCOL 6**    **Assembly of human single chain Fv antibody fragments (modified from Clackson et al., 1991; Marks et al., 1991)**

1. Estimate the quantities of light-chain, heavy-chain, and linker DNA prepared in the primary PCR reactions by comparing on agarose gel electrophoresis against concentration standards. For human genes there will be two different types of light chain (kappa and lambda) and two different linkers which correspond to each, and so two separate linkage reactions are required.

Set the linkage reaction up as follows:

*Kappa assembly (mouse or human)*

| | |
|---|---|
| Purified VH fragment | (20 ng) from Protocol 4 |
| Purified V kappa fragment | (20 ng) from Protocol 4 |
| Purified kappa linker fragment | (20–50 ng) from Protocol 5 |
| 10x Taq buffer | 2.5 µl |
| dNTP (10 mM each) | 2.5 µl |
| Water to | 25 µl |

*Lambda assembly (human)*

| | |
|---|---|
| Purified VH fragment | (20 ng) from Protocol 4 |
| Purified V lambda fragment | (20 ng) from Protocol 4 |
| Purified lamda linker fragment | (20–50 ng) from Protocol 5 |
| 10x Taq buffer | 2.5 µl |
| dNTP (10 mM each) | 2.5 µl |
| Water to | 25 µl |

(See note (a) in Protocol 4 on choice of PCR buffer). For mouse antibodies the contribution of lambda chains to the repertoire is much more limited than in man and consequently we do not use lambda-specific primers for our mouse repertoires.

2. Overlay with paraffin oil, heat to 94°C for 5 min using the PCR block, and add 2.5 U Taq polymerase under the oil. Do 25 cycles of 94°C for 1 min, 60°C for 2 min, and 72°C for 2 min followed by a 10-min incubation at 72°C. One potential problem in linkage reactions involving $(gly_4ser)_3$ is staggering of the codons for the $gly_4ser$ repeat, giving rise to shorter linkers, and so it is important to carry out the linkage at the highest temperature possible.

3. Correctly linked products from the assembly step are selectively amplified in a "pull-through" reaction using primers which anneal at the ends of the assembled product (Fig. 3, primer sets G and H). The pull-through primers incorporate a tail with a restriction enzyme site (for *Sfi*I and *Not*I at the 5' and 3' ends, respectively) to enable cloning in frame with the gene 3 signal sequence.

4. The pull-through PCR product is often cleaner using a low input from the linkage reaction (e.g., 1 µl). The reaction is set up as follows:

*Kappa assembly*

| | |
|---|---|
| HuVHBACKSfi primers | 2.5 µl (25 pmol) |
| HuJkFORNOT primers | 2.5 µl (25 pmol) |
| 10× Taq buffer | 5 µl |
| dNTP (5 mM each) | 2.5 µl |
| Water to | 50 µl |

Add 1 µl of product from the kappa linkage reaction.

*Lambda assembly*

| | |
|---|---|
| HuVHBACKSfi primers | 2.5 µl (25 pmol) |
| HuJλFORNOT primers | 2.5 µl (25 pmol) |
| 10× Taq buffer | 5 µl |
| dNTP (5 mM each) | 2.5 µl |
| Water to | 50 µl |

Add 1 µl of product from the lambda linkage reaction.

For mouse libraries VH1BACKSfi and VK4FORNot are used for the pull-through.

5. Heat to 94°C and add 2.5 U of Taq polymerase. Then carry out 25 cycles at: 94°C for 1 min, 55°C for 1 min, 72°C for 2 min. This is followed by an elongation step at 72°C for 10 min.

6. Sufficient PCRs are set up to generate at least 1 µg of product. The linked sample from steps 1–3 can be kept to generate more product in a pull-through reaction as required.

**NOTE:** (a) It is preferable to use a high concentration of fragments to drive the hybridization and it is important to have equimolar amounts of the VH and VL. The linker fragment is 1/3 the length of the others and so the inclusion of 20 ng of each fragment represents a three fold molar excess. Thus the linker may help to drive the assembly step without compromising the subsequent pull-through (since it has no homology with the pull-through primers). (b) While higher concentrations will help drive hybridization of the ends of the fragments, it is also worth trying one or two dilutions of the mix of fragments because there may be contaminants hindering hybridization or the action of Taq DNA polymerase. Problems with PCR can often disappear on dilution of template. We routinely take 1 and 4 µl of the 25-µl reaction from step 1 and make each up to 20 µl with PCR buffer/dNTPs. The sample giving best yield from the various assemblies with differing inputs of primary PCR product is selected and scaled up. (c) The purity and yield of each primary fragment as well as the integrity of the ends of the fragments will affect the success of the linkage; it is worth double checking these by gel elctrophoresis and repeating the preparation if necessary. (d) In the event of failure despite the above modifications it may be worth trying different PCR conditions, particularly the hybridization step, since the temperature profiles may vary with different PCR blocks. (e) The protocol above has been used in the work of Marks *et al.,* (1991) and Clackson *et al.,* (1991) and in other unpublished studies. It is possible to carry out primary PCR amplification using primers with "tails" at the 5' end. Engelhardt *et al.,* (1994) have taken advantage of this by adding a tail of 36 residues onto the ends of the light-chain VKBACK sequences and a complementary tail onto the heavy-chain JHFOR sequences. These together encode the entire linker sequence (as well as an *Asc*I restriction enzyme site). This modification simplifies the assembly step by reducing the number of fragments involved from 3 to 2. This example was demonstrated with a single antibody clone and it may have potential in the generation of repertoires. The longer the tails which are added, however, the greater the potential for unwanted products caused by cross-hybridization. This may be a particular problem if a diverse repertoire of sequences is being amplified.

For cloning into pCANTAB-5 and its derivatives, the rare-cutting enzymes *Sfi*I and *Not*I are used. It is important to digest extensively to ensure efficient cloning. Digestion of restriction sites near the ends of PCR fragments can be inefficient, particularly when using enzymes with 8-bp recognition sequences (such as *Sfi*I and *Not*I), and so 10- to 15-nucleotide overhangs are built into the PCR primers. Digests are set up at 50–100 ng/µl according to manufacturer's instruction.

---

**PROTOCOL 7     Restriction digestion of assembled PCR fragments**

1. Clean up the assembled, pull-through product using "Wizard PCR Preps" (Promega) or other suitable method and elute in water according to manufacturer's instructions.

2. *Sfi*I and *Not*I digests are carried out sequentially. Set up *Sfi*I restriction digest as follows:

|  |  |
|---|---|
| 10x *Sfi*I buffer | 30 μl |
| BSA (10 mg/ml) | 3 μl |
| approx 1.0 μg DNA and water to 300 μl final volume | |
| add *Sfi*I (10 units/μl) | 5 μl |

10x *Sfi*I buffer is: 500 mM NaCl, 100 mM Tris-HCl, 100 mM MgCl$_2$, 10 mM dithiothreitol (pH 7.9).

Overlay with paraffin. Incubate at 50°C for at least 12 hr.

3. For *Not*I digestion add 6 μl of 5 M NaCl. This increases the NaCl concentration from 50 to 150 mM. Add 50 units of *Not*I and incubate at 37°C for 6–16 hr.
4. Concentrate and clean up the product with Wizard Clean Up kit (Promega), elute in 50 μl, and load on a 2% LMP gel. After electrophoresis excise band and clean up with Wizard Clean Up kit (or other suitable method).
5. Estimate the concentration of the eluted DNA by comparison on an ethidium bromide-stained agarose gel with linearized DNA concentration standards.

**NOTE:** (a) The clean-up step (step 1) is important to remove Taq polymerase and nucleotides which might otherwise "fill in" the overhanging ends generated by *Not*I and *Sfi*I digestion.

---

**PROTOCOL 8    Preparation of vector DNA**

1. Digest with *Sfi*I and *Not*I in two steps as described in Protocol 7.
2. Efficient digestion is important here because a small amount of undigested vector leads to a very large background of nonrecombinant clones. Religation of the "stuffer" fragment can also lead to a high background and there are a number of ways of getting rid of this. Treatment with phosphatase will reduce background but also reduces cloning efficiency. The stuffer can be fragmented by cleavage with an enzymes in addition to those used for cloning and this helps reduce background. The stuffer from pCANTAB vectors can be cleaved with *Pst*I or *Sac*I.
3. The vector can be separated from the stuffer fragment by gel purification followed by Geneclean or Wizard Clean Up (as described in Protocol 4) or by spin-column chromatography. The latter is our method of choice for construction of large libraries. Commercially available spin columns (CS1000 columns, Cat. No. K1304-1, Clontech) significantly reduce background without compromising cloning efficiency. These can be used for both phage and phagemid vectors, as detailed in the booklet supplied by the manufacturer. The capacity of each column is around 10 μg.

**PROTOCOL 9**   **Ligation of insert into vector**

We have had excellent results using the Amersham 30-min ligation kit, which in our hands is not only faster but more efficient than standard ligation. The key components of this kit are in two tubes labeled A and B but the components of the kit are not given.

1. Set up the following ligation reaction:

| | |
|---|---|
| Vector (50 ng/µl) | 2 µl |
| Insert (digested PCR fragment, 10–50 ng/µl) | 1 µl |
| 1 $M$ Tris, pH 7.6, 50 m$M$ MgCl$_2$ | 1 µl |
| Water | 6 µl |
| Buffer (tube A) | 40 µl |

2. Spin for a few seconds in the microfuge, then add 10 µl Enzyme (tube B)

3. This is scaled up depending on the amount of vector/insert required for electroporation. Set up a control reaction as above but omitting the insert DNA. Mix well and leave for 30–60 min at 16°C.

4. After ligation, clean up and concentrate the sample using Geneclean (Bio101), Wizard Clean Up (Promega), or other clean up procedure. Redissolve in a low volume of water (10–40 µl depending on the scale of the ligation) in readiness for electroporation.

5. Use directly in electroporation of E. coli TG1 cells (see Appendix for genotype) or store at -20°C until further use. Even on a bad day, electrocompetent cells usually give transformation efficiencies 10 times better than that of the best chemically competent cells. For library construction an electroporator is a must. Methods for preparation of competent cells and electroporation are given in the Appendix. Set up a "no DNA" control for the electroporation to ensure the TG1 cells are not contaminated.

6. Following electroporation cells are plated onto bacterial plates with the appropriate antibiotic(s) as determined by the drug resistance marker on the plasmid. For antibody-gene 3 fusions transcribed from a lac promoter (e.g., pCANTAB vectors), 2% glucose should be included in the medium, to help supress transcription. Thus for pCANTAB vectors, transformed cells are spread on 24 x 24-cm plates containing 2xYT medium, 2% glucose, 100 µg/ml ampicillin, and incubated overnight at 30°C. The actual library size can be quantitated by spreading smaller aliquots of the transformation reaction on separate plates.

7. Make a frozen glycerol stock by scraping the large plate into 10 ml of 2xYT with 15% glycerol, aliquot into freezing vial, and store at -70°C.

**NOTE:** (a) Typical transformation efficiencies for E. coli TG1 cells are as follows: pUC119, 1–10 X 10$^9$ transformants/µg; ligated vector with insert, 1–10 X 10$^7$ transformants/µg; ligated vector without insert, 2 X 10$^6$ transformants/µg or less. Thus for a library of 10$^7$–10$^8$ one might start with 1 µg of cut vector and

200 ng of insert. This is ligated in 1.0–1.2 ml of ligation mix. Following clean up, the sample is divided into 5–10 aliquots and introduced into *E. coli* by electroporation. (b) If there is a low number of colonies on the plate check the transformation efficiency (by transfecting uncut vector DNA). Also recheck the amount of vector and insert and the success of the ligation, e.g., by running ligated sample on a gel (there should be a change of mobility on ligation). If there is sufficient vector and insert but no ligation there may be a problem with restriction digest of the insert. If available, a *SfiI/NotI* fragment from an existing plasmid will be useful to check out the vector and ligation mixes. (Digestion of restriction sites near the ends of fragment is more difficult than digestion of internal sites within a longer piece of DNA.) (c) If there is a sufficient number of clones but a high proportion without insert, this suggests either that the removal of the stuffer fragment during vector preparation was incomplete or that the double digestion of the vector has not worked. If there is a significant amount of vector, or vector cut at a single site, this will lead to high background. (d) It is worth checking the diversity of the library by taking 10–20 clones and sequencing or amplifying the inserts by PCR and digesting with a common cutting enzyme such as *Bst*N1 (see Clackson *et al.,* 1991).

## MANIPULATION OF LIBRARIES

Following generation of a library, the next step is to prepare phage particles expressing antibody–gene 3 fusions and select these repertoires against an appropriate antigen. The initial bacterial stock of the library is generated from transformed cells, plated out on TY agar plates with the appropriate antibiotic. There is no formal objection to growth in liquid culture, but we believe that growth on solid medium is probably safer as there is less direct competition between the clones. The lawn of cells is scraped from the plates and frozen glycerol stocks are prepared for long-term storage. As a precaution it may be desirable to prepare plasmid DNA from these cells for "back-up" storage.

Preparation of phage particles from libraries created in phage vectors (e.g., fd-tet and its derivatives) simply requires growth of bacteria in medium with the appropriate antibiotic. Generation of phage particles from phagemid libraries requires the use of a helper virus. Phagemid vectors such as pCANTAB5-E and pCANTAB6 use a lac promoter to drive gene 3 expression. With phagemid vectors it is important to repress gene 3 expression at all times, except when expression is specifically needed, i.e., after helper phage infection to create the phage antibody. The host cell needs to express lac $I^q$, which, together with catabolite repression by glucose, negatively regulates the lac promoter controlling gene 3 transcription in the phagemid. Removal of glucose is used to induce expression of the fusion gene protein for incorporation into phage particles. Cells expressing an exported product are generally happier at 30°C than at 37°C and so this temperature is used for phagemid growth. When performing the infection, cells should be grown at 37°C as phage do not infect efficiently at 30°C.

**PROTOCOL 10    Rescuing phagemid libraries**

1. Measure the $OD_{600}$ of a diluted aliquot of the library glycerol stock. An $OD_{600}$ of 1.00 represents approximately $8 \times 10^8$ cells. It should be noted, however, that not all of the frozen cells are viable and that viability is reduced with repeated rounds of freezing/thawing. Optical density measures both live and dead cells; hence, when using frozen aliquots of your library, it is a good idea to thaw an aliquot and measure the number of viable cells in each by plating dilutions on 2x agar containing 100 μg/ml ampicillin and 2% glucose (2x /glu/amp plates) the first few times you do this. The number of viable cells in the inoculum should be in excess of the number of different clones in the library by at least 10-fold to ensure that diversity of the library is not reduced.

2. Inoculate 50 ml 2x YT containing 100 μg/ml ampicillin and 2% glucose (2x YT/ glu/amp) in a 250-ml flask with sufficient cells to ensure that the representation of the library is not compromised; e.g., for libraries of $10^8$ use $10^9$ cells. If all cells in the inoculum were alive this would represent 1.25 OD of cells, giving 0.025 OD's per ml in the initial culture. The OD will be higher if there are significant numbers of dead cells.

3. Grow at 30°C for 2–2.5 hr or longer, until the $OD_{600}$ is about 0.5. This is to get the cells into mid-log phase so that they express the F pilus. If the OD is over 0.5 in 2–2.5 hr, dilute to 0.5 ($4 \times 10^8$ cells/ml) with prewarmed 2x YT/glu/amp.

4. For infection add sufficient helper phage to give a 1:1 ratio of helper phage:bacteria, i.e., add $2 \times 10^{10}$ pfu M13-KO7 or VCS-M13 helper phage to a 50-ml culture and incubate at 37°C for 60 min.

5. Spin at 3000g for 10 min and resuspend the cells in 500 ml 2x YT containing 100 μg/ml ampicillin and 50 μg/ml kanamycin (no glucose) in a 2-liter flask. This step induces expression of intact fusion by removing glucose. Smaller volumes of induced culture can be prepared if required, but we usually still dilute the infected culture 10-fold into the glucose-free medium (containing 100 μg/ml ampicillin, and 50 μg/ml kanamycin).

6. Incubate the culture at 30°C overnight with vigorous shaking. Next day, harvest the phage-antibody particles by centrifugation and PEG precipitation. Add $^1/_5$th volume of 20% PEG, 2.5 M NaCl, stand for 30 min at 4°C, and collect precipitation by centrifugation at 4000g per 10 min. Resuspend in PBS or T.E. ($^1/_{100}$ starting volume). Spin out any residual bacterial debris, in a microcentrifuge.

**NOTE:** (a) This protocol usually gives rise to $1–10 \times 10^{11}$ infectious particles (TU)/ml.

**PROTOCOL 11    Selection of phage-antibody libraries by panning in "immunotubes"**

1. Maxisorb immunotubes (Nunc; 75x12 mm, Cat. No. 4-44202) are coated overnight by addition of 4 ml of an appropriate concentration of antigen in PBS.

If antigen is limiting it is possible to use only 1 ml of antigen to coat. Proteins are usually coated at 10–100 µg/ml but concentrations as high as 1 mg/ml are used in the case of certain proteins (e.g., lysozyme). For some proteins coating may be better in 50 mM sodium hydrogen carbonate, pH 9.6, than in PBS. Different people have different preferences for coating temperatures between 4 and 37°C. ELISAs performed with the antigen on Nunc Maxisorb 96-well plates (Cat. No. 439454) are usually a good guide as to the best temperature, buffer, and antigen concentration to use for coating.

2. Next day, rinse the panning tube three times with PBS and block by filling to the brim with PBS containing 2% skimmed milk protein (PBSM). Cover with Parafilm/Nescofilm and incubate at 37°C for 2 hr to block.

3. Rinse tube three times with PBS.

4. Add $10^{12}$ to $10^{13}$ phage in 4 ml of 2% PBSM and incubate for 1–2 hr at room temperature, with inversion of the panning tube.

5. Wash tubes with 10–20 rinses in PBS, 0.1% Tween 20, then 10–20 rinses with PBS. Each rinsing step is performed by pouring buffer in and out immediately. This is best achieved using a wash bottle.

6. Elute phage from tube by adding 1 ml 100 mM triethylamine or 100 mM glycine (pH 3.0) to the tube for 10 min. If more than 1 ml of antigen has been used to coat the tube, then inversion of the tube during elution may be advisable to achieve maximum recovery. Neutralize immediately after elution by adding 0.5 ml 1.0 M Tris-HCl, pH 7.4. Store eluted viral particles at 4°C.

7. It is advisable to infect only $^1/_3$–$^1/_2$ of the eluate in case of problems or in case you subsequently decide to screen colonies from the eluate for soluble expression in nonsuppressor HB2151 cells (genotype in Appendix). Infect phage/phagemids particles into *E. coli* TG1 cells by incubating for 30 min with 5 ml of exponentially growing cells and spread on 24 x 24-cm 2x TY/2%glu/amp plates (Nunc Bio-Assay dish). This plate is scraped next day and the rescue/panning process repeated as required. The titre of the eluted phage can be determined from dilutions of the large-scale infection, or more accurately by diluting the eluted phage and mixing each dilution with exponentially growing TG1 cells for 30 min. The infected cells are then grown overnight on selective plates and counted next day to determine the phage titer. Numbers eluting will increase in later rounds of panning.

**NOTE:** (a) It is important in step 7 to plate out a control of uninfected cells to ensure that the host *E. coli* cells are not already infected. Discard and repeat the infection step if the host cells are already found to be infected. (b) It is also worth plating out this control not just on selective media associated with the library (e.g., ampicillin for most phagemid libraries) but also on tetracycline (on fd-tet-based vectors) or kanamycin (on helper viruses) if phage carrying these markers are used in the lab. Presence of the latter will reduce the apparent titer of the selected phage on ampicillin. (c) TG1 cells should be grown first on minimal plates to select for the F pilus, otherwise their infectability will be reduced. The infection step should be carried out at 37°C. (d) In some but not all cases, a successful panning can be monitored by the increasing output of phage after each round of selection. We generally carry out two to five rounds of selection. Greatest diversity (but lower proportion of binders) will be found in the earlier rounds of selection. We generally work from rounds two and three of the selection. (e) ELISAs may be carried out using populations of phage from each round of the selection. These polyclonal ELISAs provide a convenient insight into the success of a panning scheme. (f) If a panning does not seem to be working, it is worth trying to devise some means of checking out antigen coating conditions, e.g., if available, using an existing monoclonal or polyclonal antibody which recognizes the antigen. (g) The library itself may be checked out against a different antigen, e.g., BSA, thyroglobulin. The nonimmunized human library of $2 \times 10^7$ clones described by Marks et al., (1991) yielded binders to a large number of antigens. In the case of thyroglobulin 45% of binders after three rounds were specific (Griffiths et al., 1993). (h) See Chapter 15 for a discussion on factors affecting selection.

### Selection using biotinylated antigen

The ability to use soluble antigen for selections imparts a greater degree of control over the selection process. Soluble selections can be achieved using biotinylated antigen. Phage are allowed to bind the biotinylated antigen and are recovered using streptavidin-coated magnetic beads (e.g., Dynabeads M-280 Sreptavidin, Cat. No. 112.05/112.06 Dynal). The Dynabeads are pulled out of solution by a magnet (Dynal MPC-E, Dynal, Cat. No 120.04) and washed. The recovered phage can be introduced into bacteria by eluting with triethylamine or glycine (as in Protocol 11, step 6) or by adding the magnetic beads directly to bacteria. Alternatively, specific elution can be carried out using competing soluble antigen.

Biotinylation may be carried out using NHS esters of biotin (Pierce, NHS-LC-Biotin Cat. No. 21335) to modify primary amine groups. Alternatively, biotin LC hydrazide (Pierce, Cat. No. 21340) can be coupled to carbohydrates via the activator EDC (Pierce, Cat. No. 22980). For protocols see manufacturer's instruction.

A guide to getting started is given below, but it is at this stage that the user can be inventive and devise novel selection schemes to enrich and select for the desired binding activity (e.g., Hawkins et al., 1992). The actual amount of phage and the amount of antigen will vary according to the affinities present in the library and the stage of panning. The final antigen concentration should be around or below the expected $K_d$ as discussed in Chapter 15.

---

**PROTOCOL 12    Selection of phage-antibody libraries by biotin selection**

1. Block the phage by making up a solution of approximately $10^{11}-10^{13}$ phage in 1 ml of PBS/2% milk powder (PBSM) and leave at room temperature for 30 min. The streptavidin magnetic beads should also be separately blocked in the same way by removing from the storage buffer using the magnet and resuspending in PBSM.

2. After blocking, add a range of concentrations of biotinylated antigen, e.g., 0, 1, 10, and 100 n*M*, to tubes containing the phage. The numbers of recovered phage should increase with increasing input of antigen. If stringent selection is required one should then focus on the minimum concentration of antigen giving output numbers significantly above background (0 n*M* selection). If greater diversity of binders is sought it is advisable to use higher concentrations of antigen.

3. After incubation of phage with antigen for 1–2 hr, add an appropriate volume of blocked streptavidin coated magnetic beads for 15 min with inversion to keep the beads mixed. Dynal beads have a capacity for biotinylated protein of approximately 50 pmol/100 μl. Thus 200 μl of the original bead suspension will be required to recover all antigen in a 1-ml solution of 100 n*M* biotinylated antigen.

4. Beads with bound antigen:phage complex are recovered by placing in a magnetic rack for 2 min, removing the supernatant and washing the beads by resuspending in 1 ml of PBS containing 0.1% Tween 20, and transferring to a new tube. This procedure is repeated until the beads have been washed three times with PBS/Tween and three times with PBS. Finally the beads are resuspended in 100 μl of PBS and stored at 4°C.

5. Phage are reintroduced into *E. coli* by mixing half the recovered beads with 5 ml of exponentially growing TG1 cells for 30 min and plating out on a large 2x YT/ 2%glu/Amp plates as described in Protocol 11. This plate is scraped the next day and the rescue/panning process repeated as required. The titre can be determined from dilutions of the large-scale infection or, more accurately, by diluting the eluted phage and using the titration procedure described in Protocol 11 (step 7).

**NOTE:** (a) See notes for Protocol 11.

### Introducing diversity by chain shuffling

Having isolated binders by phage display it may be desirable to increase the affinity of the resultant clones. This first requires diversification of the clone followed by selection of the best variants from the resultant libraries. Diversification can be carried out by:

Mutagenesis of the original clone by PCR amplification using primers with randomized residues, Oligonucleotide site-directed mutagenesis (e.g., *in vitro* mutagenesis kit, Amersham International), "Chain shuffling."

Chain shuffling is an approach which has been used successfully for antibodies where either the heavy or light chains of a clone are recombined with a whole repertoire of complementary chains (Fig. 4). The method below describes how to do this for antibodies where the antibody to be shuffled and the repertoire of complementary chains are already in the form of scFvs. Primers based in the vector and in the linker are used in PCR to generate the V gene DNA fragments. The linker based primers introduce complementarity into the termini of heavy and light chains. This complementarity drives assembly. The procedure can be adopted for shuffling FAb fragments and also for shuffling smaller domains of individual heavy- and light-chain V genes.

**1. Primary PCR (on starting clone or complementary repertoire)**

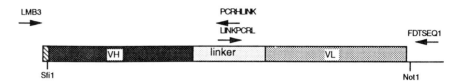

**2. Assembly**

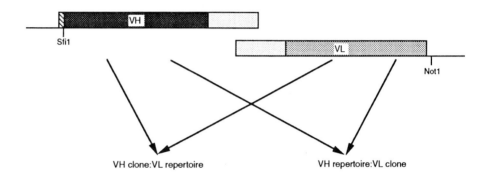

**3. Pull through PCR**

**FIGURE 4**          Chain shuffling scFv clones.

---

**PROTOCOL 13    Introducing diversity by chain shuffling**

The following primers are used and their location is shown in Fig. 4:

| | |
|---|---|
| FDTSEQ1 | 5′GTC GTC TTT CCA GAC GTT AGT 3′ |
| LMB3 | 5′CAG GAA ACA GCT ATG AC 3′ |
| PCRHLINK | 5′ACC GCC AGA GCC ACC TCC GCC 3′ |
| LINKPCRL | 5′GGC GGA GGT GGC TCT GGC GGT 3′ |

The primers PCRHLINK and LINKPCRL are complementary to each other and located in the linker sequence. The primer FDTSEQ1 is located downstream of the antibody gene in gene 3. The primer LMB3 is located in the vector sequences upstream of the gene 3 signal sequence (Fig. 4). If the clones to be chain shuffled are in a phage rather than a phagemid vector the primer LMB3 should be replaced with the primer KSJ28. This is also located upstream of the gene III signal sequence.

KSJ28   5′GTC ATT GTC GGC GCA ACT ATC GGT ATC 3′

1. Prepare primary heavy- and light-chain products in the following reactions:

| HEAVY | | LIGHT | |
|---|---|---|---|
| LMB3 primer (10 pmol/μl) | 2.5 μl | FDTSEQ1 primer (10 pmol/μl) | 2.5 μl |
| PCRHLINK primer (10 pmol/μl) | 2.5 μl | LINKPCRL primer (10 pmol/μl) | 2.5 μl |
| 10X PCR reaction buffer | 5.0 μl | 10X PCR reaction buffer | 5.0 μl |
| 10 mM each dNTP's | 5.0 μl | 10 mM each dNTP | 5.0 μl |
| Water to | 50 μl | Water to | 50 μl |
| Taq polmerase (5 U/μl) | 0.5 μl | Taq polmerase (5 U/μl) | 0.5 μl |

   For isolated clones, template can be provided most simply as a toothpick inoculum from a bacterial colony. For library material, DNA is prepared (e.g., from a frozen bacterial stock) and 2–10ng of this template added to the reaction. See note (a) in Protocol 4 on choice of PCR reaction buffer. PCR conditions are 25 cycles of: 94°C for 1 min, 60°C for 1 min, 72°C for 2 min, with a final 10 min at 72°C.

2. Primary PCR products are purified on agarose gels as described in Protocol 4, step 6. Check the recovery of the DNA by analysis on an ethidium bromide-stained agarose gel and estimate the concentration against linearized DNA concentration standards. It is possible to see 5 μl of the resultant product easily on a gel. Slightly more product can be obtained with a 55°C rather than a 60°C step in the PCR cycles, but it is preferable to use the higher temperature in this and in subsequent PCRs to avoid primer hybridization to closely related sequences in the $(gly_4ser)_3$ peptide linker.

3. Assembly is carried out using the same reaction conditions as used in the primary PCR amplification but without primers:

| | |
|---|---|
| Purified heavy chain (approximately 20 ng), e.g., from clone | 2.5 μl |
| Purified light chain (approximately 20 ng), e.g., from repertoire | 2.5 μl |
| 10X reaction buffer | 2.0 μl |
| 10 mM each dNTP | 2.0 μl |
| Water to | 20 μl |
| Taq DNA polymerase | 0.2 μl |

   The PCR conditions given for the primary PCR are used. It is advisable to shuffle both heavy and light chains with repertoires of complementary chains unless it is known which chain is most important to binding. The two resultant libraries can be mixed and selected as one.

4. For pull-through PCRs use 1 µl of the linked material as template. Set the reaction up as follows:

| | |
|---|---|
| Linked PCR product | 1 µl |
| LMB3 primer (10 pmol/µl) | 2.5 µl |
| FDTSEQ1 primer (10 pmol/µl) | 2.5 µl |
| 10X PCR reaction buffer | 5.0 µl |
| 10 mM each dNTP | 5 µl |
| Water to | 5 µl |
| Taq DNA polymerase | 0.2 µl |

PCR conditions are 25 cycles of 94°C for 1 min, 60°C for 1 min, 72°C for 2 min with a final 10 min at 72°C.

Five microliters can easily be seen on a gel as a band of approximately 1000 bp. (The absolute size will vary depending on CDR lengths within the VH and VL genes).

5. Digestion, purification, ligation, and transformation are as described in Protocols 7–9. In this case the *Sfi*I and *Not*I restriction sites are not introduced by the primer but are present within the amplified region. This arrangement permits a more efficient digestion due to the larger overhang. Another consequence is that the result of digestion is detectable by a change in mobility on agarose gels. It is possible this way to reduce each digestion step to 2 hr and confirm completeness of the digest on agarose gels.

---

## SCREENING AND EXPRESSION OF SELECTED CLONES

Following panning, individual colonies can be assayed directly for the ability to bind specific antigens. These may be screened as phage particles or as soluble fragments, by immunoassay techniques such as ELISA. Phage that have bound an immobilized ligand via the expressed antibody are detected using an antiserum raised against bacteriophage fd. The principal component recognized by this antiserum is gene VIII protein of which there are approximately 2800 copies, making up the bulk of the capsid. The structure of the phage itself can thereby be exploited as a convenient signal amplification system.

For detailed study of proteins, it is often necessary to express them in soluble form. The phagemid vectors described here carry an amber codon (TAG) between the protein of interest and gene 3. In supE bearing strains of *E. coli*, such as TG1 this codon is read as glutamic acid and a fusion between antibody and gene 3 is produced and incorporated onto phage particles. It is possible in TG1 to overexpress

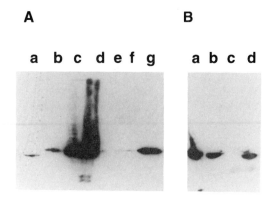

**FIGURE 5**    Expression of soluble single-chain Fv in supernatant and periplasm of *E. coli*. Samples were run on a 12% polyacrylamide gel, blotted, and detected with the 9E10 monoclonal antibody and ECL substrates (Amersham). All samples except (A, a) are ScFv from pOX16his-11 expressed in HB2151 (nonsuppressor) cells, which gives rise to the 28-kDa product:

$$\text{OX16 scFv} — \text{his}_6 — \text{myc tag.}$$

**A:** (a) Control antibody: D1.3 scFv with myc tag. (b) Total cell protein from 50 μl of uninduced cells. (c) Total cell protein from 50 μl of cells induced for 3 hr with 1 m*M* IPTG. (d) Total cell protein from 50 μl of cells induced for 18 hr with 1 m*M* IPTG. (e) Culture supernatant from 10 μl of uninduced cells. (f) Culture supernatant from 10 μl of cells induced for 3 hr with 1 m*M* IPTG. (g) Culture supernatant from 10 μl of cells induced for 18 hr with 1 m*M* IPTG.
**B:** All samples in B are from cells induced for 3 hr with 1 m*M* IPTG. (a) Total cell protein from 10 μl of cells. (b) Periplasmic proteins from 10 μl of cells. (c) Wash fraction of spheroblast pellet equivalent to 10 μl of cells. (d) Spheroblast proteins from 10 μl of cells.

this fusion by induction of the lac promoter with IPTG and also achieve significant production of soluble fragment. This is usually done in the absence of helper phage. Alternatively the phagemid can be infected into nonsupressor strains such as HB2151 which reads the amber codon as a stop signal. Soluble expression levels are usually higher in this strain (Fig. 5).

## PROTOCOL 14    Induction of soluble antibody fragments from phagemid vectors

1. Pick colony into 2x YT medium with 100 μg/ml ampicillin and 2% glucose and incubate overnight, shaking, at 30°C.
2. Inoculate 50 μl of the overnight culture into 5 ml 2x YT broth containing 100 μg/ml ampicillin and 0.1% glucose. Grow shaking at 30°C until OD$_{600}$ is 0.9 (approx 2–3 hr).
3. Add IPTG to a final concentration of 1 mM and continue shaking at 30°C for a further 4–16 hr.
4. Spin at 8000 rpm for 15 min to remove bacteria and debris. For overnight inductions the secreted antibody fragment can be found in the supernatant. For

shorter incubations the antibody fragment can be found in the periplasmic space (Fig. 5). Protocol 15 (steps 1–4) describes the isolation of the periplasmic contents.

**NOTE:** (a) This induction method can be adapted to growth of cultures in 96-well plates. Overnight cultures are grown in 96-well polysytrene plates (Corning Cell Wells) and used to inoculate a fresh 96-well plate next day. (Five microliters into 150 µl of YT/100 µg/ml Amp/ 0.1% glucose. We use a 96-position multiprong device for these transfers). When the cells have grown to approximately OD 0.5–0.9, 50 µl of 4 mM IPTG (in the same medium) is added (i.e., final conc. 1 mM). This is grown at 30°C overnight and the supernatant collected after centrifugation. Whatever centrifuge is used, ensure that the plates are not spun so fast that they crack (e.g., use 1800 rpm for a Centra 8R bench centrifuge). The above method reduces the number of manipulations by growing the cells in 0.1% glucose and adding IPTG directly to the culture. If there are a small number of samples and one is trying to maximize yield, it may be preferable to grow and infect in 2% glucose and then change to glucose-free medium for the IPTG induction step (e.g., protocol 15). (b) The expression level varies from one clone to another and can range from 2 to 1000 µg/ml. Western blotting of culture supernatants using the anti-tag antibody (e.g., 9E10 for myc tag) can be used to check if scFv is being produced (see Fig. 5).

A number of different expression vectors have been devised which introduce polyhistidine tags, thereby permitting rapid and simple protein purification by immobilized metal affinity chromatography (Fig. 5). pCANTAB6 (McCafferty *et al.*, 1994) carries such a tag (see vector in Appendix).

---

**PROTOCOL 15**   **Preparation of periplasmic fraction and purification of antibodies by immobilized metal affinity chromatography (IMAC)**

1. *E. coli* cells carrying the plasmid of interest are grown to 0.7–1.0 OD/ml in 50 ml of 2x YT medium supplemented with 2% glucose, 100 µg/ml ampicillin.
2. The culture is centrifuged in a 50-ml Falcon tube at 3000g for 10 min at room temperature, resuspended in 50 ml of 2x YT/100 µg/ml ampicillin/1 mM IPTG, and grown at 30°C for 3–6 hr.
3. The culture is centrifuged in a 50-ml Falcon tube at 3000g for 15 min at a temperature of 4°C and is resuspended in 1 ml of cold periplasmic buffer (PBS/1 M NaCl/1 mM EDTA) and left on ice for 15 min.
4. The sample is centrifuged 10 min, the supernatant carrying the periplasmic contents collected, and $MgCl_2$ added to 1–2 mM (Fig. 5B).
5. Four hundred microliters of a 1:1 slurry of Ni-NTA agarose:periplasmic buffer (Qiagen) is added to the periplasmic preparation and incubated for 10 min on an inverting platform.
6. The mixture is centrifuged at low speed on a microfuge for 10–15 sec and the pellet washed by resuspending in 10 ml of periplasmic buffer. This wash process is repeated another two times before eluting in 100 µl of either periplasmic

buffer or PBS carrying 250 mM imidazole (pH readjusted to 7.0–7.4). After 10 min the supernatant carrying the purified antibody is collected, the pellet is re-extracted with another 100 µl of the same buffer, and the eluates are pooled.

**NOTE:** (a) All volumes are for an initial culture volume of 50 ml but this can be scaled up or down as appropriate. Quiagen have introduced a spin column version of their IMAC matrix which is convenient and works well.

---

**PROTOCOL 16    Small-scale rescue of phagemid particles using helper virus**

Protocol 16 relates to the rescue of display particles from phagemid vectors. (For clones in phage display vectors a simple inoculum into medium with the appropriate selective antibiodic is sufficient.) The protocol is for 5-ml cultures; the volumes can be adjusted proportionately for different scale preparations. It is essential that the cultures are vigorously aerated. If the volumes are increased larger vessels will be needed.

1. Grow cells containing the phagemid in 5 ml 2x YT containing 100 µg/ml ampicillin, 2% glucose at 30°C to mid-log phase ($A_{600}$= 0.5). Cultures can be initiated from an overnight culture or directly from a plate.
2. For infection add an equal number of infectious phage to the bacteria, i.e., add $2.0 \times 10^9$ pfu/ml M13KO7 and grow at 37°C with moderate shaking (e.g., 200 rpm) for 1 hr.
3. Transfer 1/10th vol (0.5 ml) into 5 ml 2x YT containing 100 µg/ml ampicillin and 50 µg/ml kanamycin without glucose (TY/K/A) or add all 5 ml into 50 ml TY/K/A and grow overnight at 30°C. Pellet cells and precipitate phage from the supernatant by adding $^1/_5$th volume of 20% PEG, 2•5 M NaCl (Protocol 10 step 6). For ELISA it will often be possible to obtain good signals without precipitation, using 80 µl supernatant directly, diluted with 20 µl 5x PBS containing 10% skimmed milk powder.

**NOTE:** (a) This method can be adapted to growth of cultures in 96-well plates. Overnight cultures are grown in 96-well plates containing 2x YT, 100 µg/ml Amp, and 2% glu and used to innoculate a fresh 96-well plate the next day containing 100 µl of the same medium. For phage rescues we prefer to use polypropylene plates (e.g., Greiner) These plates are placed in an appropriate container such as a Boehringer Enzyme box fixed into the shaker. When the cells have grown to approximately OD 0.5, $4 \times 10^7$ helper phage are added in 100 µl of same medium and incubated at 37°C for 1–2 hr. After this 5 µl is transferred to a fresh 96-well plate containing 200 µl 2x YT, 100 µg/ml Amp, 50 µg/ml kanamycin (no glucose). Reproducibility and yield may be improved by incubating the final plate without a lid. The culture is grown at 30°C overnight and the supernatant collected. Centrifugation should be at a speed which pellets cells firmly but does not crack plates.

## PROTOCOL 17  ELISA using either soluble fragments or phage particles

1. Different antigens have different optimal coating conditions with respect to buffer (PBS or 50 mM $NaHCO_3$, pH 9.6), temperature (4°C, room temperature, 37°C), and concentration (see Protocol 11, point 1). Optimal coating concentration can be determined empirically if an antibody already exists. If not, start by coating the ELISA plate (e.g., Falcon 3912 plates) with 200 μl of 10 mg/ml antigen in 50 mM $NaHCO_3$, pH 9.6, or PBS overnight at 37°C.
2. Rinse wells three times with PBS and block with 200 μl per well of 2% (w/v) skimmed milk powder in PBS (PBSM) for 2 hr at 37°C.
3. Prepare supernatants from phage rescue or soluble expression. Phage can be concentrated using PEG precipitation (Protocol 10, step 6) although it should be possible to get signal directly from culture supernatants made up to 1 x PBSM (80 μl supernatant, plus 20 ml 5xPBS, 10% skimmed milk protein). Incubate sample for 1 hr at room temperature.
4. Wash wells three times for 2 min each, using PBS/0.1% Tween 20. Repeat using PBS.
5. Add 150 μl of an appropriate dilution of secondary antibody in 2% marvel/PBS to each well;. e.g., for soluble fragments carrying a myc tag (see Appendix) add an appropriate dilution of 9E10 antibody (from Cambridge Research Biochemicals). For antibodies with the E tag, add anti-E tag antibody (Pharmacia). For phage ELISAs an antibody against the phage coat protein is used. The Pharmacia Phage Detection module (27-9402-01) provides an anti-fd antibody which is already conjugated to peroxidase. Incubate for 1 hr at room temperature.
6. Wash as in step 4.
7. Add 150 μl peroxidase-conjugated antibody, e.g., for soluble fragments detected with mouse anti tag antibodies, use peroxidase-conjugated anti-mouse (Sigma). Incubate for 1 hr at room temperature.
8. Wash as in step 4.
9. Add one 10-mg ABTS (Sigma: 2,2′-azino bis(3-ethylbenzthiazoline-6-sulfonic acid, diammonium salt) tablet to 20 ml 50 mM citrate buffer (50 mM citrate buffer is made by mixing equal volumes 50 mM trisodium citrate and 50 mM citric acid).
10. Add 2 μl 30% hydrogen peroxide to the above solution immediately before dispensing.
11. Add 150 μl of the above solution to each well. Leave at room temp. until sufficiently developed (5 min to overnight). Read at 405 nm periodically.

### Acknowledgments

We acknowledge the numerous colleagues at Cambridge Antibody Technology and in Greg Winters' laboratory at the MRC, Cambridge, for their contributions to the methods presented here.

## References

Abrol, S., Sampath, A., Arora, K., and Chaudhary, V. J. (1994). Construction and characterization of M13 bacteriophages displaying gp120 binding domains of human CD4. *Indian J. of Biochem. and Biophys.* **33,** 5689–5695.

Ausubel, F. M., Brent, R., Kingston, R. E., Moore, D. D., Seidman, J. G., Struhl, K., and Smith, J. A. (1995). "Current Protocols in Molecular Biology." Wiley, New York.

Barbas, C. F., III, Bain, J. D., Hoekstra, D. M., and Lerner, R. A. (1992). Semisynthetic combinatorial antibody libraries: A chemical solution to the diversity problem. *Proc. Natl. Acad. Sci. U. S. A.* **89,** 4457– 4461.

Bass, S. H., Greene, R., and Wells, J. A. (1990). Hormone phage: An enrichment method for variant proteins with altered binding properties. *Proteins* **8,** 309–314.

Clackson, T., Hoogenboom, H. R., Griffiths, A. D., and Winter, G. (1991). Making antibody fragments using phage display libraries. *Nature (London)* **352,** 624–628.

Eckert, K. A., and Kunkel, T. A. (1990). High fidelity DNA synthesis by the Thermus aquaticus DNA polymerase. *Nucleic Acids Res.* **18,** 3739–3744.

Engelhardt, O., Grabherr, R., Himmler, G., and Ruker, F. (1994). Two step cloning of antibody variable domains in a phage display vector. *BioTechniques* **17,** 45–46.

Griffiths, A. D., Malmqvist, M., Marks, J. D., Bye, J. M., Embleton, M. J., McCafferty, J., Gorick, B. D., Hughes-Jones, N. C., Hoogenboom, H. R., and Winter, G. P. (1993). Human anti-self antibodies with high specificity from phage display libraries. *EMBO J.* **12,** 725–734.

Griffiths, A. D., Williams, S. C., Hartley, O., Tomlinson, I. M., Waterhouse, P., Crosby, W. L., Kontermann, R., Jones, P. T., Low, N., Allison, T. J., Prospero, T., Hoogenboom, H. R., Nissim, A., Cox, J. P. L., Harrison, J. L., Zaccolo, M., Gherardi, E., and Winter, G. (1994). Isolation of high affinity human antibodies directly from large synthetic repertoires. *EMBO J.* **13,** 3245–3260.

Hawkins, R. E., Russell, S. J., and Winter, G. P. (1992). Selection of phage antibodies by binding affinity-mimicking affinity maturation. *J. Mol. Biol.* **226,** 889–896.

Hoogenboom, H. R., and Winter, G. (1992). By-passing immunization: Human antibodies from synthetic repertoires of germline VH gene segments rearranged in vitro. *J. Mol. Biol.* **227,** 381–388.

Huston, J. S., Levinson, D., Mudgett, H. M., Tai, M. S., Novotny, J., Margolies, M. N., Ridge, R. J., Bruccoleri, R. E., Haber, E., Crea, R., and Opperman, H. (1988). Protein engineering of antibody binding sites: Recovery of specific activity in an anti-digoxin single-chain Fv analogue produced in *Escherichia coli. Proc. Natl. Acad. Sci. U.S.A.* **85,** 5879–5883.

Jamieson, A.C., Kim, S., and Wells, J. A. (1994). In vitro selection of zinc fingers with altered DNA binding specificity. *Biochemistry* **33,** 5689–5695.

Kushwaha, A., Chowdhury, P. S., Arora, K., Abrol, S., and Chaudhary, V. J. (1994). Construction and characterization of M13 bacteriophages displaying functional IgG binding domains of Staphylococal protein A. *Gene* **151,** 45–51.

Marks, J. D., Hoogenboom, H. R., Bonnert, T. P., McCafferty, J., Griffiths, A. D., and Winter, G. (1991). By-passing immunization: Human antibodies from V-gene libraries displayed on phage. *J. Mol. Biol.* **222,** 581–597.

McCafferty, J., Griffiths, A. D., Winter, G., and Chiswell, D. J. (1990). Phage antibodies: Filamentous phage displaying antibody variable domains. *Nature (London)* **348,** 552–554.

McCafferty, J., Jackson, R. H., and Chiswell, D. J. (1992). Phage-enzymes: Expression and affinity chromatography of functional alkaline phosphatase on the surface of bacteriophage. *Protein Eng.* **8,** 955–961.

McCafferty, J., FitzGerald, K. J., Earnshaw, J., Chiswell, D. J., Link, J., Smith, R., and Kenten, J. (1994). Selection and rapid purification of murine antibody fragments that bind a transition-state analog by phage display. *Appl. Biochem. Biotechnol.* **47,** 157–173.

Orlandi, R., Gussow, D. H., Jones, P. T., and Winter, G. (1989). Cloning immunoglobulin variable domains for expression by the polymerase chain reaction. *Proc. Natl. Acad. Sci . U.S.A.* **86,** 3833–3837.

Parmley, S. F., and Smith, G. P. (1988). Antibody-selectable filamentous fd phage vectors: Affinity purification of target genes. *Gene* **73,** 305–318.

Rebar E. J., and Pabo, C. O. (1994). Zinc finger phage: Affinity selection of fingers with new DNA binding specificities. *Science* **263,** 671–673.

Roberts, B. L., Markland, W., Ley, A. C., Kent, R. B., White, D. W., Guterman, S. K., and Ladner, R. C. (1992). Directed evolution of a protein: Selection of potent neutrophil elastase inhibitors displayed on M13 fusion phage. *Proc. Natl. Acad. Sci. U.S.A.* **89,** 2429–2433.

Sambrook, J., Fritsch, E. F., and Maniatis, T. (1989). *"Molecular Cloning: A Laboratory Manual,"* 2nd ed. Cold Spring Harbor Lab., Cold Spring Harbor, NY.

Soumillion, P., Jespers, L., Bouchet, M., Marchand-Brynaert, J., Winter, G., and Fastrez, J. (1994). Selection of β-lactamase on filamentous bacteriophage phage by catalytic activity. *J. Mol. Biol.* **237,** 415–422.

Swimmer, C., Lehar, S. M., McCafferty, J., Chiswell, D. J., Blattler, W. A., and Guild, B. C. (1992). Phage display of ricin B chain and its single binding domains: System for screening galactose-binding mutants. *Proc. Natl. Acad. Sci. U.S.A.* **89,** 3756–3760.

Whitlow, M., Bell, B. A., Feng, S. L., Filpula, D., Hardman, K. D., Hubert, S. L., Rollence, M. L., Wood, J. F., Schott, M. E., Milenic, D. E., Yokota, T., and Schlom, J. (1993). An improved lialzer for single chain Fv with reduced aggregation and enhanced proteolytic stability. *Protein Eng.* **6,** 989–995.

# 7

# Phagemid-Displayed Peptide Libraries

## Diane Dottavio

## INTRODUCTION

The most salient features of a standard *Escherichia coli* cloning vector are the origin of replication, the antibiotic resistance gene, and the cloning sites. When a filamentous bacteriophage origin of replication (f1 ori) is added to such a standard vector, the vector can be packaged as single-stranded DNA in a bacteriophage coat. Single-stranded plasmid DNA that is packaged in a bacteriophage coat is referred to as a phagemid. All of the functions for replication and packaging of the phagemid DNA are supplied by a "helper phage." Cells are transformed with phagemid DNA and are infected with helper phage. The helper is impaired in its own replication and self-packaging functions so that only a small fraction of helper phage is produced relative to phagemids. Once the helper has made single-stranded phagemid DNA and packaged it in a bacteriophage coat, the phagemid is as infectious as a bacteriophage; however, it cannot replicate and package to form infectious units without the aid of the helper phage. For the pIII phagemid library, the surface of each phagemid particle can display both wild-type (encoded by the helper) and modified pIII (encoded by the phagemid DNA).

The phagemid system has been used in bacteriophage display libraries for the purpose of presenting a modified bacteriophage protein (Bass *et al.,* 1990; Barbas and Lerner, 1991) and the wild-type version of this coat protein on the same bacteriophage particle. The phagemid system allows one to present, and at the same time compensate for, a nonfunctional bacteriophage protein. This is essential when the modification of the pIII or the pVIII protein results in its inactivation. In some cases it is not essential, but desirable, to decrease the copy number of the displayed protein on the surface of the phage. This decrease in the copy number of the modified protein avoids artifactually tight binding events that are the result of polyvalent rather than high-affinity binding.

Another way of obtaining bacteriophage that display wild-type and modified surface proteins is to clone the modified pIII or pVIII into the bacteriophage genome, such that one bacteriophage genome encodes both the wild-type and modified genes. Differential transcriptional regulation is used to control the ratio of modified to wild-type protein. Theoretically, this "double-gene" system gives the same display as the helper phage system without some of the technical difficulties. To date, several proteins (Huse *et al.,* 1992; Corey *et al.,* 1993; Markland *et al.,* 1991) have been displayed via the double gene approach. (To avoid recombinational events in the double-gene system, the second copy of the gene is often synthesized, taking advantage of the degeneracy of the genetic code, to avoid repeating the same nucleotide sequence twice in the same genome.) Thus, when large insertions are desirable for either pVIII or pIII, helper phage or the double-gene system are employed. For example, at times it is useful to exploit the bacteriophage system for multiple mutagenesis of a protein (Barrett *et al.,* 1992; Lowman *et al.,* 1991; Swimmer *et al.,* 1992; Garrard *et al.,* 1991; Gram *et al.,* 1993). In these experiments a protein domain, or an entire protein, is expressed in place of the entire N terminal domain of pIII. This can only be accomplished in a system that coexpresses the wild-type and the modified pIII. Since pVIII is inactivated by most modifications, most pVIII modifications utilize double-gene or phagemid vectors.

Employing the phagemid approach, we designed a phage display system that utilized phagemid technology and *in vitro* mutagenesis (Kunkel *et al.,* 1987) to generate libraries of random peptides on the surface of bacteriophage. Expression of a modified gene III protein was placed under control of the *lac* promoter on the phagemid vector, pGEM-3Zf+. The advantages of this system (Wright *et al.,* 1995) include the (i) ability to control the expression level of peptides on the phage surface through induction of the *lac* promoter; (ii) facile, large-scale production of ssDNA, which serves as a "universal template" from which many different peptide libraries can be prepared; and (iii) elimination of restriction digests during preparation of libraries.

# MATERIALS

    1. pGEM-3Zf+ plasmid vector (Promega, Madison, WI).

    2. Supercompetent *E. coli* strain DH10B (Gibco/BRL, Grand Island, NY).

    3. Electrocompetent XL1-Blue No. 200236 (Stratagene, La Jolla, CA). Store at -80°C.

    4. CJ236 [*dut⁻,ung⁻*] cells in agar stab (Stratagene). (Store in the dark at room temperature.)

    5. Interference resistant helper bacteriophage R408 (Invitrogen Corp., San Diego, CA). Store at -20°C.

6. Restriction enzymes and T4 DNA ligase (Gibco/BRL).

7. Bio-Rad (Hercules, CA) Muta-Gene Phagemid *In Vitro* mutagenesis kit No. 1703582 with T7 DNA polymerase. Store at -20°C.

8. Oligonucleotide synthesizer Model 392 (Applied Biosystems).

9. OPC Sepak columns, No. 400771.

10. Perkin Elmer thermocycler and PCR kit (Perkin Elmer).

11. Small scale CPG columns (0.2 µm, 1000 Å pore size) for oligonucleotide synthesis (Applied Biosystems). Bio-Rad Electroporator and 0.2-cm cuvettes (Gibco/BRL). Store cuvettes at -20°C.

12. Applied Biosystems 373A DNA sequencer and PRISM Dyedeoxy terminator sequencing kit.

13. Wizard PCR Clean Up kit No. A7170 (Promega).

14. Prep-a-gene kit (Bio-Rad).

15. LB broth (10 g bacto-tryptone, 5 g yeast extract, and 5 g NaCl, pH 7.5) supplemented with 30 µg tetracycline/ml.

16. TE buffer (10 m$M$ Tris-HCl, pH 8.0, 5 m$M$ EDTA).

17. 2x YT broth (16 g bacto-tryptone, 10 g yeast extract, 5 g NaCl/liter), pH 7.5, containing antibiotics (70 µg/ml chloramphenicol or 50 µg/ml ampicillin).

18. TBS buffer (50 m$M$ Tris-HCl, pH 7.4, 150 m$M$ NaCl) is used to resuspend phagemids.

19. TBS buffer + 0.5 % Tween 20 is used as wash buffer.

20. Elution buffer for phage, 0.1 $M$ glycine buffer, pH 2.2.

21. LB plates contain ampicillin (100 µg/ml).

22. Antibody coating buffer (50 m$M$ NaHCO$_3$ buffer, pH 9.6).

## CONSTRUCTION OF PHAGEMID VECTOR PIII.PEP

This section deals with the construction of the modified M13 gene III (Fig. 1) as it appears in the pGEM-3Zf+ phagemid vector. Production of single-stranded DNA from this vector in CJ236 cells will provide template DNA for synthesis of the peptide libraries.

---

**PROTOCOL I**    **Generation of the two fragments needed to construct the gene III minigene**

Two separate PCR reactions are set up as follows (final volume, 100 µl):

> 100 ng M13 RF template (Gibco/BRL)
> 10x PCR buffer
> Taq DNA polymerase (0.5 units)
> 1.5 mM MgCl$_2$
> 1.25 m$M$ each dNTP (final concentration)
> 10 pmol oligonucleotides 1/2 to generate
>   the leader sequence of gene III, and
>   3/4 for the gene III second domain.

Diane Dottavio

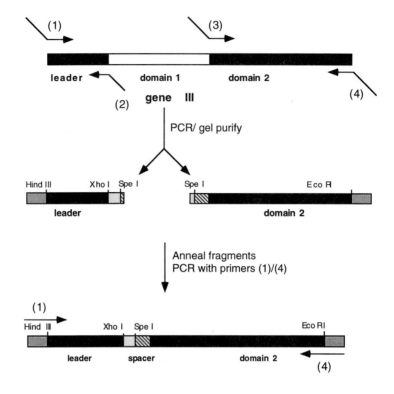

Oligonucleotide primers:
(1) 5'-(X)₁₂AAGCTTTAAGAAGGAGATATACATGTGAAAAAATTATTATTCGCA-3'
(2) 5'-GCCACCACTAGTGAAATCAGGACTCGAGAGCGGAGTGAGAATAGA-3'

(3) 5'-TCCTGATTTACTAGTGGT GGCGGTGGCTCTCCATTCGTTTGTGAATATCAA-3'
(4) 5'-(X)₁₂GAATTCTTATTAAGACTCCTTATTACGCAG-3'

GGT GGCGGTGGCTCT= flexible linker ▨

AAGCTT= Hind III;  CTCGAG=Xho I;  ACTAGT= Spe I;  GAATTC= Eco RI

---

**FIGURE I**  Construction of gene III fusion. Gene III of bacteriophage M13 is engineered according to the scheme presented. The signal sequence and the C-terminal portion (domain 2) of gene III are amplified by two separate polymerase chain reactions (PCR). The two products are purified from an agarose gel and then used in another PCR, with primers (1) and (2). The final product contains a short spacer between the two segments of protein III and has the sites *Xho*I and *Spe*I. The sequences of the primers and spacer region are shown. Underlined sequences correspond to restriction sites of *Hind*III, *Xho*I, *Spe*I, and *Eco*RI.

The PCR program for both reactions (35 cycles) is 94°C for 30 sec, 55°C for 30 sec, and 72°C for 1 min. Using the M13 RF as a template, the oligonucleotide pairs 1/2 and 3/4 are used to PCR-amplify DNA encoding the gene III leader sequence

(110 bp) and gene III second domain (651 bp), respectively. The 5′ and 3′-overhangs incorporate the restriction sites and additional features (10-bp spacer and GGGGS linker). The two PCR products are isolated on agarose gels (2.0 and 0.6%, respectively) and purified using the Wizard DNA clean up kit. The DNA is eluted from the matrix in 25 μl of TE, pH 7.5, previously heated to 55°C.

---

**PROTOCOL 2**  **Construction of the gene III fusion gene by splice overlap extension (SOE) and PCR amplification**

Equimolar amounts (0.5 pmol) of each fragment are mixed together in an initial volume of 50 μl containing the following:

> M13 leader fragment (18 ng)
> Gene III second domain fragment (100 ng)
> 10X PCR buffer (Perkin Elmer)
> 1.5 m$M$ MgCl$_2$
> 1.25 m$M$ each dNTP (final concentration)
> *Taq* DNA polymerase (0.5 units)

The PCR program links three different reactions: for the first 5 cycles, the program is set at 94°C for 30 sec, 42°C for 30 sec, 72°C for 1 min; a 5-min hold at 40°C follows, during which 50 μl, containing the same salts, dNTP concentrations, and 10–20 pmol each of oligonucleotides 1 and 4 is added to the reaction. The PCR program for the next 40 cycles is as follows: 94°C for 30 sec, 65°C for 30 sec, and 72°C for 1 min.

No oligonucleotides are added to the initial reaction. During the first five cycles, the 3′ end of the M13 leader fragment (110 bp) anneals to the complementary overlap (22 bp) added to the 5′ end of the DNA fragment encoding the gene III second domain (651 bp), and primes the extension of the overlapping ends. Addition of oligonucleotides 1 and 4 during the subsequent cycles amplifies a 743-bp fragment. This fragment is isolated on 0.6% low melt agarose in TAE buffer, and purified using Wizard DNA Clean Up kit. The fragment is restricted with *Hind*III and *Eco*RI, and ligated into pGEM-3Zf+ (which had been similarly digested), to position the gene III fusion gene under control of the *lac* promoter.

---

**PROTOCOL 3**  **Introduction of *Not*I linker in place of the gene III spacer**

A final modification in the pIII.PEP vector was introduced by the insertion of a *Not*I linker, which replaces the spacer and adds a unique *Not*I restriction site between the *Xho*I and *Spe*I sites (Fig. 2). In this way, phagemids which do not undergo mutagenesis could be identified by *Not*I digestion. This protocol demonstrates the utility of the restriction sites *Xho*I and *Spec*I for the introduction of a specific

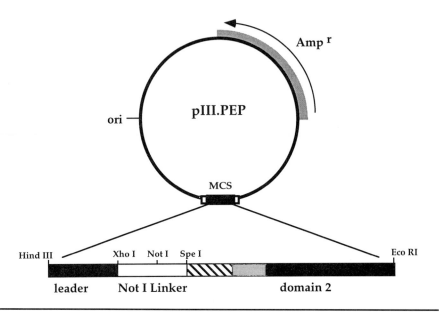

**FIGURE 2**    Diagram of the pIII.PEP phagemid vector with the selectable NotI linker. The final PCR product, shown in Fig. 1, is cleaved with *Hin*dIII and *Eco*RI and cloned into pGEM-3Zf+, to generate the vector pIII.PEP. This vector has been modified by cloning a double-stranded oligonucleotide with a *Not*I site between the *Xho*1 and *Spe*I sites of pIII.PEP.

(known) coding sequence into gene III. Alternatively, the *in vitro* mutagenesis protocol can be used to replace the spacer with a sequence encoding a specific peptide.

1. The oligonucleotide, 5′-G TCC TGA CGC TCG **GCG GCC GCT** AAT ACT AG-3', and its complement, 5′-T ATT A**GC GGC CGC** CGA GCG TCA GGA CTC GA-3′ which encode the *Not*I linker were synthesized, using an Applied Biosystems Model 392 sequencer.
2. The oligonucleotides were purified on an OPC Sepak column equilibrated in tri-ethanolamine bicarbonate buffer, according to the manufacturer's protocol.
3. The primers (10 pmol each) were annealed in 10 µl annealing buffer by placing the tube in a small heating block at 94°C, and allowing it to cool to 37°C, such that the ends formed 5′ overhangs complementary to *Xho*I- and *Spe*I-restricted sites.
4. The *Not*I linker was then ligated into the pIII.PEP vector, which had been restricted with *Xho*I and *Spe*I.
5. One microliter of the ligation mix was used to transform electrocompetent XL1-Blue cells, using standard protocols (Sambrook *et al.*, 1989).

## PREPARATION OF PHAGEMID PEPTIDE LIBRARIES

The oligonucleotides consist of random nucleotides encoding five or more amino acids, flanked by 30 fixed residues, which are complementary to the template on either side of the *Not*I linker. For example, libraries of (-) strand oligonucleotides were synthesized containing the sequence 5' TGG AGA GCC ACC GCC ACC ACT AGT **(SNN)**$_x$CTC GAG AGC GGA GTG AGA ATA 3' (where **S** is G or C, and **N** is A, G, T, or C) for linear peptides and 5' TGG AGA GCC ACC GCC ACC ACT AGT **TGT (SNN)**$_x$ **TGT** CTC GAG AGC GGA GTG AGA ATA 3' for cyclic peptides. Varying the number of nucleotides encoding the random peptide region **(SNN)**$_x$ from 18 to 30 bases (*X*=6–10 amino acids) has not affected the yields of dsDNA, or transformation efficiencies. The partial duplex is then converted into double-stranded DNA *in vitro* and used to generate a library of phage-displayed random peptides (Fig. 3).

---

**PROTOCOL 1**   **Synthesis and purification of random oligonucleotides for library synthesis**

1. Oligonucleotides bearing random sequences are synthesized on an Applied Biosystems Model 392, using the 0.2 μ*M* synthesis scale and the 1000 Å pore size ABI columns.
2. The oligonucleotides are cleaved from the synthesis column and deprotected in ammonium hydroxide (55°C for 8 hr).
3. They are then isolated on base-resistant OPC column in TEAB buffer, according to the manufacturer's protocol, and aliquots (8 x 500 μl) are dried down in a Savant speed-vac. At this point, the oligonucleotide preparation is approximately 80% pure.
4. It is necessary to further purify the oligos on polyacrylamide (20%)/urea gels to remove contaminants which interfere with the second-strand priming and synthesis. The oligonucleotides are resuspended in 50 μl TE, an equal volume of sample buffer is added, and aliquots are electrophoresed at 150 V until the bromphenol blue runs about two-thirds of the way into the gel.
5. The oligonucleotide band is visualized by UV absorption, cut out, and eluted overnight in TE buffer at 4°C.
6. After precipitation, using 1/10 vol 7.8 *M* ammonium acetate and 2 vol ethanol, the pellet is washed in 70% ethanol, dried, and resuspended in water (50 μl), and the concentration adjusted to 10 pmol/μl, based on the absorbance determined at 260 nm.
7. The oligonucleotides (200 pmol) are phosphorylated according to the protocol outlined in the Bio-Rad Muta-Gene kit, and used as primers for *in vitro* mutagenesis with (+) single-stranded, uracil-containing phagemid template.
8. After heat-inactivation of T4 polynucleotide kinase, the reaction mix is diluted to 10 pmol/μl and stored at -20°C.

---

**PROTOCOL 2**   **Preparation and purification of template ssDNA**

The single-stranded, uracil-containing DNA packaged in the phagemid particles serves as the template for second-strand synthesis. The random oligonucleotide

Diane Dottavio

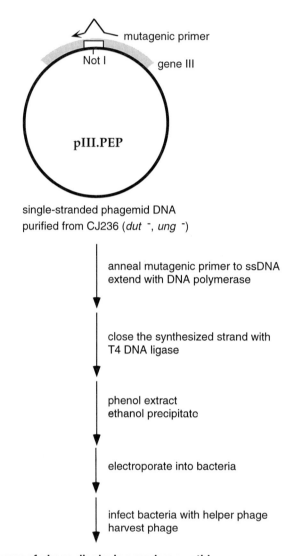

single-stranded phagemid DNA
purified from CJ236 (*dut* ⁻, *ung* ⁻)

anneal mutagenic primer to ssDNA
extend with DNA polymerase

close the synthesized strand with
T4 DNA ligase

phenol extract
ethanol precipitate

electroporate into bacteria

infect bacteria with helper phage
harvest phage

**library of phage displaying random peptides**

**FIGURE 3**   Flow-diagram of the generation of a phage-displayed library through site-directed, oligonucleotide-mediated mutagenesis. The mutagenic primer, encoding random codons, is annealed to single-stranded template DNA, corresponding to the pIII.PEP phagemid vector. The second strand of the partial duplex is synthesized and closed with DNA polymerase and DNA ligase, respectively. After removal of proteins and salt, the DNA is introduced into bacteria by electroporation. The transfected cells are then infected with helper phage and the secreted viral particles are harvested. Recombinant and nonrecombinant genomes can be distinguished by *Not*I restriction enzyme digestion.

pool described above is used collectively to prime second-strand synthesis. The DNA template is isolated as outlined in the Bio-Rad M13 mutagenesis protocol using the bacterial strains CJ236 [dut⁻, ung⁻]. To improve the efficiency of second-strand synthesis and reduce spurious priming, the uracil–DNA is further purified by ion exchange chromatography using the Bio-Rad Prep-a-Gene kit. The additional purity achieved by the Prep-a-Gene (Bio-Rad) purification step dramatically improves dsDNA synthesis and subsequent transformation efficiency. Promega (Wizard) and Bio-Rad (Prep-a-Gene) kits no longer contain the same DNA-binding resin; the recovery of ssDNA from the Promega resin is not recommended by the manufacturer. If another kit is used, it is important to check whether ssDNA can be recovered from the resin supplied.

1. Inoculate 4 x 125-ml flasks containing 25 ml 2x YT broth (50 µg/ml ampicillin and 70 µg/ml chloramphenicol) with 2.5 ml CJ236 [dut⁻, ung⁻] grown to approximately 0.5 $OD_{600}$.
2. Culture the cells for 8 or 16 hr (overnight).
3. Centrifuge twice at 12,000 rpm at 4°C for 30 min to remove cells and debris.
4. Precipitate the phagemid by adding 1/4 vol 20% PEG 8000/3.5 $M$ ammonium acetate to the clarified supernatant and let stand on ice for 1 hr.
5. Collect phagemid by centrifugation at 12,000 rpm at 4°C for 30 min.
6. Carefully pore off the supernatant from the pellet and drain dry on paper towels.
7. Resuspend the pellet in 0.2–1.0 ml TE buffer and extract twice with equal volumes of TE-saturated phenol, followed by equal volumes chloroform until the interface is clear.
8. Precipitate with 3 $M$ sodium acetate and ethanol as previously described.
9. The DNA is diluted to 200 ng/µl and stored at -20°C in 100-µl aliquots. The yield of purified ssDNA is approximately 20–100 µg, depending on the growth interval.

---

**PROTOCOL 3       Preparation and amplification of phagemid libraries**

1. Just prior to use, purify 2 µg ssDNA (10 µl) on 10 µl resin according to the Bio-Rad protocol in the Prep-a-Gene kit.
2. Elute the template in 25 µl TE buffer and use directly in the annealing reaction.
3. Using the Bio-Rad in vitro mutagenesis kit, mix 1.5–2 µg (approximately 1 pmol) of template ssDNA, 20–30 pmol of primers, and 10 µl annealing buffer (10x) in 100 µl annealing reaction.
4. Upon completion of the synthesis reaction [200 µl containing 20 µl synthesis buffer (10x), 10 units T7 DNA polymerase, and 30 units T4 DNA ligase], run 10 µl on an agarose gel to determine if the reaction is complete.
5. The double-stranded DNA is extracted twice with an equal volume of phenol (equilibrated with 10 mM Tris-HCl, pH 8; 1 mM EDTA), extracted twice with chloroform, and then precipitated with 2 vol ethanol and 0.3 $M$ NaCl.
6. The pellet is rinsed with 70% ethanol, vacuum dried, and resuspended in water (10 µl).

7. This DNA is then electroporated into ten aliquots of electrocompetent XL1-Blue (500 ng DNA, 80 μl cells, 0.2-cm cuvette, 2.5 kV/cm, 200 Ω, 25 μF) using a Bio-Rad Gene Pulser apparatus.

8. The transformed cells are immediately diluted into 2 ml of ice-cold SOC medium and shaken (200 rpm) at 37°C for 1 hr to let the cells recover.

9. After 1 hr, the cultures are combined, 10 μl is removed, and dilutions are plated onto LB plates containing ampicillin (100 μg/ml) to score the total number of independent transformants.

10. R408 helper bacteriophage is added to the remaining culture at a multiplicity of infection (MOI) of 10 and incubated for 1 hr at 37°C without shaking. Alternatively, to generate multivalent phagemid libraries, add IPTG (1 mM) to the culture for 1 hr, prior to the addition of helper phage.

11. The culture is then diluted into 1 liter (2 × 500 ml) of superbroth (in 2-liter flasks) containing ampicillin (100 μg/ml) and tetracycline (30 μg/ml) and shaken at 200 rpm for 15 hr (Sambrook et al., 1989).

12. Cells are removed from the culture by centrifugation (twice at 10,000g for 10 min) and the cleared supernatant is transferred to sterile flasks.

13. To precipitate the phage, 1/4 vol of 3.5 M ammonium acetate, 20% polyethylene glycol (PEG) 8000 is added to the supernatant and the mixture chilled on ice for 1 hr.

14. The phage are harvested by centrifugation at 17,000g for 20 min and the supernatant is discarded. The pellet is resuspended in a final volume of 5 ml TBS (50 mM Tris-HCl, pH 7.4, 150 mM NaCl).

15. The phage (transducing units) are titered by infecting E. coli XL1-Blue cells (0.5 $OD_{600}$) and plating on LB plates containing ampicillin (100 μg/ml).

---

**PROTOCOL 4     Preparation of electrocompetent cells**

Electrocompetent XL1-Blue cells can be made in the laboratory, using the following protocol. All steps should be carried out in a cold room, using prechilled pipettes and tubes, and the cells transported on ice for centrifugation.

1. Prepare an overnight culture from frozen XL1-Blue cells in 20 ml LB broth and tetracycline (35 μg/ml) at 37°C.

2. Dilute 20 ml overnight culture (1/100) into 2 liters LB broth containing tetracycline.

3. When the culture reaches an $OD_{600}$ equivalent to 1.0, set culture on ice for a minimum of 1 hr.

4. Centrifuge in 4 × 500-ml or 8 × 250-ml flasks at 4000 rpm for 15 min.

5. Wash by resuspending cells in 1 vol (2 liter) ice-cold water.

6. Collect cells by centrifugation at 4000 rpm for 10 min.

7. Resuspend pellet in 1/2 vol (1 liter) water (1 mM Hepes, pH 7.5, may be used in place of $H_2O$).

8. Centrifuge as before and resuspend in 1/50 vol (40 ml) 10% glycerol/water (v/v).

9. Collect cells by centrifugation and resuspend in 1/100 volume (20 ml) glycerol / water (10% v/v).

10. Aliquot cells (100 µl) and freeze in liquid nitrogen (Quick-freezing in liquid nitrogen is important for long term stability). Store at -80°C freezer. Check efficiency by electroporation the following day.

---

**PROTOCOL 5a**   **Panning libraries on mAb using acid elution**

Several strategies and rationales for screening phage-displayed libraries will be discussed throughout this text. In this system, a receptor which displays high binding affinity for a ligand may be used effectively to screen a library displaying a low copy number of peptides on the phage surface. However, a receptor which exhibits a weak interaction with its cognate ligand might first be used to screen an IPTG-induced, multivalent-display library where selection would be dependent upon avidity. In a subsequent pan, higher-affinity motifs might be distinguished from the avidity-dependent, lower affinity peptides by panning the amplified subset, grown without induction.

1. Microtiter plates (96 well) are coated with 100 µl of the antibody (20 µg/ml in 50 mM NaHCO$_3$, pH 9.6) for 2 hr at 37°C.

2. The plates are washed twice with TBS/0.5% Tween 20, rinsed twice with TBS, and blocked with bovine serum albumin (3% in 50 mM NaHCO$_3$, pH 9.6) for 2 hr at 37°C.

3. The plates are washed and rinsed as before and then incubated with the mixture of phage (1 x 10$^{10}$ transducing units in 100 µl) for 4 hr at room temperature.

4. The wells are washed 10 times with 250 µl of TBS/0.5% Tween 20 over a period of 1 hr to remove unbound and nonspecifically bound phage, then rinsed twice with 250 µl of TBS.

5. Bound phage are eluted from the antibody with 100 µl of 0.1 M glycine buffer, pH 2.2, for 10 min at room temperature. The eluate is transferred to a fresh microfuge tube and neutralized with 0.1 vol of 2 M Tris-HCl, pH 8.

6. Each 100 µl eluate is incubated with 1 ml of E. coli XL1-Blue cells (A$_{600}$=0.5 in LB broth supplemented with 30 µg tetracycline/ml) for 30 min at 37°C.

7. Samples (10 µl) of infected cells were then plated onto LB plates containing ampicillin (100 µg/ml) to determine the number of phage eluted.

8. The remaining phage are amplified and harvested as described above, but in 10 ml instead of 1 liter.

---

**PROTOCOL 5b**   **Panning phage libraries by competitive elution**

When more information about the ligand is known, alternative strategies may be employed to determine the epitope more quickly or efficiently. For example, phage libraries were panned by competitive elution with somatostatin-14 (Sigma), the

$NH_2$-terminal 14 amino acids of the ligand known to cross-react with this mouse anti-human somatostatin antibody (SOM-14; BiosPacific, Emeryville, CA), as follows:

1. Microtiter plates (96 well) are coated with 100 μl of the same antibody (20 μg/ml in 50 mM $NaHCO_3$, pH 9.6) for 2 hr at 37°C.
2. The plates are washed twice with TBS/0.5% Tween 20, rinsed twice with TBS, and blocked with bovine serum albumin (3% in 50 mM $NaHCO_3$, pH 9.6) for 2 hr at 37°C.
3. The plates are washed and rinsed as before and then incubated overnight with the library phage ($10^{10}$ transducing units in 100 μl) at 4°C.
4. The wells are washed 10 times with 250 μl of TBS/0.5% Tween 20 to remove unbound and nonspecifically bound phage, then rinsed twice with 250 μl of TBS.
5. Bound phage are first eluted for 1 hr at room temperature with 100 μl of 100 μM somatostatin 14.
6. The eluates are transferred to microfuge tubes and a second elution is carried out for 24 hr at room temperature with an additional 100 μl of 100 μM somatostatin 14.
7. Following the second elution, each 100 μl eluate is incubated with 1 ml of *E. coli* XL1-Blue cells ($A_{600}$=1.0 in LB broth with 30 μg tetracycline/ml) for 30 min at 37°C.
8. Samples (10 μl) of infected cells are then plated onto LB plates containing ampicillin (100 μg/ml) to determine the number of phage eluted at each step. The remaining phage are amplified and harvested as described above.

The amplified phage recovered from the 24-hr elution are used for the next round of panning and the entire process is repeated for a third round. After the third pan, 50 colonies from each library are picked from the final 24-hr elution plates and the phagemid DNAs are sequenced. Several of the deduced peptides are synthesized for binding evaluation.

---

## TROUBLESHOOTING PHAGEMID PEPTIDE LIBRARY SYNTHESIS

Conversion of template ssDNA to covalently closed dsDNA is required for high-efficiency transformation of bacteria and generation of vast repertoires of peptide sequences. *In vitro* mutagenesis protocols generally predict a 50–80% conversion of "wild-type" template to mutant sequence. That suggests that up to 50% of the transformants carry the *Not*I template sequence. To reduce this background, the synthesis reaction is purified as described above and restricted with *Not*I in the appropriate digestion buffer. The material which remains undigested is gel-purified and used in the electroporation protocol.

Heating the annealing mixture of random primers and ssDNA template to 95°C in a water bath (or heating block) and then allowing the bath to come to 37°C increases the efficiency of priming.

## References

Barbas, C. F., III, and Lerner, R. A. (1991). Combinatorial immunoglobulin libraries on the surface of phage (Phabs): Rapid selection of antigen-specific Fabs. *Methods: Companion Methods Enzymol.* **2,** 119–124.

Barrett, R. W., Cwirla, S. E., Ackerman, M. S., Olson, A. M., Peters, E. A., and Dower., W. J. (1992). Selective enrichment and characterization of high affinity ligands from collections of random peptides on filamentous phage. *Anal. Biochem.* **204,** 357–364

Bass, S. H., Greene, R., and Wells, J. A. (1990). Hormone phage: An enrichment method for variant proteins with altered binding properties. *Proteins: Struct., Funct., Genet.* **8,** 309–314.

Corey, D. R., Shiau, A. K., Yang, Q., Janowski, B. A., amd Craik, C. S. (1993). Trypsin display on the surface of bacteriophage. *Gene.* **129,** 129–134

Garrard, L. J., Yang, M., O'Connell, M. P., Kelley, R. F., and Henner, D. J. (1991). Fab assembly and enrichment in a monovalent phage display system. *Bio/Technology* **9,** 1373–1377.

Gram, H., Strittmatter, U., Lorenz, M., Glueck, D., and Zenke, G. (1993). Phage display as a rapid gene expression system: Production of bioactive cytokine-phage and generation of neutralizing monoclonal antibodies. *J. Immunol. Methods* **161,** 169–176.

Huse, W. D., Stinchcombe, T. J., Glaser, S. M., Starr, L., MacLean, M., Hellström, K. E., Hellström, I., and Yelton, D. E. (1992). Application of a filamentous phage pVIII fusion protein system suitable for efficient production, screening, and mutagenesis of F(ab) antibody fragments. *J. Immunol.* **149,** 3914–3920.

Kunkel, T. A., Roberts, J.D., and Zakour, R. A. (1987). Rapid and efficient site-specific mutagenesis without phenotypic selection. *In* "Methods in Enzymology" (R. Wu and L. Grossman, eds.), Vol. **154,** pp. 367–382.

Lowman, H. B., Bass, S. H., Simpson, N., and Wells. J. A. (1991). Selecting high-affinity binding proteins by monovalent phage display. *Biochemistry* **30,** 10832–10838.

Markland, W., Roberts, B. L., Saxena, M. J., Guterman, S. K., and Ladner, R. C. (1991). Design, construction and function of a multicopy display vector using fusions to the major coat protein of bacteriophage M13. *Gene* **109,** 13–19.

Sambrook, J., Fritsch E. F., and Maniatis, T. (1989). "Molecular Cloning: A Laboratory Manual," 2nd ed. Cold Spring Harbor Lab., Cold Spring Harbor, NY.

Swimmer, C., Lehar, S. M., McCafferty, J., Chiswell, D. J., Blättler, W. A., and Guild, B. C. (1992). Phage display of ricin B chain and its single binding domains: System for screening galactose-binding mutants. *Proc. Natl. Acad. Sci. U.S.A.* **89,** 3756–3760.

Wright, R., Gram, H., Vattay, A., Byrne, S., Lake, P., and Dottavio, D. (1995). Binding epitope of somatostatin defined by phage-displayed peptide libraries. *Bio/Technology* **13,** 165–169.

# 8

# Multiple Display of Foreign Peptide Epitopes on Filamentous Bacteriophage Virions

*Pratap Malik, Tamsin D. Terry,*
*and Richard N. Perham*

## INTRODUCTION

Display of foreign peptides on the surface of filamentous bacteriophage virions (fd, M13, etc.) has become an important technique for generating specific anti-peptide antibodies and exploring vaccine design (Greenwood *et al.,* 1991; Minenkova *et al.,* 1993; Veronese *et al.,* 1994; Cortese *et al.,* 1994; Meola *et al.,* 1995). Display on gVIIIp, the major viral coat protein, is especially useful in raising an immune response against the displayed peptide since a much higher frequency of peptide display can be achieved when peptides are incorporated into gVIIIp (2700 copies per virion), compared with gIIIp, a minor coat protein (up to 5 copies per virion) (Willis *et al.,* 1993; Meola *et al.,* 1995; Perham *et al.,* 1995).

To display a peptide on gVIIIp, a fragment of DNA encoding the peptide is inserted at an engineered restriction site in the bacteriophage gene VIII. The new amino acid sequence appears, for example, between residues 3 and 4 of the mature coat protein that is generated by cleavage of the precursor pro-coat by leader peptidase during the assembly process (Greenwood *et al.,* 1991). In such a recombinant virion, all 2700 copies of the major coat protein display the peptide. However, there is a limit of about 6–8 amino acids on the size of the peptide that can be inserted into the major coat protein in a recombinant phage (Greenwood *et al.,* 1991; Iannolo *et al.,* 1995). To display larger peptides, a hybrid virion must be constructed in which

*Phage Display of Peptides and Proteins*

**Wild-type**

**Recombinant**

**Hybrid**

**FIGURE I**     Schematic diagram of wild-type, recombinant, and hybrid bacteriophage virions. (a) Wild-type filamentous bacteriophage fd; (b) recombinant virion in which each copy of the major coat protein has a small peptide insert in the N-terminal region; and (c) hybrid virion containing a mixture of the wild-type major coat protein and a modified coat protein with a large peptide insert in the N-terminal region. (⬭) gVIIIp, the major coat protein; (●—) gIIIp; (⬭) gVIp; (▬) gVIIp and gIXp, all minor coat proteins.

the modified coat protein subunits are interspersed with copies of the wild-type coat protein (Greenwood *et al.,* 1991; Felici *et al.,* 1991; Markland *et al.,* 1991).

Here we describe various vectors and techniques for generating recombinant and hybrid virions (Fig. 1) displaying peptide epitopes. Such virions have many potential applications, prominent among them being their use as immunogens for eliciting an immune response.

## VECTORS

The following vectors are based on Greenwood *et al.* (1991), Malik and Perham (1996), and unpublished work (Malik, P., and Perham, R. N.).

### Vectors (fdH, fdISPLAY8 and fdAMPLAY8) for Making Recombinant Virions

In the bacteriophage fdH, a blunt-ended fragment of DNA encoding the peptide is inserted at an engineered *Hpa*I restriction site in the bacteriophage gene VIII, such that the new amino acid sequence appears between residues 3 and 4 of the mature coat protein. Generating the *Hpa*I restriction site in fdH causes two amino acid replacements, G3V and D4N, in the N-terminal region of the mature gVIIIp (Greenwood *et al.,* 1991).

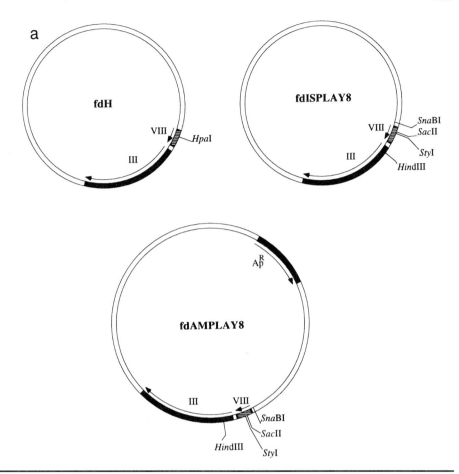

**FIGURE 2**   Schematic diagrams of the genomes of bacteriophage vectors. (a) Genomes of bacteriophages fdH, fdISPLAY8, and fdAMPLAY8 used for making recombinant bacteriophage virions; (b) genomes of the plasmid pKfdH, the pfd8 (pfd8SHU, pfd8SU, and pfd8SY) series of vectors, and bacteriophage fdAMPLAY88 used for making hybrid bacteriophage virions. Ap$^R$, gene encoding ampicillin resistance; P*tac*, *tac* promoter; VIII, gene VIII of bacteriophage fd; III, gene III of bacteriophage fd. (c) Amino acid sequence of the pro-coat of bacteriophage fd plus the DNA and amino acid sequences corresponding to residues -1 to +10 in wild-type fd, fdH, fdISPLAY8, fdAMPLAY8, fdAMPLAY88, and the vectors pfd8SHU, pfd8SU and pfd8SY. The cloning sites are underlined and indicated in italics. *continues*

The bacteriophage fdISPLAY8 has engineered *Sac*II and *Sty*I sites to permit directional cloning of DNA fragments into the bacteriophage gene VIII (Malik and Perham, 1996). In spite of these mutations, the amino acid sequence of the resulting gVIIIp is identical to that of the wild type. Bacteriophage fdISPLAY8 also has a 3' *Hind*III site, in its gene III, which makes it possible for the modified gene VIII to be shuttled easily between the bacteriophage and plasmids (see Fig. 2), if required. The bacteriophage fdAMPLAY8 has cloning sites similar to fdISPLAY8 but in addition has a β-lactamase gene enabling the phage-infected cells to grow under ampicillin selection. These vectors are depicted schematically in Fig. 2a.

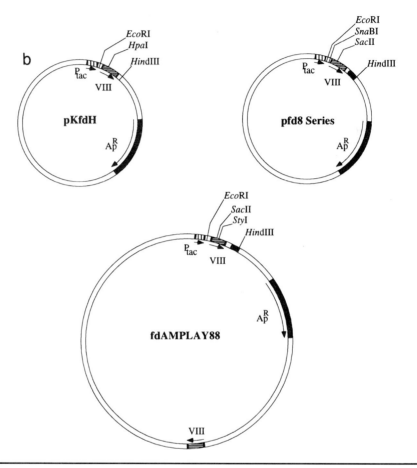

**FIGURE 2**      *Continued*

## Vectors (pKfdH, pfd8SHU, pfd8SU, pfd8SY, fdAMPLAY88) for Making Hybrid Virions

In these plasmid vectors, the modified gene VIII is placed under the control of the IPTG-inducible *tac* promoter when grown in *lac*I^q strains of *Esherichia coli*. In the vector pKfdH, a blunt-ended DNA fragment encoding the desired peptide insert is cloned into the engineered *Hpa*I site, as for bacteriophage fdH (Greenwood *et al.,* 1991). The introduction of vectors pfd8SHU, pfd8SU, and pfd8SY permits more efficient directional DNA cloning (Malik and Perham, 1996). Appropriate DNA fragments, can be cloned into plasmids pfd8SHU and pfd8SU digested with *Sac*II and *Stu*I, or into the plasmid pfd8SY digested with *Sac*II and *Sty*I (this site has been made unique in pfd8SY). Alternatively, fragments of DNA encoding the insert can be generated by carrying out PCR reactions downstream from the *Eco*RI site or upstream from the *Hin*dIII site. These can then be cloned into the appropriate sites.

C

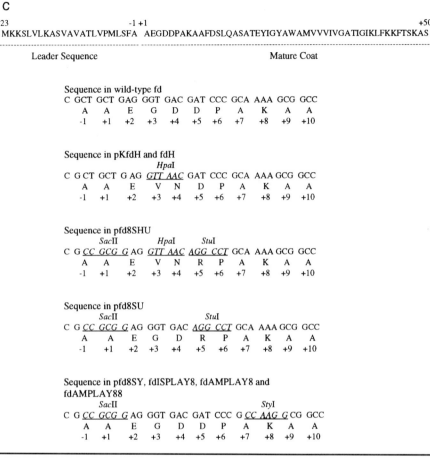

-23                                                -1 +1                                                                                +50
MKKSLVLKASVAVATLVPMLSFA  AEGDDPAKAAFDSLQASATEYIGYAWAMVVVIVGATIGIKLFKKFTSKAS
-----------------------------------------------------------------------------------------------------------------------------------------------------------------------------
                Leader Sequence                                                                    Mature Coat


                    Sequence in wild-type fd
                    C GCT GCT GAG GGT GAC GAT CCC GCA AAA GCG GCC
                       A    A    E    G    D    D    P    A    K    A    A
                       -1   +1   +2   +3   +4   +5   +6   +7   +8   +9   +10


                    Sequence in pKfdH and fdH
                                         *HpaI*
                    C G CT GCT G AG *GTT AAC* GAT CCC GCA AAA GCG GCC
                       A    A    E    V    N    D    P    A    K    A    A
                       -1   +1   +2   +3   +4   +5   +6   +7   +8   +9   +10


                    Sequence in pfd8SHU
                        *SacII*           *HpaI*    *StuI*
                    C G *CC GCG G* AG *GTT AAC* *AGG CCT* GCA AAA GCG GCC
                       A    A    E    V    N    R    P    A    K    A    A
                       -1   +1   +2   +3   +4   +5   +6   +7   +8   +9   +10


                    Sequence in pfd8SU
                        *SacII*                         *StuI*
                    C G *CC GCG G* AG GGT GAC *AGG CCT* GCA AAA GCG GCC
                       A    A    E    G    D    R    P    A    K    A    A
                       -1   +1   +2   +3   +4   +5   +6   +7   +8   +9   +10


                    Sequence in pfd8SY, fdISPLAY8, fdAMPLAY8 and
                    fdAMPLAY88
                        *SacII*                              *StyI*
                    C G *CC GCG G* AG GGT GAC GAT CCC G *CC AAG G* CG GCC
                       A    A    E    G    D    D    P    A    K    A    A
                       -1   +1   +2   +3   +4   +5   +6   +7   +8   +9   +10


**FIGURE 2**        *Continued*

---

The bacteriophage fdAMPLAY88 carries two versions of gene VIII. One of these is present as in the wild-type phage fd, whereas the other is placed under the control of the inducible *tac* promoter. A β-lactamase gene has also been incorporated to provide ampicillin resistance to *E. coli* cells infected with the modified phage. These were introduced into an engineered *Xba*I site in the intergenic region of the bacteriophage genome (P. Malik and R. N. Perham, unpublished work). The unique *Sac*II and *Sty*I restriction sites used for cloning are present on the gene VIII segment that comes under the control of the *tac* promoter.

In addition to offering the advantage of directional cloning, these vectors make it possible to place the N-terminus of the peptide insert anywhere between residues 2 and 6 in the mature coat protein in the case of pfd8SHU and pfd8SU, and between residues 2 and 7 in the case of pfd8SY and fdISPLAY8. Changes or insertions can also conveniently be made upstream of the leader peptidase site.

---

**PROTOCOL 1**     **Cloning protocols**

We have found the *E. coli* strain TG1recO (K12, Δ*(lac-proAB)*, *supE*, *thi*, *hsd*Δ5, *recO*::Tn5 Kan$^r$/F′ *traD36*, *proA$^+$B$^+$*, *lacI$^q$*, *lacZ*ΔM15 ) to be the most effective in successful cloning and preparation of filamentous bacteriophages. One may also use *E. coli* strain JM109 (K12, Δ*(lac-proAB)*, *supE44 thi*, *gyrA96* (Nal$^r$), *endA1*, *recA1*, *hsdR17* ($r_k^-m_k^+$), *relA1*, /F′ *traD36*, *proA$^+$B$^+$*, *lacI$^q$*, *lacZ*ΔM15 ). The general recombinant DNA techniques used in the following protocols are fully described in Ausubel *et al.* (1994).

### Cloning in vector pKfdH

1. Linearize the plasmid (0.5–1 μg) with the restriction enzyme *Hpa*I. We prefer to perform a several-fold overdigestion. Extract with phenol–chloroform and precipitate the DNA with ethanol.
2. Anneal the synthetic complementary DNA strands encoding the peptide insert.
3. Ligate the annealed insert into the linearized plasmid. We usually take at least a 10-fold molar excess of insert to plasmid. The final reaction volume should be around 50 μl.
4. Heat-inactivate the ligase and treat again with *Hpa*I in order to reduce the background (omit this step in cases where a *Hpa*I site is present in the final plasmid).
5. Transform into *E. coli* TG1recO cells and select on ampicillin plates. Sequence the DNA from several colonies.

**NOTE:** In our experience *Hpa*I does not cut to the full extent to give a zero background. Cloning into the blunt-ended *Hpa*I site of pKfdH is not very efficient and several clones may have to be sequenced to find the right one. Moreover, clones with the insert in the reverse orientation will also obtained because of the blunt-end nature of the cloning procedure; these can be discarded.

### Cloning in vectors pfd8SHU and pfd8SU

1. Digest the vector (0.5–1 μg) with the restriction enzyme *Sac*II. Extract with phenol–chloroform and precipitate the DNA with ethanol. Digest the DNA with the restriction enzyme *Stu*I and purify the linear plasmid after gel electrophoresis.
2. Anneal the synthetic complementary DNA strands encoding the peptide insert.
3. Ligate the annealed insert (at least a 10-fold molar excess of insert to plasmid) into the linearized plasmid. The final reaction volume should be around 50 μl.
4. Heat-inactivate the ligase and treat again with *Stu*I (for pfd8SHU and pfd8SU) or *Hpa*I (for pfd8SHU) in order to reduce the background (except when a *Stu*I or a *Hpa*I site is present in the final plasmid).
5. Transform into *E. coli* TG1recO cells (one may also use *E. coli* JM109 cells although we prefer the TG1recO strain) and select on ampicillin plates.

**NOTE:** It is important to make the first digestion with *Sac*II since the enzyme is inhibited by the presence of excess salt. In designing the oligonucleotide inserts, bases that regenerate the sequence flanking the insert as in the wild-type should be included, and appropriate overhangs created where required. For example, to create an insert for the vector pfd8SHU, the oligonucleotide sequences should read as

5′-GGAGGGTX... XXXXXX...XGACGAT-3′
3′-CGCCTCCCAX... XXXXXX...XCTGCTA-5′

### Cloning in vector pfd8SY

The protocol is identical to that for pfd8SHU, except that the restriction enzyme *Sty*I is used instead of *Stu*I.

### Cloning in the replicative forms of bacteriophages fdH, fdISPLAY8, fdAMPLAY8, and fdAMPLAY88

The protocol is similar to those described above. Use the RF form of the bacteriophage DNA for digestions and select the clones as plaques on a bacterial lawn. For fdAMPLAY8 and fdAMPLAY88, the clones can also be selected as ampicillin-resistant colonies.

---

## PROTOCOL 2    PCR protocols

In addition to the above methods of generating clones, fragments of DNA encoding the desired peptide insert can be generated by carrying out PCR reactions. These fragments can then be cloned into the appropriate sites in pfd8SHU, pfd8SU, pfd8SY, fdISPLAY8, fdAMPLAY8, and fdAMPLAY88. For PCR reactions downstream from the *Eco*RI site, we have used the oligonucleotides OLfdRIfor (5′-CGTTTTAG-GTTGGTGAATTCGTAGTGGCAT-3′) and for reactions upstream from the *Hind*III site the oligonucleotideOLfdH3rev. (5′-CGTTAGTAAATGAAGCTTCTGTATGAG-GTT-3′) as the end primers, with the vector pfd8SHU as the template. The oligonucleotide carrying the insert has at least a 15-base perfect match on the 3′ side and least a 10-base overhang from the restriction site (*Sac*II or *Sty*I) for an efficient digestion of the PCR fragments.

In fdISPLAY8 and fdAMPLAY8 where there is no *Eco*RI site, the unique *Sna*BI site can be used for cloning in the DNA fragment generated by PCR using the oligonucleotide OLfdRIfor.

### PCR reactions

1. A typical PCR reaction is set up as follows:

>      200 ng of plasmid DNA in TE buffer or water
>      100 pmol of each primer
>      8 µl of stock solution of deoxynucleotides
>          (2.5 m$M$ with respect to each of the four, deoxynucleotides)
>      10 µl of 10x native *Pfu* buffer (Stratagene) (200 mM Tris-HCl,
>          pH 8.2, 100 mM KCl, 60 m$M$ (NH$_4$)$_2$SO$_4$, 20 mM MgCl$_2$, 1%
>          TritonX-100, 100 µg/ml nuclease-free bovine serum albumin)
>      2.5 units of native *Pfu* polymerase (Stratagene).

2. Overlay the reaction with 100 µl of paraffin oil.
3. Denaturation: 1 min at 96°C, followed by 2 min at 95°C.
4. Amplification: 25 cycles of 1 min at 94°C, 2 min at 50°C, and 3 min at 72°C.
5. Final extension: 10 min at 72°C.
6. Maintain the amplified DNA at 25°C until used. Remove the paraffin oil layer by chloroform extraction and analyze the PCR products (5 µl) by agarose gel electrophoresis.
7. Purify the reaction product by phenol:chloroform extraction, followed by ethanol-precipitation of the DNA or by using Promega Wizard PCR preps.

**NOTE:** In some instances when the above amplification program did not work, the following alternative amplification cycle was successful: 10 cycles of 1 min at 94°C, 2 min at 65°C, and 3 min at 72°C; 10 cycles of 1 min at 94°C, 2 min at 60°C, and 3 min at 72°C; 10 cycles of 1 min at 94°C, 2 min at 55°C, and 3 min at 72°C; and 15 cycles of 1 min at 94°C, 2 min at 50°C, and 3 min at 72°C.

---

**PROTOCOL 3**    **Preparation and purification of bacteriophages**

### Preparation of recombinant bacteriophages

The preparation of recombinant bacteriophages on a small scale is carried out as follows:

1. Inoculate a 1:200 dilution of an overnight culture of *E. coli* TG1recO cells in 2x TY medium (1.5 ml) with a single bacteriophage plaque and grow for 8–12 hr with shaking at 37°C.
2. Separate the cells by centrifugation at 12–15,000 rpm in a microfuge.
3. Remove 1 ml of the supernatant and add to 200 µl of 20% PEG 6000, containing 2.5 *M* NaCl.
4. Incubate at room temperature for 15 min, and then harvest the virions by centrifugation at 12,000 rpm in a microfuge for 10 min. Remove the supernatant and centrifuge again at 12,000 rpm for 5 min. Remove any traces of supernatant with a micropipette and resuspend the bacteriophage pellet in 100 µl TE buffer.

The large-scale purification of recombinant virions is carried out as follows:

1. Inoculate 1 liter of 2x TY medium in a 2-liter Erlenmeyer flask with a 1-ml overnight culture of infected *E. coli* TG1recO cells. Grow the cells for 14–24 hr with shaking at 37°C.
2. Separate the cells by centrifugation for 10 min at 4500*g* (Beckman JA-10 rotor, 5000 rpm) at 4°C.
3. Add PEG 6000 and NaCl to final concentrations of 4% and 0.5 *M*, respectively, to the supernatant to precipitate the virions.
4. Incubate on ice for 4 hr and then collect the precipitate by centrifugation for 30 min at 15,000*g* (Beckman JA-14 rotor, 10,000 rpm) at 4°C. Resuspend the pellet

in a total volume of 30 ml of TE buffer, pH 8.0, and clarify by centrifugation for 20 min at 45000$g$ (Beckman 42.1 rotor, 20,000 rpm). Precipitate the bacterio-phage virions from the supernatant by adding $^1/_4$ vol of 20% PEG 6000 containing 2.5 $M$ NaCl, and collect by centrifugation for 20 min at 12000$g$ (Beckman JA-20 rotor, 10,000 rpm).

5. Resuspend the pellet in 8–10 ml of TE buffer, pH 8.0. The volume is adjusted to ensure complete resuspension. Add 0.5 g/ml cesium chloride and purify by density gradient centrifugation for 17 hr at 200,000$g$ (Beckman SW 50.1 rotor, 40,000 rpm) at 15°C. After centrifugation the virions appear as translucent bands.

6. Collect these bands using a Pasteur pipette and dialyze against 0.5 $M$ NaCl in TE buffer, pH 8.0, for 12 hr, followed by dialysis twice against TE buffer, pH 8.0. Add 0.02% sodium azide for long-term storage at 4°C.

7. To quantify the yield of virions, measure the absorbance of the bacteriophage solution at 269 nm (1 mg/ml of bacteriophage solution = 3.8 absorbance units). The pfu depends on the nature of the insert; usually a 1 mg/ml bacteriophage solution has a pfu of the order of $10^{11}$ per ml. The $A_{260}/A_{280}$ ratio should be approximately 1.1.

**NOTE:** For fdAMPLAY8 and derivatives, add ampicillin to a final concentration of 100 µg/ml to the growth medium. The yield from a 1-liter preparation of a recom-binant bacteriophage displaying a six amino acid insert is about 5 mg (as com-pared with about 100 mg of wild-type fd). If required, a solution of bacteriophage particles may be filter-sterilized by passage through a 0.2-µm filter.

### Preparation of hybrid bacteriophages

The preparation of hybrid bacteriophages on a small scale is carried out as follows:

1. Inoculate 20 ml of 2x TY medium containing 100 µg/ml ampicillin in a 100-ml Er-lenmeyer flask with 100 µl of E. coli TG1recO cells carrying the appropriate ex-pression plasmid.

2. When the culture reaches an $A_{600}$ of 0.2, infect with wild-type bacteriophage fd at a multiplicity of infection of 10:1.

3. After 20 min, induce (see note below) by adding IPTG.

4. Continue to grow the cells for a further 8–12 hr before separating them by cen-trifugtion, and harvest the hybrid virions from the supernatant as described above for the recombinant bacteriophages.

The large-scale purification of hybrid virions is carried out as follows:

1. Inoculate 1 liter of 2x TY medium containing 100 µg/ml ampicillin with 1 ml of an overnight culture of E. coli TG1recO cells carrying the plasmid of interest. Grow to an $A_{600}$ of 0.2.

2. Infect with wild-type bacteriophage fd at a multiplicity of infection of 10:1.

3. After 20 min, induce by adding IPTG.

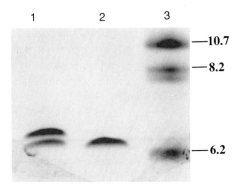

**FIGURE 3**          An SDS–polyacrylamide gel electrophoresis to resolve the modified and wild-type major coat proteins in a hybrid virion. Lane 1, hybrid bacteriophage; Lane 2, wild-type bacteriophage; Lane 3, low-molecular-weight markers. The peptide insert in the hybrid bacterioophage contained six amino acids. Note that the electrophoretic mobility of the coat protein (wild-type $M_r$ 5240) is slightly anomalous. The effect of the insert on electrophoretic mobility will depend on its size and charge.

4. Continue to grow the cells for a further 14–24 hr and harvest the hybrid virions as described above for the recombinant bacteriophages.

**NOTE:** The above protocol for making hybrid virions is quite general and should be optimized for each insert. The amount of IPTG required for induction may vary from 0 to 1 m$M$. We have found that overinduction can decrease both the yield of virions and the density of display, *i.e.*, the amount of modified coat protein compared with wild-type coat protein in the virion. In some instances, growing large-scale (1-liter cultures) from cells already infected with wild-type fd followed by induction at an $A_{600}$ of about 0.2 gave a much better yield of virions and a higher density of display.

When using the fdAMPLAY88 system, grow the cells to an $A_{600}$ of 0.2–0.5, and then induce by adding IPTG. After about 30 min, harvest the cells by centrifugation and resuspend in the original volume of fresh 2x TY medium containing ampicillin and IPTG. Grow the cells for another 12–18 hr before removing them by centrifugation. Purify the hybrid virions as before. The density of peptide display varies from 5 to 50%, depending on the insert.

## ANALYSIS OF BACTERIOPHAGE PARTICLES

Before the recombinant or hybrid virions are used for any purpose, the fidelity of the displayed peptide should be verified (in rare instances, mutations and deletions may occur). This may readily be done by means of polyacrylamide gel electrophoresis and direct N-terminal sequence analysis of the coat proteins.

## Polyacrylamide Gel Electrophoresis of Coat Proteins

The wild-type and modified gene VIII coat proteins can normally be resolved using SDS–polyacrylamide gel electrophoresis in a tricine buffer (Schägger and von Jagow, 1987). We have obtained the best results with 16.5%T, 1%C SDS-polyacrylamide gels (14 x 15 x 0.1 cm). A typical gel is shown in Fig. 3. The compositions of the gels, sample buffer, and running buffer are listed below.

|  | *Resolving gel* | *Stacking gel* |
|---|---|---|
| Acrylamide (49.5% acrylamide, 0.5% bisacrylamide) | 7.5 ml | 1.0 ml |
| Distilled $H_2O$ | 3.40 ml | 6.8 ml |
| 3 *M* Tris-HCl buffer, pH 8.3 | 7.5 ml | 2.5 ml |
| Glycerol | 4.2 ml | — |
| 10% Sodium dodecyl sulfate | 200 µl | 75 µl |
| 10% Ammonium persulfate | 100 µl | 50 µl |
| TEMED | 10 µl | 10 µl |

Running buffer (per liter): 17.9 g Tricine, 12.1 g Tris base, 1.0 g SDS.

Loading buffer (2x): 2 ml 10% sodium dodecyl sulfate, 1 ml glycerol, 170 µl 1 *M* Tris-HCl buffer, pH 6.8, 1.63 ml double-distilled $H_2O$, 200 µl bromophenol blue (0.2% in ethanol), 154 mg dithiothreitol.

Mix the bacteriophage (0.05 to 1 µg ) with an equal volume of sample buffer and immediately denature by boiling for 5 min. Load and run the gel at a constant voltage of 130 V for approximately 16 hr until the bromophenol blue dye front is about 1 cm from the bottom of the gel.

Visualize the proteins by silver staining.

## Silver Staining

Incubate the gels in a glass dish in 200 ml of the following solutions, made up in water:

| | | |
|---|---|---|
| 1. | 50% methanol, 5% acetic acid | 15 min |
| 2. | 7% methanol, 5% acetic acid | 15 min |
| 3. | 10% glutaraldehyde | 15 min |
| 4. | Water | 10 min |
| 5. | Repeat step 4 a total of five times | |
| 6. | 5 µg/ml dithiothreitol | 10 min |
| 7. | 1.0 g/liter silver nitrate | 15 min |
| 8. | Water | 1 min |
| 9. | Developer[a] | 30 sec |
| 10. | Developer | until bands appear (about 5 min) |
| 11. | Wash briefly with water | |
| 12. | Stop solution, 10% acetic acid | 30 min |

*a*Composition of developer: 8.16 g anhydrous $Na_2CO_3$ + 125 µl formaldehyde in 200 ml water.

**NOTE:** It is important to use double-distilled or MilliQ grade water throughout the above protocol for silver-staining.

## N-Terminal Sequence Analysis of Bacteriophage Coat Proteins

Spin the bacteriophage particles in TE buffer through a Prospin column (Perkin Elmer/Applied Biosystems) and apply directly to an Applied Biosystems 477A protein sequencer. The density of peptide display in a hybrid virion can be calculated from the relative yields of the two sequences (derived from the wild-type and modified coat proteins) that are obtained.

## Mass Spectrometry

Good mass spectra of the bacteriophage coat protein can be obtained by electrospray mass spectrometry of the denatured protein. However, more conveniently, the mass of the coat protein can be estimated directly by submitting intact virions to analysis in a MALDI-TOF spectrometer. Apply the matrix, α-cyano-4-hydroxycinnamic acid, to the sample slide in acetone (Vorm *et al.,* 1994). Spot 1 µl of phage containing 6–60 ng of phage (1–10 pmol) onto the matrix and dry. Wash the samples twice by placing a large droplet of double-distilled $H_2O$ or 0.1% TFA on the sample spot. Remove the liquid after 10 sec using a tissue. Dry the slide thoroughly and collect spectra (Kratos Kompact MALDI III) using positive ion mode and a laser power slightly above the threshold for ionization. Insulin and the matrix peaks are normally used as external calibrants.

## IMMUNIZATION AND ANTIBODY PREPARATION

The bacteriophage virions, both recombinant and hybrid forms, may be used as conventional immunogens. It may not be necessary to include an adjuvant in the immunization protocol; good antibody responses have been obtained in rabbits and mice without additional adjuvants (Greenwood *et al.,* 1991; Veronese *et al.,* 1994). Antisera directed specifically against the inserted peptide can be prepared by selectively removing the anti-phage coat protein antibodies by passing the antisera over a column of immobilized wild-type virions (Greenwood *et al.,* 1991).

## CONCLUDING REMARKS

We have successfully used filamentous bacteriophage virions to display a wide variety of peptides. Inserts of up to 16 amino acids have been displayed at a density of several hundred to a thousand copies of the peptide per virion (up to 30% coverage). Fewer copies per virion are generally obtained with longer inserts. In some rare instances, peptides containing fewer than 10 amino acids could not be displayed. This may be due to a problem in the membrane insertion and processing of the modified procoat protein or in a subsequent step in the assembly of the bacteriophage virion itself (Malik *et al.,* 1996).

Peptide display on bacteriophage particles as a means of generating anti-peptide antibodies has the obvious advantage of providing a simple and inexpensive route to the immunogen. It completely obviates the need to synthesize a peptide and chemically link it to a carrier protein. The immune response generated by the virions is strong and does not generally require the use of adjuvants.

## Acknowledgements

We thank The Wellcome Trust for the award of a research grant to R. N. P. and a Prize Studentship to T. D. T. We thank Dr L. C. Packman (Protein and Nucleic Acid Chemistry Facility, Department of Biochemistry) for help with the mass spectrometry.

## References

Ausubel, F. M., Brent, R., Kingston, R. E., Moore, D. D., Seidman, J. G., Smith, J. A., and Struhl, K. (1994). "Current Protocols in Molecular Biology." Wiley, New York.

Cortese, R., Felici, F., Galfre, G., Luzzago, A., Monaci, P., and Nicosia, A. (1994). Epitope discovery using peptide libraries displayed on phage. *Trends Biotechnol.* **12**, 262–267.

Felici, F., Catagnoli, L., Musacchio, A., Jappelli, R., and Cesareni, G. (1991). Selection of antibody ligands from a large library of oligopeptides expressed on a multivalent exposition vector. *J. Mol. Biol.* **222**, 301–310.

Greenwood, J., Willis, A. E., and Perham, R. N. (1991). Multiple display of foreign peptides on a filamentous bacteriophage. Peptides from *Plasmodium falciparum* sporozoite protein as antigen. *J. Mol. Biol.* **220**, 821–827.

Iannolo, G., Minenkova, O., Petruzzelli, R., and Cesareni, G. (1995). Modifying filamentous phage capsid: Limits in the size of the major capsid protein. *J. Mol. Biol.* **248**, 835–844.

Malik, P., and Perham, R. N. (1996). New vectors for peptide display on the surface of filamentous bacteriophage. *Gene* **171**, 49–51.

Malik, P., Terry, T. D., Gowda, L. R., Langara, A., Petukhov, S. A., Symmons, M. F., Welsh, L. C., Marvin, D. A., and Perham, R. N. (1996). Role of capsid structure and membrane protein processing in determining the size and copy number of peptides displayed on the major coat protein of filamentous bacteriophage. *J. Mol. Biol.* **260**, 9–21.

Markland, W., Roberts, B. L., Saxena, M. J., Guterman, S. K., and Ladner, R. C. (1991). Design, construction and function of a multicopy display vector using fusions to the major coat protein of bacteriophage M13. *Gene* **109**, 13–19.

Meola, A., Delmastro, P., Monaci, P., Luzzago, A., Nicosia, A., Felici, F., Cortese, R., and Galfé, G. (1995). Derivation of vaccines from mimotopes. Immunological properties of HBsAg mimotopes displayed on filamentous phage. *J . Immunol.* **154**, 3162–3172.

Minenkova, O. O., Ilyichev, A. A., Kishchenko, G. P., and Petrenko, V. A. (1993). Design of specific immunogens using filamentous phage as the carrier. *Gene* **128**, 85–88.

Perham, R. N., Terry, T. D., Willis, A. E., Greenwood, J., di Marzo Veronese, F., and Appella, E. (1995). Engineering a peptide epitope display system on filamentous bacteriophage. *FEMS Microbiol. Rev.* **17**, 25–31.

Schägger, H., and von Jagow, G. (1987). Tricine-sodium dodecyl sulfate–polyacrylamide gel electrophoresis for the separation of proteins in the range from 1 to 100 kDa. *Anal. Biochem.* **166**, 368–379.

Veronese, F. D. M., Willis, A. E., Boyer-Thompson, C., Appella, E., and Perham, R. N. (1994). Structural mimicry and enhanced immunogenicity of peptide epitopes displayed on filamentous bacteriophage. *J. Mol. Biol.* **243**, 167–172.

Vorm, O., Roepstorff, P., and Mann, M. (1994). Improved resolution and very high sensitivity in MALDI-TOF of matrix surfaces made by fast evaporation. *Anal. Chem.* **66**, 3281–3287.

Willis, A. E., Perham, R. N., and Wraith, D. (1993). Immunological properties of foreign peptides in multiple display on a filamentous bacteriophage. *Gene* **128**, 79–83.

# Phage Libraries Displaying Random Peptides Derived from a Target Sequence

*Dion H. du Plessis and Frances Jordaan*

## INTRODUCTION

A phage library displaying peptides encoded by arbitrary fragments of a target gene can be used to identify antigenic determinants or other binding domains on proteins. Although a gene-targeted library is not as diverse as a universal random peptide library (Chapter 6), the localization of continuous epitopes in particular is facilitated by the fact that the fusion phages express cognate peptides. In a typical application (Fig. 1), a target gene is randomly fragmented by partial digestion with DNase I (Stanley and Herz, 1987). Next, appropriate modifications are made to the termini of the DNA fragments to allow insertion into the replicative form (RF) of the phage vector. After affinity selection from the resulting expression library, the amino acid sequences of peptides recognized by the antibody or other ligate are aligned with the original sequence to locate the binding region.

This chapter describes the fragmentation of a target gene, the addition of linkers, and the construction and characterization of a peptide library (Wang *et al.,* 1995). A suitable display vector is fUSE 2 (Parmley and Smith, 1988). It can be propagated like a plasmid in the presence of tetracycline and is conveniently titred as antibiotic-resistant bacterial colonies rather than phage plaques. The DNA fragments are spliced into a unique *Bgl*II site and the expressed peptides are fused to the minor capsid protein, pIII.

*Phage Display of Peptides and Proteins*
Copyright © 1996 by Academic Press, Inc. All rights of reproduction in any form reserved.

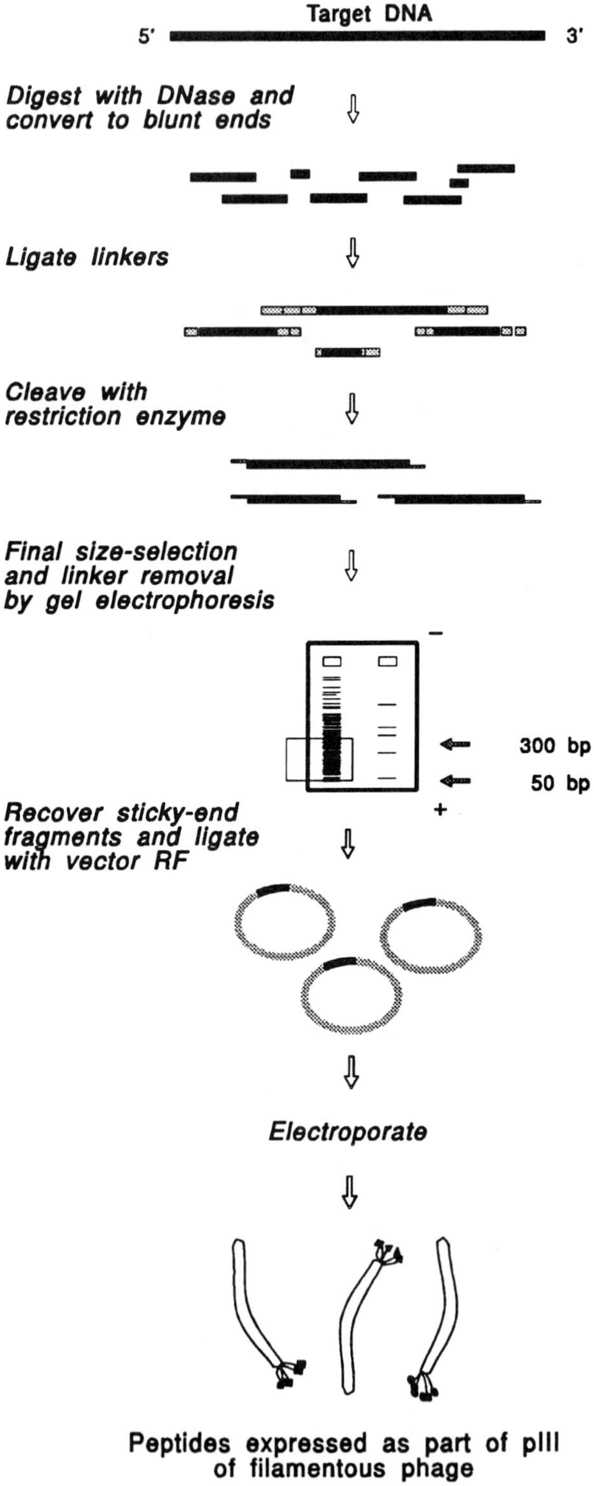

**Target DNA**

5'  3'

*Digest with DNase and convert to blunt ends*

*Ligate linkers*

*Cleave with restriction enzyme*

*Final size-selection and linker removal by gel electrophoresis*

300 bp

50 bp

*Recover sticky-end fragments and ligate with vector RF*

*Electroporate*

**Peptides expressed as part of pIII of filamentous phage**

**FIGURE I**      Flow diagram illustrating construction of a peptide library based on random DNase fragmentation of a target gene.

## PROTOCOL 1   Fragmentation of the target gene

1. First, do a pilot digestion to determine conditions which will yield suitably sized DNA fragments. For epitope mapping, this is usually 50–300 bp. Divide an amount of target plasmid or insert dissolved in DNase I buffer (50 mM Tris buffer, pH 7.6, containing 1 mM $MnCl_2$ and 0.1 mg/ml BSA) into 10-μl aliquots, each containing approximately 1.0 μg DNA. Keep on ice.
2. Make a twofold dilution series of DNase I from 10 units/ml to 0.3125 units/ml in cold DNase I buffer. Add 3.5 μl of each dilution of enzyme to the individual DNA-containing tubes.
3. Transfer all tubes simultaneously to a 15°C water bath and incubate for 10 min. Stop the digestions by adding 2.0 μl of stop solution (70% glycerol, 75 mM EDTA, 0.3% bromophenol blue) to each tube.
4. Load the DNA in each tube onto a 2.5% agarose (SepRate; Amersham) gel for electrophoresis. Include appropriate size markers. After electrophoresis, stain the gel with ethidium bromide and examine with transmitted UV light to determine the resulting size distribution (Fig. 2).

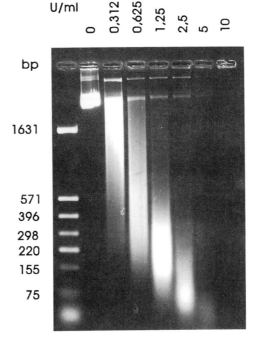

**FIGURE 2**   Partial digestion of a target plasmid by different concentrations of DNase I in the presence of manganese ions. Digestion products were analyzed by electrophoresis in a 2.5% agarose gel.

5. Using conditions determined in step 4, scale-up the digestion proportionally. Approximately 10–15 µg DNA is required in a volume of 100 µl. Add DNase to the required concentration and incubate for 10 min at 15°C. Stop the reaction with 10 µl of 250 mM EDTA. Remove a 5-µl sample and check by agarose gel electrophoresis that the scale-up has yielded fragments of the correct size.

6. After confirming that the size distribution is acceptable, dilute the remaining DNA fragments to 500 µl in TE buffer and extract once with a 50:50 mixture of phenol and chloroform. Precipitate the DNA overnight at -20°C with 2 vol of ethanol in the presence of 0.3 M sodium acetate.

**NOTE:** (a) Since many steps follow during which losses can occur, it is important to begin the procedure with sufficient DNA. A quantity of 20–25 µg is usually sufficient to allow several attempts at library construction. (b) If the gene of interest is cloned into a plasmid, it is usually acceptable to fragment the total DNA. Alternatively, the target insert may be isolated using preparative gel electrophoresis. (c) The manganese-containing DNase I buffer should be made up just before use since it discolors during storage. (d) To reduce the probability that the scale-up will yield fragments with a different size distribution, use the same DNase preparation and do the scale-up soon after the test reactions. Alternatively, do several pilot scale reactions and pool the digests. (e) Some degree of bias in the representation of peptide sequences is probably inevitable. Even if both digestion and cloning are truly random, it is unlikely that all fusion phages in a library will be equally well secreted or have equivalent infectivity. We have nevertheless used gene-targeted display libraries to successfully map continuous epitopes on several viral antigens. A more defined peptide library can be made by ligating restriction enzyme-generated gene fragments into the display vector (Tsunetsugu-Yokota et al., 1991).

---

## PROTOCOL 2    Modifying termini of DNase-generated fragments for cloning

1. Centrifuge to recover the DNA, wash the pellet with 70% ethanol, allow to dry (use Speedivac or air-dry at 37°C), and redissolve in 100 µl repair buffer (40 mM Tris-HCl, pH 8.0, 10 mM ammonium sulfate, 10 mM 2-mercaptoethanol, 5 mM magnesium chloride, 0.5 mM EDTA). Add dNTPs to a final concentration of 100 µM each.

2. Add 30 units of T4 DNA polymerase (Promega, 10 U/µl) and incubate for 1 hr at 15°C. Then add 2 µl (8–10 units) of the Klenow fragment of *Escherichia coli* DNA polymerase I (Boehringer, Mannheim) and incubate for a further 30 min at 37°C.

3. Stop the reaction with 5 µl of 250 mM EDTA and add 10 µl Strataclean (Stratagene) resin. Vortex, incubate at 20°C for 1 min, centrifuge (2000g for 1 min), and remove the original volume of supernatant fluid. Repeat the extraction, and precipitate and wash the fragments with ethanol as described above (Protocol 1,

step 6; Protocol 2, step 1). Dilution to 750 µl with TE followed by phenol/CHCl$_3$ extraction is an alternative to the use of Strataclean resin.

4. Redissolve the blunt-ended DNA fragments in 50 µl of water. Next, add 3 µl of 10x T4 DNA ligase buffer (supplied by manufacturer of the ligase) to a 24-µl sample of the fragments. Keep the rest of the DNA as a backup. For ligation into fUSE2, add 2 µl (1 µg/µl, anneal before use) of 5′ phosphorylated *Bgl*II linkers and 1 µl of T4 DNA ligase (Promega, 15 U/µl).

5. Allow the ligation to proceed overnight at 16°C. Inactivate the ligase by heating at 65°C for 10 min and after cooling on ice, add 95 µl water and 16 µl of 10X *Bgl*II buffer (provided by manufacturer). Next, add a total of 200 U of *Bgl*II (Amersham, 10 U/µl) and incubate at 37°C for 4 hr. Add another 60 U of restriction enzyme and incubate at 37°C for a further 60 min.

6. Inactivate the restriction enzyme by heating at 65°C for 10 min. Extract the DNA twice with Strataclean resin (or once with phenol/ChCl$_3$). Ethanol precipitate as described.

**NOTE:** (a) Alternative cloning strategies can be used. These include blunt-end ligation, the addition of adenosine with *Taq* DNA polymerase for ligation into a vector with a 3′ thymidine overhang (Marchuk *et al.,* 1991), and the use of nonphosphorylated adapters (Haymerle *et al.,* 1986). (b) Use of Strataclean resin in place of phenol extraction may help reduce losses of DNA.

---

## PROTOCOL 3    Final size selection and removal of linkers

1. Redissolve the DNA pellet in 20 µl gel loading buffer (1/7 dilution of stop buffer; Protocol 1, step 3) and load into a single well of a 5% vertical polyacrylamide minigel. Include appropriate size markers in an adjacent well. Run the gel in TBE buffer at 120 V until the dye has migrated 3/4 of the gel length.

2. Stain the gel for 30 min in a solution of 0.5 µg/ml ethidium bromide in TBE buffer. Using as short an exposure as possible to transmitted long-wave *UV* light, locate the DNA smear, and cut out the region containing fragments of the desired size.

3. To recover the DNA, weigh the excised gel slice and cut into small cubes with a scalpel. Crush the gel in an Eppendorf tube using a minipestle (Eppendorf) or glass rod and add 2 vol of extraction buffer (0.5 *M* ammonium acetate, 10 m*M* magnesium acetate, 1 m*M* EDTA, 0.1% SDS). Vortex and shake vigorously overnight at 37°C.

4. Remove gel fragments by centrifugation in a microfuge and filter the supernatant fluid through a Millipore or equivalent 0.2-µm filter. Reextract the polyacrylamide by vortexing with a further 500 µl of extraction buffer. Centrifuge and filter as above. Pool the supernatant fluids and precipitate the DNA fragments by adding 2 vol of ethanol. No sodium acetate is required. Store overnight at -20°C.

5. Centrifuge to recover the DNA and redissolve the pellet in 450 µl TE. Add 45 µl of 3 *M* sodium acetate and reprecipitate with 2 vol of ethanol. Recover by

centrifugation, wash the pellet with 70% ethanol, allow to dry, and redissolve in 20 µl TE. Check recovery and size distribution by agarose gel electrophoresis of 1–2 µl of the DNA.

**NOTE:** (a) Before cloning, free linkers must be removed by agarose or polyacrylamide gel electrophoresis. This step also allows final size selection. While agarose gel electrophoresis is more convenient, it is not always possible to consistently recover the DNA fragments using, e.g., Gene-clean or Qiaex. Furthermore, perhaps due to impurities in some agaroses, recovered fragments are not always efficiently cloned. (b) To make it easier to locate small fragments after electrophoresis, omit the bromophenol blue from the gel-loading buffer. Instead, run the dye in an adjacent well.

---

**PROTOCOL 4     Ligation, electroporation, and harvesting the library**

1. Set up two ligation reactions according to the table; column a gives volumes (in µl) for the ligation proper, and b is the control.

|                                                    | a  | b  |
|----------------------------------------------------|----|----|
| Dephosphorylated *Bg*III-cut fUSE2 RF (0.5 µg DNA) | 5  | 5  |
| Linkered DNA fragments                             | 2  | —  |
| T4 DNA ligase (1 U/µl)                             | 3  | 3  |
| 10X ligase buffer                                  | 3  | 3  |
| Distilled water                                    | 17 | 19 |
| Total                                              | 30 | 30 |

Dephosphorylated vector DNA is prepared by standard methods (Sambrook *et al.*, 1989).

2. Incubate at 16°C overnight. Extract twice with Strataclean resin and precipitate the DNA with ethanol. It is important to wash the pellet with 70% ethanol to remove traces of salt. Redissolve in 5 µl of water. Electroporate (Chapter 5) 1-µl aliquots into electrocompetent (MC1061 for fUSE2) *E. coli* cells.

3. After gene expression in the presence of the appropriate antibiotic (shake at 37°C for 60 min in LB containing 0.2 µg/ml tetracycline for fUSE2), titre the transformants by plating 200 µl of electroporated cells onto an antibiotic-containing (40 µg/ml tetracycline for fUSE2) LB agar plate (Appendix). Incubate overnight at 37°C.

4. Inoculate the remainder of the electroporated cells into a flask containing 250 ml LB broth (Appendix) and antibiotic (20 µg/ml tetracylcine for fUSE2). Shake overnight at 37°C at 200–250 rpm.

5. Confirm that ligation of fragments with the dephosphorylated vector RF has taken place by comparing the number of colonies resulting from the ligation reaction with the control. Remove bacterial cells from the liquid culture by

centrifugation and harvest the library from the supernatant medium by PEG/NaCl precipitation (Chapter 4). Resuspend the phage library in 30 ml Tris-buffered saline, pH 7.5 (TBS), containing 0.02% sodium azide (TBS/azide). Repeat the PEG/NaCl precipitation and resuspend the final pellet in 1 ml TBS/azide.

6. The library may be further purified by CsCl density-gradient centrifugation (Chapter 4). Fusion phage are titred either as colony-forming (fUSE2) or plaque-forming units (Chapter 4). Store the library at 4°C in TBS/azide or with glycerol as described in Chapter 6.

**NOTE:** (a) Ligation can be checked by agarose gel electrophoresis of a small fraction of the ligation mix prior to electroporation. Successful ligation may cause the vector band to shift or become fuzzy. Sometimes faint new bands appear above the vector band. If these changes are not seen, it is usually still worthwhile proceeding with electroporation. (b) It is necessary to optimize the ratio of insert to plasmid in order to get efficient incorporation. Since it is difficult to accurately quantify fragment ends, several ligations with different fragment/vector ratios may be necessary. (c) In liquid cultures, competition could conceivably lead to certain clones becoming dominant. To reduce this possibility, the total amount of electroporated cells can be plated onto 24 X 24-cm LB agar plates containing antibiotic. The library is then recovered by scraping the colonies off into TBS.

---

## PROTOCOL 5   Characterization of the library

1. Using a sterile toothpick for each transformant colony, pick 10–12 randomly chosen clones from the agar plate used to titre the transformants. Transfer the bacteria to an Eppendorf tube containing 15 µl of water. Vortex briefly and heat to 100°C for 5 min in a water bath.

2. Chill immediately on ice and add a cocktail consisting of 10 µl of 10X PCR buffer (166 mM ammonium sulfate, 670 mM Tris-HCl, pH 8.8, 67 mM magnesium chloride, 100 mM 2-mercaptoethanol), 10 µl of 2 mM dNTPs, 1 µl of 1 mg/ml BSA, and 1 µl of a 100µM solution of suitable oligonucleotide primers. Add 62.5 µl water and 0.5 µl thermostable DNA polymerase (e.g., *Taq* DNA polymerase, 5 U/µl; Promega).

3. Subject the samples to PCR for 35 cycles at 94°C for 20 sec, 45°C for 20 sec, and 72°C for 90 sec (conditions may need to be be varied to suit primer sequences). Analyze 5-µl samples by agarose gel electrophoresis (Fig. 3).

4. Calculate the average size and number of productive clones in the library. Take into account that three reading frames and two orientations are possible for each insert. The number of clones required to achieve a 99% probability that a sequence is present is $N = \ln(1-P)/\ln[1-(I/G)]$ (Clarke and Carbon, 1976), where $N$ = number of clones required, $P$ = the probability desired (0.99), $I$ = the average size of the cloned fragment, and $G$ = the size of the target sequence (all sizes in base pairs).

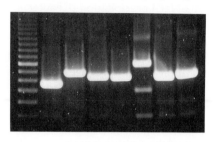

**FIGURE 3** Electrophoresis in a 1.5% agarose gel of PCR products illustrating successful library construction. DNA size markers in lane 1 represent a 100 bp-ladder. The no-insert control is in lane 2. Lanes 3–8 show products resulting from the amplification of DNA fragments inserted into the RFs of six randomly selected phage clones.

**NOTE:** (a) Oligonucleotide primers should be selected to amplify a 400- to 600-bp region of the vector DNA which includes the cloning site. (b) If the PCR indicates that at least some clones contain inserts, proceed directly to affinity selection. Even a suboptimal library can yield useful information.

---

**PROTOCOL 6** **Affinity selection of phage-displayed peptides**

Protocols for panning and identification of binding peptides are described in Chapter 13. Figure 4 shows an example of the identification of an antigenic region on a virus structural protein. The gene-targeted library was screened by affinity selection with a monoclonal antibody (Wang *et al.*, 1995).

**NOTE:** (a) In epitope mapping studies, either monoclonal antibodies or IgG pre-pared from an immune serum can be used to screen the library. With antiserum, phages are sometimes isolated nonspecifically. In this case, try panning with affinity-purified antibodies. Proteins can be blotted from an SDS–polyacrylamide gel onto a nitrocellulose membrane for use as an affinity matrix (Harlow and Lane, 1984). A potential advantage of this approach is that the purified antibodies are likely to recognize antigenic determinants which do not have strict confor-mational requirements (du Plessis *et al.*, 1995). (b) Representative phage clones expressing *all* the different sequences obtained after one, two, or three rounds of affinity selection should be tested for binding using ELISA, Western, or dot blots. In our experience, the isolation of a particular clone is no guarantee that its displayed peptide will be recognized in other assays.

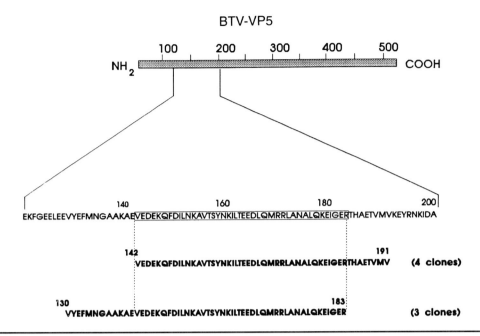

**FIGURE 4**

Identification of an antigenic region on VP5, a structural protein of bluetongue virus (BTV). Two classes of peptide (shown in **bold**) which were affinity-purified by a monoclonal antibody from a gene-targeted phage display library are aligned with the amino acid sequence of the cognate protein. Overlapping residues represent the putative antigenic region. Modified from *J. Immunol. Methods* **178,** Wang *et al.,* use of a gene-targeted phage display random epitope library to map an antigenic determinant on the bluetongue virus outer capsid protein VP5 (1995) with kind permission of Elsevier Science-NL, Sara Burgerhartstraat 25, 1055 KV Amsterdam, The Netherlands.

## Acknowledgments

We are indebted to Professor George Smith of the University of Missouri at Columbia for providing us with phage display vectors and his comprehensive laboratory protocols. We thank Marco Romito and Frank Vreede for assistance with computer graphics and electronic mail.

## References

Clarke, L., and Carbon, J. (1976). A colony bank containing synthetic Col E1 hybrid plasmids representative of the entire *E. coli* genome. *Cell (Cambridge, Mass.)* **9,** 91–99.

Du Plessis, D. H., Romito, M., and Jordaan, F. (1995). Identification of an antigenic peptide specific for bluetongue virus using phage display expression of NS 1 sequences. *Immunotechnology* **1,** 221–230.

Harlow, E., and Lane, D. (1988). "Antibodies: A Laboratory Manual." Cold Spring Harbor Lab., Cold Spring Harbor, NY.

Haymerle, H., Herz, J., Bressan, G. M., Frank, R., and Stanley, K. K. (1986). Efficient construction of cDNA libraries in plasmid expression vectors using an adapter strategy. *Nucleic Acids Res.* **14,** 8615–8624.

Marchuk, D., Drumm, M., Saulino, A., and Collins, F. S. (1991). Construction of T vectors, a rapid and general system for cloning of unmodified PCR products. *Nucleic Acids Res.* **19,** 1154.

Parmley, S. F., and Smith, G. P. (1988). Antibody-selectable filamentous fd phage, vectors: Affinity purification of target genes. *Gene* **73,** 305–318.

Sambrook, M., Fritsch, E. F., and Maniatis, T. (1989). "Molecular Cloning: A Laboratory Manual," 2nd ed. Cold Spring Harbor Lab., Cold Spring Harbor, NY.

Stanley, K. K., and Herz, J. (1987). Topological mapping of complement C9 by recombinant DNA techniques suggests a novel mechanism for its insertion into target membranes. *EMBO J.* **6,** 1951–1957.

Tsunetsugu-Yokata, Y., Tatsumi, M., Robert, V., Devaux, C., Spire, B., Chermann, J.-C., and Hirsch, I. (1991). Expression of an immunogenic region of HIV by a filamentous bacteriophage vector. *Gene* **97,** 261–265.

Wang, L.-F., du Plessis, D. H., White, J. R., Hyatt, A. R., and Eaton, B. T. (1995). Use of a gene-targeted phage display library to map an antigenic determinant on the bluetongue virus outer capsid protein VP5. *J. Immunol. Methods* **178,** 1–12.

# *10*

# Display and Selection of Proteins on Genetic Packages

## *Robert Charles Ladner*

## INTRODUCTION

The development of selection methods of binding molecules from immense libraries of peptidyl compounds displayed on genetic packages began in the late 1980s (Parmley and Smith, 1988; Ladner, 1990). I refer to the method, as applied to proteins, as DSPGP (Display and Selection of Proteins on Genetic Packages; Table 1 contains abbreviations used in this chapter).

With the development of phage display, it is possible to obtain peptidyl compounds from as few as three or four amino acids up to proteins of several hundred amino acids that bind to almost any chosen target with affinities (quantified as $K_D$) that range from picomolar (p$M$) up to micromolar ($\mu M$). The selection process can make huge changes in the binding properties while only minor changes occur in other properties. While $K_D$ has been known to change (in one round of variegation and several rounds of selection) by as much as six logs, other properties (such as size, solubility, stability, toxicity, and antigenicity) may change little or not at all from that of the parent. Even the smallest protein framework can support a diversity of more than $10^7$ distinct proteins (six variable positions) and most protein frameworks can support much larger numbers of structures. Thus, one is no longer limited to antibodies as the framework on which binding proteins can be devised.

**TABLE I**   Abbreviations

| Abbreviation | Meaning |
|---|---|
| DSPGP | Display and Selection of Proteins on Genetic Packages |
| $K_D$ | Dissociation constant: $K_D(A,B) = [A][B]/[AB]$ |
| GP | Genetic Package |
| BD | Binding Domain |
| PBD | Potential Binding Domain |
| IPBD | Initial Potential Binding Domain |
| SBD | Successful Binding Domain |
| PPBD | Parental Potential Binding Domain |
| AfM($X$) | Affinity Molecule that binds $X$ |
| Ab | Antibody |
| scAb | single-chain Ab |
| w.t. | wild-type |
| Ambiguous | N,ACGT; B,CGT; D,AGT; H,ACT; V,ACG; |
| Bases | K,GT; M,AC; R,AG; S,CG; W,AT; Y,AC |
| SAR | Structure–Activity Relationship |
| BPTI | Bovine Pancreatic Tryspin Inhibitor |
| CMTI | *Cucurbid$^a$ maxima* Trypsin Inhibitor |
| CM | Conditioned Medium |
| CDR | Complementarity Determining Region (of an Ab) |
| NMP | Nuclear Magnetic Resonance |
| PCR | Polymerase Chain Reaction |

One must ask, "what physicochemical properties do I want my selected binding molecule to have and what known domains have these properties?" A high-affinity binding protein may be used in several ways. Four important possible uses are: (1) as a drug, (2) as a binding reagent *in vitro,* (3) as a targeting agent *in vivo,* and (4) as a structural probe of what is required to effect high-affinity binding to a particular target. Which of these uses is desired will greatly affect the choice of parental protein. This chapter is a guide to different strategies for DSPGP, of which phage display is one example. Many of the laboratory procedures used for phage display are applicable with minimal modification to other forms of DSPGP.

## SUMMARY OF THE DSPGP METHOD

Display and Selection of Protein on Genetic Packages and phage display involve the construction and expression of vast collections of potentially useful proteins and selection of those mutated genes that specify novel proteins with actual desirable binding properties. The substances bound by these proteins (referred to as "targets") may be, but need not be, proteins. Targets may be other macromolecules as well as other organic and inorganic substances.

The fundamental principle of the DSPGP method is one of "forced" or "directed" evolution. In nature, evolution of binding proteins results from the accidental combination of genetic variation, selection for advantageous traits, and reproduction of the selected individuals, thereby enriching the population for the trait. This process is fairly slow in nature because almost all mutated proteins are less fit for their current function than is the parental protein. Hence, organisms are highly conservative in copying their genomes. Most novel binding proteins that do appear in nature are likely to be rejected because the new binding that the protein exhibits is of no use to the organism that makes it.

The DSPGP method overcomes these "shortcomings" of nature by bringing a highly diverse set of proteins together with a target for which a binding agent is wanted. We force a huge amount of diversity into a small genetic space that is picked to be nonessential to the organism carrying it. Because we can see the connection between the gene and the gene product, we can variegate proteins at the codon level. Furthermore, any protein can be used as a framework.

DSPGP selects for mutated genes that specify novel proteins with desirable binding properties by (1) arranging that the product of each mutated gene be displayed on the outer surface of a replicable Genetic Package (GP) [a cell, a spore, a virus, a complex of ribosome:mRNA:nascent peptide (Kawasaki, 1991; Mattheakis *et al.,* 1994), or a plasmid:protein complex (Cull *et al.,* 1992)] that contains coding sequences (gene, mRNA), and (2) using affinity selection to enrich the population for those packages that display a protein that binds the target. When the mutations are focused on residues the side groups of which are on the surface of a parental protein domain, a very large number of surfaces is generated and the evolution can be highly efficient.

The "display" vector is developed by modifying a genetic package to display a stable, structured domain (the "initial potential binding domain," IPBD) for which an affinity molecule (*e.g.,* an antibody) is available. To date, all display vectors involve a fusion of the sequence to be displayed to an outer-surface protein or protein fragment of the GP, the "anchor protein." The success of the modifications is readily measured by determining whether the modified genetic package can be separated from unmodified packages based on binding to the affinity molecule. Two display genetic packages have been widely used, *viz.* M13 (f1 and fd are essentially identical to M13) and *Escherichia coli.* For M13, both whole phage and phagemids using either proteins III or VIII as anchor have been shown to work (Smith, 1985; Il'ichev *et al.,* 1989). For *E. coli,* LamB (Brown, 1992) and OmpA (Charbit *et al.,* 1987, 1988a,b) have been used as anchors. Once a display vector has been shown to work for (1) a variety of unrelated domains and (2) a variety of closely related variants of each of several unrelated domains, it is likely that the vector will work for other domains.

The IPBD is chosen with a view to its tolerance for extensive mutagenesis. The structure of a protein is determined to a very large extent (sometimes exclusively) by its amino acid sequence. Typically, only a small percentage of the information of the sequence is needed to define the 3D structure. As Rossmann (1987) observed several years ago, structure is more persistent than amino acid sequence. It is thus expected that most residues in most known proteins can be altered without destroying the structure of the protein. Optimally, the residues within the IPBD have a well-defined spatial relationship to other residues in the IPBD, but the IPBD is not restrained any more than necessary in its connection to the anchor protein.

The gene encoding the IPBD in the display vector is subjected to variegation, which leads to the production of a population of GPs, each of which displays a single Potential Binding Domain (PBD) (*i.e.,* a mutant of the IPBD). The population collectively displays a multitude of different, though structurally related, PBDs. Each GP carries the version of the *pbd* gene that encodes the PBD displayed on the surface of that particular GP; one might view most of the GP as a very complicated bit of string that ties the gene product to the gene.

Affinity selection and amplification of binders is used to enrich the population for GPs bearing PBDs with the desired binding characteristics. After one or more cycles of enrichment, the DNA encoding the "Successful Binding Domains" (SBDs) may then be recovered from selected packages. If need be, the DNA from the SBD-bearing packages may then be further variegated, using an SBD (or a pool of SBDs) of the last round of variegation as the "Parental Potential Binding Domain" (PPBD) to the next generation of PBDs, and the process continued until you are satisfied with the result. At that point, the SBD may be produced by any conventional means, including (for small proteins) chemical synthesis.

Usually the number of different amino acid sequences obtainable by mutation of the PBD is very large compared to the number of different domains which are actually present in the population of GPs. The efficiency of the forced evolution is greatly enhanced by careful choice of which residues are to be varied and the range of amino acid types allowed at each varied position. Residues on the surface of a protein have the greatest potential to alter the binding properties of the domain and the least potential to greatly disrupt the structure of the framework. The efficiency of the forced evolution is particularly high if the side groups of the variegated residues are close together on the surface of the framework molecule. (See the web page for additional information.)

In its most basic form, the DSPGP method involves:

1. preparing a variegated population of replicable GPs, each GP including a genetic construct encoding an outer-surface-displayed potential binding protein comprising (a) a signal directing the display of the protein on the outer surface of the GP, and (b) a PBD for binding the target, where the population collectively displays a multitude of different PBDs having a vast range of variation in sequence;

2. causing expression of the protein and display of the protein on the surface of the GPs;

3. contacting the population of GPs with target material so that the PBDs of the proteins and the target material may interact, and separating GPs bearing PBDs that succeeds in binding the target material from GPs that do not bind; and

4. recovering and replicating GPs which bear an SBD.

After step 4, one has a number of options, including: (a) repeat the fractionation steps (2–4); (b) sequence the DNA of several packages to get implied amino acid sequences of binders; and (c) introduce additional variegation, *i.e.,* repeat steps 1–4 but include the information obtained.

Indications of successful screening include: (1) increase in the fraction of phage that adhere to the target, (2) appearance of a consensus sequence, (3) high levels of binding of clonally pure phage selected for binding, and (4) nonbinding of selected phage to a matrix that differs from the selective matrix only by absence of the target material. The most important measure of success is an improved $K_D$ measured under the conditions of intended use.

Key questions in using DSPGP include:

1.  What parental protein to use,
2.  What genetic package to use,
3.  What anchor protein to use,
4.  Where to attach the displayed protein to the anchor protein,
5.  Whether to have a linker between displayed domain and anchor,
6.  Which amino acids to variegate,
7.  Which amino acids to allow at the variegated positions,
8.  How many transformed GPs are needed,
9.  How to fractionate the population of GPs,
10. How much target to use,
11. How many rounds of fractionation,
12. How to express the selected proteins,
13. How small a component to expect to extract from a library, and
14. What to do if DSPGP does not "work? "

## NOMENCLATURE

A "protein" is a polypeptide that has a measurable structure in solution as determined by NMR, X-ray diffraction, or other physical or chemical measurement. Although proteins are thought of as being larger than peptides, this is not a very useful distinction. Heat-stable enterotoxin IA (ST-IA), for example, is a bacterial toxin protein related to guanylin and having only 19 amino acids, 6 of which are cysteines (Dallas, 1990). Guanylin has only 15 amino acids, 4 of which are cysteine. ST-IA can be boiled without loss of activity and can withstand passage through mouth, stomach, and small intestine. ST-IA has a very well-defined 3D structure.

I will discuss binding on the assumption that it is noncovalent, reversible, and bimolecular between species "A" and "B." Affinity between A and B is discussed by reference to the dissociation constant ($K_D = [A][B] \div [AB]$). $K_D$ is related to the kinetic "on" and "off" rates: ($K_D = k_{off} \div k_{on}$). The maximum value of $k_{on}$ is limited by diffusion. In water for protein molecules of 5000 Da or more, $k_{on} = 10^7/M$/sec is characteristic of a fairly fast reaction. A binding pair having $K_D = 1$ n$M$ and $k_{on} = 10^7/M$/sec has $k_{off} = 10^{-2}$/sec so that $t_{1/2} = 69$ sec.

The term "variegated DNA" (vgDNA) refers to a mixture of DNA molecules of the same or similar length which, when aligned, vary at some codons so as to encode at each of these codons a number of different amino acids, but which encode only a single amino acid at other codons. Some workers prefer to have a large fraction of

**TABLE 2** Poisson Sampling Statistics[a]

| $X$ | $(1-e^{-X})$ |
| --- | --- |
| 0.01 | 0.00995 |
| 0.1 | 0.09516 |
| 0.25 | 0.2212 |
| 0.5 | 0.39346 |
| 0.693 | 0.5 |
| 1 | 0.6321 |
| 2 | 0.865 |
| 2.302 | 0.9 |
| 3. | 0.95021 |
| 4 | 0.98168 |
| 4.605 | 0.99 |
| 5 | 0.993 |
| 10 | 0.99995 |
| 100 | $1.000 - 3.7 \times 10^{-44}$ |

[a]$X$ is the ratio (independent transformants)/(possible DNA sequences). $(1-e^{-X})$ is the fraction of possible DNA sequences that should be represented in an $X$-fold sampling of the possible DNA sequences. For example, if 1,000,000 DNA sequences were possible, then a sample of 500,000 independent transformants should have 0.39346∗1,000,000 = 393,460 sequences actually present.

the possible encoded amino acid sequences present in the population of variegated GPs. This approach is termed "Majority Sampling." Majority Sampling requires the production of a large number of transformed GPs. Table 2 shows the fraction of possible DNA sequences expected in a population obtained by $X$-fold sampling. Here, $X$ is the ratio of (transformed GPs)÷(possible DNA sequences). Sampling that gives twice as many transformants as DNA sequences should give 0.865 of the sequences actually available. Threefold sampling is better, but higher levels can give only very little benefit. Obtaining $10^8$ transformants is feasible, so one can use Majority Sampling on a library having a DNA diversity of about 3 x $10^7$. To allow Majority Sampling, the number of designated variable codons in the variegated DNA can be no more than about 20 codons, usually no more than 5–10 codons. The mix of amino acids encoded at each variable codon may differ from codon to codon.

Other workers prefer to cast a very wide net and accept that only a small fraction of possible proteins or peptides will be present for selection. This approach is termed "Minority Sampling." Both approaches are useful, but it is important to understand what to expect from each.

## Types of Display

Roughly following Smith's suggested nomenclature (Parmley and Smith, 1989), I denote the display systems in M13, fd, and f1 as follows:

| Designation | Anchor protein | Genetic Package |
|---|---|---|
| D3 | Mature pIII | M13 phage |
| 3+D3 | Mature pIII (w.t. pIII also present) | M13 phage |
| 3+3stump | Carboxy terminal portion of pIII, wt. pIII present | M13 phage |
| 3mid | Mature pIII | M13-derived phagemid |
| 3stump | Carboxy terminal portion of pIII | M13-derived phagemid |
| D8 | Mature pVIII, no w.t. pVIII present (works for peptides) | M13 phage |
| 8+D8 | Mature pVIII, w.t. pVIII also present | M13 phage |
| 8mid | Mature pVIII | M13-derived phagemid |

## GENERAL CONSIDERATIONS ON PROTEINS

It is important to remember that molecules that interact with proteins do not "see" residues or groups; molecules "see" only a large, covalently linked collection of atoms. It is the constraints imposed by protein folding which hold groups of atoms (often from residues that are widely separated in sequence) in highly similar relative orientations for quite long times to allow proteins to exhibit high affinities and specificities.

Because each residue in a protein sits in a more-or-less constant relationship to the other residues, it is meaningless to slide selected protein sequences around to find consensus sequences. For example, two BPTI derivatives that differ only in that one has Ile $_{15}$-Asp $_{16}$-Ala $_{17}$-Trp $_{18}$-Lys $_{19}$ and the other has Asp $_{15}$-Ala $_{16}$-Trp $_{17}$-Lys $_{18}$-Ile $_{19}$ are very different. The first protein is more closely related to a protein having Val $_{15}$-Glu $_{16}$-Gly $_{17}$-Phe $_{18}$-Arg $_{19}$ than it is to the second protein.

In a protein, a particular amino acid type may be required (as judged by ubiquity in sequences selected in nature or artificially) for structural reason, although it may be outside the binding site. In a peptide, requirement of a particular amino acid type implies that the residue in question is directly involved in binding because most peptidse lack stable structures.

### Protein Binding

Protein binding usually involves exclusion of bulk water in the interface between the protein and its binding mate. It is widely thought that the freeing of solvent is a major driving force to hold proteins to their binding partners. Although X-ray diffraction and NMR can show where the atoms are in a protein:target complex, one cannot see which residues contribute most to the binding energy. Cunningham and Wells (1993) reported that 85% of the binding energy of human growth hormone (hGH) to its receptor (hGHBP) is contributed by 8 of the 31 hGH residues that contact hGHBP. Recently, Clackson and Wells (1995) reported that only 11 of 33 contact residues of the receptor make appreciable energetic contributions, with 4 or 5 residues dominating. These results support the view of Chothia and Janin (1975) that a protein–protein interface is highly similar to a cross-section through a folded protein.

# REASONS TO USE PROTEINS AS BINDING MOLECULES

This section discusses why one might select a non-antibody protein as a parental protein for display and selection to obtain a high-affinity binder to a known molecular target. The alternatives are antibodies (*via* immunization or selection from phage-displayed Abs), peptides (*via* phage display or synthesis), and synthetic molecules *via* synthetic "diversimer"'s or trial and error. Proteins have the advantage over peptides, that, due to their internal structure, they are very likely to show similar binding properties when: (a) displayed on phage, (b) expressed as soluble domains, or (c) fused to other protein domains. Until the development of DSPGP, Abs were essentially the only broad class of proteins from which one might obtain a binder for any given substance. The preeminence of Abs arose early in vertebrate evolution with the conjunction of a mechanism for substituting the DNA that encodes the complementarity determining regions (CDR) into the DNA that encodes the Ab framework with machinery to: (1) amplify cells expressing useful Abs, and (2) eliminate Abs that bind "self." One might hypothesize that once antibodies developed during evolution, there was no incentive for nature to need hypervariation of other protein frameworks as part of an immune system.

## Affinity

Generally, proteins are capable of higher affinity and specificity in binding than are peptides. Proteins can bind other molecules (ranging from single ions such as $Ca^{2+}$ to macromolecules such as peptides, other proteins, glycoproteins, and polysaccharides) with affinities in the picomolar range (and below) while peptides that lack their own structure in solution usually have affinities for other molecules in the range 1 m$M$ to 1 n$M$. When a peptide has an appreciable affinity for another molecule, it is because the other molecule has a site that is highly complementary to the peptide in one of its more stable conformations. A protein can achieve binding to a second macromolecule by having appropriate side groups present at appropriate positions on its surface. The side groups and exposed main-chain atoms fit into the target less deeply than is thought necessary for peptide binding.

Abs represent a class of proteins which have historically been a rich source of affinity reagents. Abs can be found that bind almost any molecule that is stable in water. Abs typically have affinities for antigen in the range 1 n$M$ to 1 μ$M$. Furthermore, Abs have been selected through phage display that have higher affinities (Vaughan *et al.,* 1996), and there is no reason to think that there is any barrier that limits the affinity of an Ab for an antigen; in theory, Abs could have $K_D$ for a target of $10^{-15}$ $M$.

In addition to the general ability of proteins to show high affinity for molecular targets, an important consideration is whether a protein or protein domain is known to have affinity for <u>your</u> target or for a molecule closely related to your target. For example, Kunitz domains are thought of as having affinity for serine proteases (in general) and the binding of Kunitz domains to proteases has been shown to be alterable by changing the amino acid sequence. Other small domains are known to bind to particular molecular targets. The affinity of such binding can be greatly improved (by up to $10^6$-fold) by displaying a few (such as 10 million) variants and selecting the ones with the highest affinity.

I have adopted as a working hypothesis that most natural proteins have not been selected to have maximal affinity for the molecules they bind. Rather, proteins have evolved to fulfill one or more biological functions. This means they must have, among other properties, appropriate affinity for their ligands and appropriate susceptibility to proteolysis.  Let T be a molecular target and let X be a natural protein ligand for T. We and others (particularly Wells and colleagues) have shown that mutants of X can be found that have affinity for T substantially increased above that shown by X. Starting with molecules having micromolar to nanomolar affinity, mutants have been found, via DSPGP, that have affinities down to 0.9 p$M$ (Lowman and Wells, 1993). To change a Kunitz domain, so that it has $10^6$ greater affinity for hNE than its parent, required only four amino acid sequence changes which were found in one round of variegation and three rounds of selection. To improve hGH from about 0.34 n$M$ to about 0.9 p$M$ took 15 changes, accumulated from selections from several libraries (Lowman and Wells, 1993). Wells and co-workers have also selected peptide linkers for susceptibility and for resistance to specific proteases (Matthews and Wells, 1993). There is no reason that such a selection could not be applied to a binding domain to obtain a domain that has a particular binding property and is resistant to a particular protease.

## Stability

Proteins are preferable to peptides as binding molecules because they can be more stable than peptides. That is not to say that all proteins are more stable than all peptides, but rather that particular proteins are more stable than almost all peptides. There are proteins that can survive various hostile environments (such as low pH, high pH, high temperature, chaotropes, the gut, the blood stream, and other body fluids) where peptides quickly vanish.

## Size

Although proteins are thought of as being (and often are) larger than peptides, there are proteins of many different sizes; quite small proteins can bind with affinities and specificities typical of Abs, and somtimes better. Thus, I have proposed that non-antibody proteins may offer some advantages compared to Abs as DSPGP proteins. These advantages include:

1. smaller display gene leading to: (a) possible to synthesize variegated DNA as one (or two) piece(s), (b) excellent chance of getting display;

2. smaller displayed protein, hence smaller final protein, leading to: (a) lower molecular weight, (b) possible medical (*in vivo*) advantages: (1) better organ penetration, (2) faster elimination, (3) lower antigenicity, (c) possibility of chemical synthesis for proteins having about 30 or fewer amino acids.

The listed advantages are not of equal importance. The most important are probably 2b and 2c. The increase in diffusion of small proteins gives them only a small advantage over Abs, but the effect is in the right direction.

Medical advantages could be important. It has been observed that whole Abs do not penetrate tumors very well, while single-chain Abs (scAb) penetrate better. Small protein domains, such as a Kunitz domain, should diffuse even better; the relative sizes of an Ab to a scAb to a Kunitz domain are approximately 150:30:6.8 kDa.

The structure of a small protein bound to the target may be easier to determine by X-ray diffraction or NMR than is the complex of the target with an Ab. Furthermore, once the structure of a small-protein:target-molecule complex is found, the interactions may be limited to fewer residues, making interpretation of the interactions easier. For example, the CDRs of a typical Ab comprise about 60 residues which are held in place by approximately 200 to almost 1500 residues. The whole of BPTI (a typical Kunitz domain) is only 58 amino acids and most binding partners of Kunitz domains touch only 10–20 amino acids. The interpretation of these interactions is further simplified by the availability of several closely related sequences that may have high but different affinities for the target. Proteins of 20 to 30 amino acids may be even better as scaffolds because the binding must be focused into a small number of residues. For exampe, the analysis of Wells and co-workers (Cunnigham and Wells, 1993; Clackson and Wells, 1995) has shown that the energetics of HGH binding by its receptor are not at all evenly distributed. Using a small parental protein avoids the possibilities of (a) getting binding that requires 20 or more amino acids in particular places, or (b) having to make 20 or more mutants to find which of the contact residues are giving the binding. For proteins of about 30 amino acids or less, it is possible chemically to synthesize the protein and variants. The advantages of such an approach are discussed below.

## Pharmacokinetics

Proteins have some advantages over small-molecule drugs which are often overlooked. Proteins usually do not cross lipid bilayers. Unless special delivery techniques are used, intact proteins will stay in the gut if ingested, in the blood stream if injected, on the surface of the skin if applied topically, or in the lung if inhaled. Thus, the toxicology of small proteins can be simpler than the toxicology of small organic molecules that cross a variety of biological barriers.

Another advantage is that while proteins are typically antigenic, small binding proteins can be engineered on a human framework such that the final protein differs from a human protein by only a few residues. Thus, such molecules may escape immune surveillance due to tolerance. Futhermore, if the parental and engineered derivative proteins are highly resistant to proteolysis they may fail to present and elicit an immune respone. Moreover, while many therapeutic proteins, such as human insulin, elicit an immune response, the resulting Abs are not neutralizing.

## CHOICE OF PROTEIN DOMAIN TO BE DISPLAYED

The following section assumes that you are going to display, variegate, and select a non-Ab protein and covers the considerations needed to pick the best parental domain.

### Having a Domain That Already Binds the Target

If you have a domain that is known to bind the target (with some level of specificity and affinity of at least 1 $\mu M$), and if the domain meets all your other requirements of stability, size, producibility, *etc.,* then this domain is your prime candidate as IPBD. The binding of a parental protein having affinity of 1 n$M$ can be improved, perhaps by 10- to 1000-fold. Even parental proteins with lower affinities can most likely be improved. Parental proteins having less affinity (higher $K_D$) can most likely be improved by much larger factors. Even if you have such a prime candidate, you should go through the following sections to verify all relevant criteria are met. A domain that binds the target may also be very useful in fractionation of libraries even if you do not use it as IPBD.

### Knowledge of 3D Structure

Knowing the 3D structure of the PBD allows you to see which side groups are on the surface, how the residues are arranged in space (*viz.* which pairs of residues have side groups that could contact one molecule of the target simultaneously), and which residues are involved in determining the structure of the framework. Examination of complexes of proteins with other molecules shows that the complexes are highly idiosyncratic but that the contacts between proteins and other macromolecules usually form a contiguous and irregular surface. Viewed in linear amino acid-sequence space, the interface often appears discontinuous, but not in 3D space. Furthermore, the residues in the interface (and particularly residues having side groups that contact the binding partner) have the greatest potential to affect the binding if altered by mutation.

Knowing the 3D structure of the target is less important than knowing the 3D structure of the IPBD. The 3D structure of the target may be useful and could suggest that certain IPBDs are too big to fit into well defined cavities on the target.

### Size

The following patterns are frequently observed in high-affinity complexes between proteins and other molecules:

1. A small molecule (from atomic ions up to covalent compounds of about 500–1000 Da) binds strongly to a protein by sitting in an internal cavity so that the protein covers substantial parts or all of the small molecule (*e.g.,* streptavidin binding biotin, subtilisin binding $Ca^{2+}$, and hemoglobin binding heme)

**TABLE 3**    Characteristic Binding by Various Sizes of Molecules

| | Size of binder 1 | | |
| --- | --- | --- | --- |
| Size of binder 2 | Small molecule (up to about 500 Da) | Small protein (<80 amino acids) | Large protein (≥80 amino acids) |
| Small molecule | Usually weak | Very high affinity possible, modest affinity known | Strong binding in known cases |
| Small protein | Very high affinity possible, modest affinity known | Very high affinity possible, modest affinity known | Examples of very strong binding |
| Large protein | Strong binding in known cases | Examples of very strong binding | Examples of very strong binding |

2. A molecule in the range of 500 to 10,000 Da binds tightly to a protein of 20 kDa or more by the smaller molecule fitting into a concavity in the larger molecule (*e.g.*, BPTI binding trypsin), with some portion of the smaller molecule exposed.

3. A protein of 15 kDa or more binds tightly to another protein of 15 kDa or more having an interface that is flat at low resolution and interdigitated at the level of side groups.

Table 3 shows typical binding strength between classes of molecules: small molecules, small proteins, and large proteins. Here, "very strong binding" means $K_D < 100$ p$M$. "Modest affinity" means $1\ \mu M > K_D > 1$ n$M$. If the target is less than about 2000 Da, then a protein of 80 or more amino acids may be a better IPBD than is a smaller protein. Theoretically, small proteins (about 80 or fewer amino acids) could bind small-molecule targets very tightly, but this has not been demonstrated yet.

## Knowledge of Homologous Proteins

The amino acid sequences of homologous proteins gives several kinds of information: (1) which amino acids are essential, (2) what is allowed at the variable positions, and (3) definition of binding sites. Below, I discuss Kunitz domains because they are perhaps the most fully documented class. Many other protein domains, however, are potentially as useful as or superior to Kunitz domains. DSPGP can be used to generate the kind of diversity data that nature has provided for Kunitz domains.

When homologous sequences from widely separated sources are known (as for Kunitz domains) we can see which amino acids are essential to maintain the structure. The sequences of >70 naturally occurring Kunitz domains are known; some of the sequences are identical, though they come from different sources. Kunitz domains have been found in mammals, insects, crustaceans, reptiles, birds, coelenterates, amphibians, and mollusks. (To examine the amino acid sequences of selected Kunitz domains, consult the web site.) Thus, if these proteins arose from a common ancestral gene, they must have done so hundreds of millions of years ago. Even so, certain amino acids predominate at certain amino acid positions of Kunitz domains. (To examine the frequency of amino acids at each position in a collection of Kunitz domains, consult the web site.) For example, at position 12 of a typical Kunitz domain only four different amino acids are found; 68 of the sequences have Gly, 1 sequence has Asp, 1 has Ile, and 1 has Lys. If we assume that all the positions are independently substitutable, then we could make about $10^{47}$ amino acid sequences using only the observed types at each position.[1] Of course, this type of anlayis of Kunitz domains may be highly biased, due to the focus on those domains that have trypsin inhibition activity. Nevertheless, in such a small sample size, we have already seen 16 amino acid types at three positions (26, 34, and 46) and 15 at 5 positions (1, 2, 3, 9, and 17). Note that residues 34 and 17 of BPTI contact trypsin in the BPTI:trypsin complex.

One can divide, the positions of the Kunitz domain into six categories: (1) Cysteine, (2) Glycine, (3) Aromatic, (4) Conserved, (5) Proline excluded, and (6) Anything. (A complete listing of Kunitz domains by amino acid residues is available at the web site. A table of proline frequencies is also available at the web site.) There are 17 positions where any amino acid type is allowed, 20 where only proline is excluded, and 11 positions allowing at least twofold diversity giving at least $10^{51}$ (i.e., $20^{17}$ x $19^{20}$ x $2^{11}$) sequences. This is, however, only about 3 x $10^{-24}$ of the $20^{58}$ (i.e., 3 x $10^{75}$) protein sequences of length 58. Note that the 6 positions listed for cysteine are those required for the standard Kunitz structure. Cysteine may occur at other positions, but these "extra" cysteines will, in most circumstances, form disulfides to some other moiety, such as a second cysteine-containing protein domain (as happens in bungarotoxin). It is not clear that the disulfides are essential; each of the proteins having one disulfide converted to a pair of alanines folds into a structure similar to BPTI, but is less stable (Eigenbrot *et al.*, 1990; Goldenberg, 1988).

With protein III[2] or VIII fusions of bacteriophage M13, there is selection against display of proteins having unpaired cysteines. For cysteines to remain unpaired in the *E. coli* periplasm, the thiol must be sheltered from exposure to oxidants or other free thiols. This can occur occasionally and has been observed in Abs (R. Lerner, personal communication).

Only two positions in Kunitz domains seem to require glycine: 12 and 37. At position 37, there are no exceptions found in nature. At position 12, each of the exceptions occurs in the context of a Kunitz domain that has an insertion or deletion that could allow or require a main-chain rearrangement. At position 36, glycine is highly preferred, but not required; the $\phi\psi$ angles are compatible with any amino acid. Thus, the preference probably arises from requirements for binding.

---

[1]This is obtained by multiplying the number of amino acid types seen at each residue 1–58 with the following exceptions: use 1 at residue 5 (anything other than Cys requires an amino terminal extension that includes a Cys) and omit residues 9a, 42a, and b.

$N = 2$ x $3^3$ x $4^5$ x $5^3$ x 6 x 7 x $8^3$ x $9^3$ x $10^5$ x $11^6$ x $12^5$ x $13^2$ x $14^4$ x $15^5$ x $16^3$

   $= 9.64$ x $10^{46}$

[2]I refer to the proteins of M13 as protein III or pIII encoded by gene III and protein VIII or pVIII for the protein encoded by gene VIII.

**TABLE 4**    Sequences of Various Cucurbida Trypsin Inhibitors

```
                                                      2
                        1111111111122222252222233
                        1234567890123456789012345a678901
EETI-II            GCPRILMRCKQDSDCLAGCVCGPN-GFCG        LENG89b
CMTI-I             RVCPRILMECKKDSDCLAECVCLEH-GYCG       LENG89b
CMTI-III           RVCPRILMKCKKDSDCLAECVCLEH-GYCG       OTLE87ᵃ
CMTI-IV          HEERVCPRILMKCKKDSDCLAECVCLEH-GYCG      OTLE87
CPTI-II            RVCPKILMECKKDSDCLAECICLEH-GYCG       LENG89b
CPTI-III         HEERVCPKILMECKKDSDCLAECICLEH-GYCG      OTLE87
CSTI-IIb           MVCPKILMKCKHDSDCLLDCVCLEDIGYCGVS     OTLE87
CSTI-IV            NMCPRILMKCKHDSDCLPGCVCLEHIEVCG       LENG89b
MRTI-CM-1          GICPRILMECKRDSDCLAQCVCKRQ-GYCG       OTLE87
MRTI-II            GCPRILMRCKQDSDCLAGCVCGPN-GFCG        LENG89b
CVTI-I             GRRCPRIYMECKRDADCLADCVCLQH-GICG      OTLE87
BDTI-II            RGCPRILMRCKRDSDCLAGCVCQKN-GYCG       OTLE87
CPGTI-I            RVCPKILMECKKDSDCLAECICLEH-GYCG       WEIC85

Parental      xxxxxCPxILMxCKxDxDCLxxCxCxxx-xxCGxx
```

[a]Mc Wherter et al. (1989) prepared 8 CMTI-III derivatives having various amino acids at position 5 (P1 *a la* Scheckter and Berger, 1968).

Some positions of Kunitz domains show a strong preference for one or two types. For example, position 16 is typically either Ala or Gly (i.e., 58 of 71 proteins). This preference most likely is rooted in the binding properties of the observed proteins; eight other types are seen, including acidic, basic, and hydrophobic amino acids.

Note that about half of the positions do not have proline, suggesting that proline is never observed at these sites because it would wreck the structure. There are, however, a few positions where the BPTI φ would indicate that proline should not be observed, but some Kunitz domains do have proline. Presumably, in these cases, the main chain rearranges to accommodate the φ requirement.

Kunitz domains are especially well known because there are sequences from so many sources. In some cases, very high diversity has been observed within a single organism. Three proteins homologous to the King Kong protein (a 29-amino acid kenotoxin) were cloned from a single conus snail (Hillyard *et al.*, 1989). Each of the secreted proteins comprises a prepro sequence and a presumed toxin domain. Within the prepro sequence these was almost identic similarity. Within the toxin domain, however, only the cysteines were conserved and at only a few positions did even two of the three have the same amino acid type.

When sequences are available only from a narrow source, one might easily overestimate the degree to which the sequence is limited by structural requirements. CMTI-type trypsin inhibitors (see Table 4) have been found only in cucurbida seeds and the degree of similarity is quite high. These proteins all inhibit trypsin and were selected for study on this basis. Had biochemists sought other inhibitory activities, they might have found CMTI-related inhibitors that do not inhibit trypsin. One would expect these to show greater sequence diversity. The sequence shown in Table 4 as "parental" is $xxxxxC_3P_4x_{I6}L_7M_8xC_{10}K_{11}xD_{13}xD_{15}C_{16}L_{17}xxC_{20}xC_{22}$ $xxxxxC_{28}G_{29}xx$. Residues that allow at least some variability are shown as "x." We have shown that $P_4$, $I_6$, $L_7$, and $M_8$ are not required for structure or binding. I find it difficult to believe that any of $K_{11}$, $D_{13}$, $D_{15}$, $L_{17}$, and $G_{29}$ are structurally required.

## Knowledge of Binding Sites

When the parental protein is known to bind a second molecule (*e.g.,* BPTI binding trypsin), knowing which residues of a parental protein make up a binding site is quite valuable. The residues of BPTI that contact trypsin or are close enough to make contact with other trypsin-homologous proteins include residues 10–20, 31, 32, and 34–42. These residues are the prime candidates to vary should one want to make an inhibitor of a different serine protease. A substantial number of known Kunitz domains come from snake venoms. Many of these proteins are potent trypsin inhibitors, but their toxic activity derives from their binding to ion channels. Dufton has identified a cluster of residues that are well removed from the residues involved in trypsin binding as being the binding site for potassium channels (Dufton, 1985). This illustrates that a single framework can support various binding sites and that having one activity does not preclude other activities. Knowing which residues are in a binding site gives us ways to endeavor to maintain, alter, or abolish the activity due to that binding.

## Stability

More rigid proteins are usually more resistant to proteolysis. BPTI, for example, is not degraded when: (1) boiled for 5 min in 8 *M* urea, (2) diluted so that urea is 2 *M,* (3) chymotrypsin is added, and (4) the reaction is incubated at 37°C overnight. BPTI does not inhibit chymotrypsin any more than does, for example, BSA. BPTI is not, however, degraded as are most proteins.

Though not rigorously demonstrated, I believe that more rigid molecules are capable of higher-affinity binding than are flexible molecules. A molecule that in isolation naturally takes on a conformation very close to the best conformation that it will need to have in a complex will have minimal tendency to leave the complex when it forms. This can lead to very small $k_{off}$ and $K_D$.

### Disulfides

In the display systems M13 III, M13 VIII and *E. coli* OSP, the displayed polypeptide is translated downstream from a functional signal sequence causing the gene product to be secreted into the periplasm. In this milieu, cysteines are oxidized to disulfides. Furthermore, the periplasm contains chaparonins that help newly secreted polypeptides fold correctly. In Table 5, the first seven classes have been displayed on phage and the expected disulfides are formed. The subsequent compounds have not, to my knowledge, been displayed, but these should work too. Disulfides are very effective in stabilizing small proteins (see Creighton and Charles, 1987).

**TABLE 5**   Sample of Displayed Proteins

| Name | Formula | Size (AAs) | Disulfides |
|------|---------|-----------|------------|
| ThioNut (TNn) | ..XC-(X)$_n$-CX..  ($n$ = 4-18) | 6 to about 25 | 1 |
| Endothelin | XCXC-(X)$n$-CXXXC..  ($n$ = 6-10) | Around 20 | 2 |
|  | XCXC-(X)$n$-CXXXC..  ($n$ = 6-10)  Both disulfide structures reported. | | |
| Cucurbida maxima Trypsin Inhibitor (CMTI) | 1   2   3   1'   2'   3'  xxCxxxxxxxCxxxxxCxxxCxCxxxxxxxCxxxCxxx | 27 to about 30 | 3 |
| Kunitz domains | 5   14   30   38   51   55  ...C.....C.........C.....C....C...C... | 51 to 60 | 3 |
| CD4 domains | C16:C84 (first domain) and C130:C159 (second domain)  (IgG superfamily) | 178 | 2 |
| Chymotrypsin-homologous serine proteases | Trypsin (C. Craik and co-workers) | 223 | 6 |
|  | Plasmin (R. B. Kent, unpublished results) | 252 | 4 |
|  | Plasmin & Kringle 5 (R. B. Kent, unpublished results) | About 300 | 7 |
| α-Conotoxins, e.g., GI, GIA, GII, MI, and SI | 12   1'   2'  ..XCCXXXCXXXXXC(XXXXX)- | 13 to 19 | 2 |
| Guanylin | PNTCEICAYAACTGC | 15 | 2 |
| ST-Ia | NSSNYCCELCCNPACTGCY | 19 | 3 |
| ω-Cono-toxins | 1   2   3 1'   2'   3'  CXXXXXXCXXXXXXC-CXX (X) CXXXX (XX) C | About 27–29 | 3 |
| Streptococcal Immunoglobulin-binding domain | MTYKLILNGKTLKGETTTEAVDAATAEKVFKQYANDNG VDGEWTYDDATKTFTVTE | 56 | 0 (T$_m$= 87°C) |
| EGF domains | xxxxCxxxxxCxxxxxxCxxxxxxxxxCxCxxxxxxCxxx | 35 to 50 | 3 |
| μ-Cono-toxins | XXCCXXXXXCXXXXCXXXXC CXX | About 23 | 3 |

### Chelation

Certain side groups on proteins have substantial tendency to bind metal ions and other molecules (*i.e.,* prosthetic groups). Proteins involving $Cu^{2+}$ or $Zn^{2+}$ chelated by a combination of histidine, methionine, and cysteine can be displayed on phage (Choo and Klug, 1994a,b). Chelation of copper and zinc though His, Met, and Cys is more stable than is the chelation of calcium by acidic side groups.

### Hydrophobic interactions

There are some small proteins, such as the IgG-binding domain of streptococcal G protein, that are small, very stable, and devoid of covalent crosslinks. These proteins are held together by hydrophobic forces and carefully positioned residues. Nevertheless, these proteins can be mutated without destroying the structure.

### Charges

It is unlikely that complementary charges can hold a small protein together.

## Antigenic Potential

Proteins are generally regarded as being antigenic. Binding proteins can be derived from human proteins or domains of human proteins to diminish this problem. It is general accepted that murine Abs are antigenic. Chimeric Abs having murine $F_V$ domains and human constant domains are decidedly less antigenic and "humanized" Abs having a human $F_V$ framework with engineered CDRs are even less antigenic. Proteins having, for example, a human Kunitz domain framework and five or six altered residues may have low antigenicity because they (1) are so much smaller, (2) are likely to be used in very small amounts, (3) are highly resistant to proteolysis (which prevents efficient presentation of fragments by antigen-presenting cells), and (4) are highly similar to self proteins.

## Example

BPTI (the prototypical Kunitz domain) is an excellent IPBD because: (1) it is highly stable, (2) it exhibits very high affinity binding for some proteins (*e.g.,* trypsin with $K_D = 6 \times 10^{-14}\,M$), (3) it shows highly specific binding (affinity for other trypsin-homologous proteins is well above micromolar), (4) it has many known homologs (each of which could also be an IPBD), (5) the binding can be altered by changing a few residues, (6) Kunitz domains can be secreted from bacterial cells, (7) the 3D structure is well known, and (8) it is one of the most thoroughly studied proteins.

## DESIGNING A DISPLAY GENETIC PACKAGE

This section covers the design of a Display Genetic Package (DGP) for display of PBDs. There are some differences between DGPs for proteins and those for peptides.

## Picking a Genetic Package

Picking a genetic package is related to the protein you intend to display. The important question is, "Do any Outer-Surface Proteins (OSPs) of the candidate GP traverse a morphogenetic pathway through which I think my PBD will go and be correctly folded?"

The most widely used GP is the filamentous phage M13 where most proteins have been displayed at the amino terminus of the proteins III or VIII. M13-based phagemids have also been widely used. The phagemids usually use a C-terminal fragment of protein III that will assemble into the phage.

### Criteria for a genetic package

The criteria for picking a DGP include:

I. Demonstration that a known GP works for your chosen IPBD or a very similar domain, or

II. High score on the following check list:
   A. Stable
      i. Long storage
      ii. Chemical insults: resistance to: (1) low pH, (2) high pH, (3) guanidinium HCl, (4) urea, (5) detergents, (6) proteases, (7) high salt, (8) organic solvents (ether, chloroform, *etc.*)
   B. Regeneration: desirable properties
      i. Genetic fidelity
      ii. Genetically alterable
   C. Morphogenesis: Desirable properties
      i. Outer-surface proteins (OSP) that can be modified
      ii. One or more OSPs that have morphogenic pathway compatible with chosen IPBD

### Bacteriophage

Certain bacteriophage are particularly suitable GPs because, among other things, they: (1) have small genomes that can be manipulated, (2) have a set of proteins on their surface that is large enough to provide suitable anchors but not so diverse as to cause adventitious binding to most targets, (3) are resistant to chemical insult, and (4) can be quickly replicated.

The filamentous coliphage M13 (f1 and fd are essentially identical for display) is the most used DGP. Makowski (1993) has reviewed the structural aspects of picking an anchor protein in M13. Table 6 shows several signal sequences and the first few amino acids of different secreted proteins. In our hands, the "signal sequence" of M13's protein VIII does not lead to secretion of proteins such as BPTI. Other signal sequences can cause BPTI or the amino-terminal portion of BPTI:mature pVIII to be secreted. Expression of a tripartite gene comprising

[secY-dependent signal sequence]:[display domain]:[mature pVIII]

**TABLE 6**   Signal Peptides[a]

| Ident | Sequence | N |
|-------|----------|---|
| PhoA | M **K** q s t i a l a l l p l l f t p v t **K** A /**R** T... | (17) |
| MalE | M **K** I **K** T G A **R** i l a l s a l t t m m f s a s a l a /**K** I... | (18) |
| OmpF | M M **K** **R** n i l a v i v p a l l v a g t a n a /a **E**... | (19) |
| Bla | M S I Q H F **R** v a l i p f f a a f c l p v f a /h ... | (>18) |
| LamB | M M I T L **R** **K** l p l a v a v a a g v m s a q a m a /v **D**... | (19) |
| Lpp | M **K** A T **K** l v l g a v i l g s t l l a g /c s... | (>17) |
| gpIII | M **K** **K** l l f a i p l v v p f y s h s /a **E** T V **E**... | (16) |
| gpIII-BPTI | M **K** **K** l l f a i p l v v p f y s g a /**R** P **D**... | (15) |
| gpVIII | M **K** **K** S L V L **K** a s v a v a t l v p m l s f a /a **E** G **D** **D**... | (16) |
| gpVIII-BPTI | M **K** **K** S L V L **K** a s v a v a t l v p m l s f a /**R** P **D**... | (15) |
| PhoA-BPTI | M **K** q s t i a l a l l p l l f t p v t **K** A /**R** P **D**... | (17) |
| mod8-BPTI | M **K** **K** s l v l l a s v a v a t l v p m l s f a /**R** P **D**... | (20) |

[a]N is the number of amino acids between the last charged residue in the amino-terminal region and the first of the mature region.

results in processing of the polypeptide to give:

a. Cleavage of the signal sequence,
b. Mature pVIII spanning the lipid bilayer, carboxy terminus in the cytoplasm, and
c. Display domain in the periplasm.

For some proteins that one might wish to display, folding in the cytoplasm might be desirable. For example, one might have a protein that contains heme or some other prosthetic group that is available in the cytoplasm. Some amino acid sequences found in proteins seem to block secretion; proteins containing such sequences cannot be displayed on M13's protein III or VIII, or on bacterial OSPs. Some proteins that have free sulfhydryls may be stable once they have folded in the reducing environment of the cytoplasm. For such cases, one might try using a phage that has cytoplasmic morphogenesis, such as ΦX174 and λ (Sternberg and Hoess, 1995).

## Bacterial Cells

PBDs can be displayed on bacterial cells attached to any of the myriad number of suface proteins. These proteins arrive at the surface in a variety of ways which have been studied extensively for several proteins. In addition to getting the displayed protein through the lipid bilayer, one must have a signal that causes the cell to transport the display protein to the outer cell surface (OSP). This has been done by fusing a protein domain or peptide to a protein that normally resides on the outer surface, *e.g.*, LamB or OmpA. The advantage of using bacteria is that propagation of the GPs is easier. A possible disadvantage is that high multiplicity of display may make separation of very-high-affinity binders from good binders difficult. I believe this can be overcome by careful choice of: (a) anchor protein, (b) regulatory features of the display gene, and (c) amount of target on matrix.

# Picking an Anchor Protein and an Attachment Site in or on the Anchor

## Morphogenic and structural considerations

To get an active collection of PBDs, one has to ensure that the PBD sequences can traverse the morphogenic pathway of the anchor protein and that the PBD as displayed has a good chance of being active. As stated, proteins that cannot be secreted cannot be displayed on M13 III or M13 VIII anchors. Also, proteins that require prosthetic groups present only in the cytoplasm are likely to function poorly if at all when expressed on III or VIII. Multimeric proteins pose special problems discussed below. Proteins that are degraded or chemically altered in the periplasm (*e.g.,* proteins that contain free thiols) may not function well if displayed on III or VIII anchors.

Some proteins have an exposed amino or carboxy terminus that is involved in binding. These proteins may be expressed and fold into essentially native form when fused to a display anchor, but the binding may be greatly altered. Proteins that have a buried terminus may not fold if this terminus is fused to a display anchor. For example, the amino terminus of chymotrypsin-homologous serine proteases is buried in the catalytically active form. It would not be surprising if display of these proteins by a fusion to the amino terminus leads to proteins that cannot obtain the native conformation. (Note, however, that tPA is about 10% active in the precursor form.) Chymotrypsin-homologous serine proteases have been displayed by fusion of the carboxy terminus of the protease to the amino terminus of an anchor (Rachel Kent, unpublished work).

One should be careful about statements that such-and-such a residue "must" be at the amino terminus. It was formerly well known that peptides having the sequence Tyr-Gly-(gly/ala)-Phe-Leu-... would bind the Ab 3-E7 and that the Tyr must be at the amino terminus. We found that the microproteins AEGXCYGgFCX... (disulfide forms six-amino acid-membered ring and the "g" can be Gly or Ala) bind 3-E7 although the Tyr is not amino terminal (McLafferty *et al.,* 1993).

## Multiplicity of display

Display of a protein on pIII when the only gene III is altered gives, in principle, five copies of the protein per phage. In practice, proteolysis may reduce the number. Alternatively, phagemids can be used in which one protein III is displayed and a second copy is provided in *trans* by helper phage. Such an arrangement has a number of interesting features. If the relative amounts of w.t. pIII and display pIII are regulated so that very few phage carry more than one display molecule, then phage with NO display are the vast majority. We have found that multiple binding can be avoided by using a suitably low amount of target and have separated phage that display a protein having $K_D$(target) $\approx$ 10 p$M$ from other phage that display a protein having $K_D$(target) $\approx$ 1 p$M$.

A second feature of having both a display and wild-type pIII is that display of a protein that interferes with infectivity is no longer lethal to the phagemid. By removing this penalty, it becomes less likely that deletion mutants will take over a population.

Since the protein VIII is not involved in infection, alteration to a few protein VIII molecules does not seem to affect the viability of the phage. On the other hand, it does not seem possible to make phage in which all copies of pVIII are altered by addition of a protein the size of BPTI. Protein VIII is the major coat protein, about 3000 copies per phage. In w.t. phage, production of pVIII does not use the SecY machinery, probably because the SecY machinery is not capable of processing this number of protein molecules rapidly enough. One can get from about 10 to perhaps 200 copies of BPTI:pVIII assembled into each phage. Presumably, these are scattered at random along the phage; BPTI does not have any known tendency to dimerize.

## Consideration for multimeric proteins

*Homomultimeric proteins*   Before the circular single-stranded DNA of M13 is extruded through the pVIII-laden cytoplasmic membrane, the mature VIII lies in the lipid bilayer with the carboxy terminus in the cytoplasm. Typically, such protein molecules enjoy a high degree of lateral mobility. Thus, we expect that two molecules of X:VIII could find each other if X has a tendency to form $X_2$ dimers. To maximize the chance of success, one would like to know that the carboxy termini of the X monomers are free or at least not buried too deeply. This might also work for trimers (such as IL-1). For higher oligomers, one should consider the symmetry. A tetramer with tetrahedral symmetry (32) or alternating square (222) symmetry would have linkers going in all directions; to get back to the phage, rather long linkers might be needed and these might interfere with binding. A protein having point-group symmetry 4 might be linked by four parallel strands to the phage as shown in Fig. 1. With sufficiently long linkers, such a tetramer could be connected to one linker in layer 1, two in layer 2 and 1 in layer 3. To my knowledge, this has not been done.

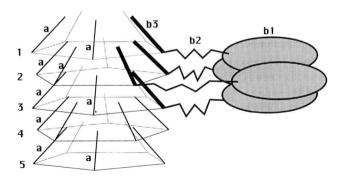

**FIGURE I**   Five layers of M13 major coat protein with a tetrameric protein attached. WT monomers of VIII are labeled "a." The fusion display protein comprises: "b1," the displayed domain; "b2," a flexible linker of from 1 to perhaps 10 amino acids; and "b3," mature VIII. The phage DNA is not shown but lies in the central tube.

***Heteromultimeric proteins***    When the components of the multimer are different, there are additional considerations. A method that has been used in display of Abs involves one component covalently anchored to III or VIII while the other component is supplied into the periplasm to allow association of the parts. It is also possible to have each component anchored; this would work best with pVIII anchoring. It is likely to be important to have essentially stoichiometric amounts of each component so that spurious homodimers do not arise.

# Linkers To Allow

## Correct post-translational processing

For PBDs fused to the amino terminus of III or VIII, an essential step is cleavage of the signal sequence. It is now known that a suitable sequence within the signal sequence allows cleavage such that any amino acid type can appear at the amino terminus of the processed polypeptide. It is likely that some amino acids will function better than others in the role of facilitating cleavage and becoming the amino-terminal residue of the secreted polypeptide. Marks *et al.,* (1986) found, however, that a long leader facilitated secretion of BPTI; with very short (or no) linkers, BPTI folds and forms disulfides within the cell and is not transported from the bacterial cell. Although BPTI and millions of its variants have been displayed at the amino terminus of proteins III and VIII, such display might be enhanced by inclusion of an unstructured amino-terminal linker. It may also be valuable to insert BPTI into one of the $(GGGS)_n$ or $(EGGGS)_n$ regions of protein III.

## Relative motion between anchor and displayed protein domain

Most groups that have displayed proteins or peptides on phage have inserted a linker or spacer between the protein or the varied peptide and the anchor protein. The reason is to allow the varied segment to bind independent of the anchor. No one has, to my knowledge, shown that any particular linker is better or worse than any other or that the linkers are actually needed. The presumed "flexible linker" from M13 III: $(GGGS)_n$ is a good choice as are sequences rich in Gly, Ser, Pro, Asp, Asn, and Thr.

## Cleavage by specific proteases

Matthews and Wells (1993) have exploited phage display of a tight-binding domain (hGH) to explore the protease resistance and susceptibility of linkers.

Introduction of a known recognition site, such as a site for bovine Factor $X_a$, offers several advantages. First, there might be members of the library that bind so tightly that they cannot be eluted without destroying the GPs. Second, removing the displayed protein from protein III should give all the phage identical infectivity, independent of the display gene carried. Having a specific cleavage site in the target might also be useful so that the target can be selectively released from the matrix.

## Picking an Amino Acid Sequence To Be Encoded

The amino acid sequence to be encoded is picked to give efficient display of the IPBD flexibly linked to a suitable anchor domain. Table 7 shows the 3′ portion of a gene to cause display of BPTI attached to M13 III. Above the sequence the codons are numbered from the first codon of the gene; the amino acid numbers of BPTI are shown as subscripts. The amino acid sequence was picked first. This gene contains a compromise: the segment labeled "§" (comprising Ala-Glu) was inserted between the signal sequence and BPTI because there is a phage (A-SHO) having an *Esp*I site in DNA that encodes this amino acid sequence. Alternatively, a *Kpn*I site could be introduced into the signal-sequence encoding part of protein III and the Ala-Glu dipeptide omitted.

## Designing a Gene

Design of a DGP for a protein is illustrated here by the display of BPTI on M13's protein III. Design of a gene for display on protein VIII or on a carboxy-terminal fragment of pIII is essentially the same. First pick a DGP (*e.g.,* a strain of M13 or a phagemid) that: (1) does not have many extraneous restriction sites; you will want all the sites you can get for your gene; (2) does carry a selectable marker gene such as ampicillin, kanamycin, or tetracycline resistance; and (3) has one or more restriction sites that bound the anchor-display junction (*e.g.,* the signal sequence of protein III and the mature protein). Alternatively, pick a DGP that has an engineered *signal-sequence:display:mature protein III* with suitable restriction sites in gene III.

Compile a list of restriction sites in the vector and of enzymes that do not cut the vector. Enzymes that leave cohesive ends of two or more bases are highly preferred for introduction of variegated DNA. Blunt-cutting enzymes are very useful for elimination of parental vectors that escape cutting when the vgDNA is introduced.

### TABLE 7

*A. Amino-acid sequence of BPTI-display protein*

| | |
|---|---|
| MKKLLFAIPLVVPFYSG | (the signal sequence) |
| AE | (an adaptor) |
| RPDFCLEPPYTGPCKARIIRYFYNAKAGLCQTFVYGGCRAKRNNFKSAEDCMRTCGGA | (BPTI) |
| YIEGRVGAE. . . | (Factor Xa cleavage site:mature III) |

*B. Designed BPTI:III display gene*

```
      1    2    3    4    5    6    7    8    9   10   11
      M    K    K    L    L    F    A    I    P    L    V
    ------  Signal sequence of M13 iii --------
    --------------------------------------------- 1 -->
    |atg|aag|aag|ctG|ctc|ttc|gct|att|cct|ctt|gtG|
              |  SapI   |                    |  KpnI
```

*continues*

**TABLE 7** *Continued*

```
 12  13  14  15  16  17  18  19  20  21  22
  V   P   F   Y   S   G   A   E   R₁  P₂  D₃
- Signal sequence -------->|<- § ->|<----BPTI---
----------------- 1 ---------------------->|
|gtA|cct|ttc|tac|tct|ggt|gcT|gaG|cgT|ccG|gat|
   KpnI      |                | EspI  | | BspEI|
```

```
 23  24  25  26  27  28  29  30  31  32  33
  F₄  C₅  L₆  E₇  P₈  P₉  Y₁₀ T₁₁ G₁₂ P₁₃ C₁₄
----------- BPTI ---------------------------
|<----- 2 ------------->|<---- 6 ----------- 
|ttc|tgt|CTC|GAG|cct|ccG|taT|acC|ggT|ccG|tgt|
        | XhoI   |          | AgeI   |
                        |         || RsrII  |
                   Bst1107I
```

```
 34  35  36  37  38  39  40  41  42  43  44
  K₁₅ A₁₆ R₁₇ I₁₈ I₁₉ R₂₀ Y₂₁ F₂₂ Y₂₃ N₂₄ A₂₅
-------- BPTI -----------------------------
------- 6 ----------------->|<--- 3 ---------
|aaa|gcG|cgC|att|att|cgc|tat|ttc|tat|aac|gcC|
    | BssHII|                          |  StyI
```

```
 45  46  47  48  49  50  51  52  53  54  55
  K₂₆ A₂₇ G₂₈ L₂₉ C₃₀ Q₃₁ T₃₂ F₃₃ V₃₄ Y₃₅ G₃₆
------------------- BPTI ------------------
---- 3 ------------->|<----- 7 ---------------
|aaG|gcA|ggC|Ctt|tgt|caa|act|ttc|gtt|tac|ggc|
StyI  | | StuI  |
```

```
 56  57  58  59  60  61  62  63  64  65  66
  G₃₇ C₃₈ R₃₉ A₄₀ K₄₁ R₄₂ N₄₃ N₄₄ F₄₅ K₄₆ S₄₇
------------- BPTI ----------------------
-------- 7 -------------------------->|<- 4 -
|ggC|tgc|AgG|gct|aaa|cgn|aac|aac|ttc|aaa|tcG|
   | PstI |                           | EagI
```

```
 67  68  69  70  71  72  73  74  75  76  77
  A₄₈ E₄₉ D₅₀ C₅₁ M₅₂ R₅₃ T₅₄ C₅₅ G₅₆ G₅₇ A₅₈
----------------- BPTI ------------------->|
------ 4 ----------------------------------->|
|gcc|gaa|gac|tgC|atg|cgt|acG|tgt|ggt|ggG|gcC|
  EagI  |        |   |    | BsiWI  |    | EcoO109I
     | BsbI  |    | SphI  |
```

```
 78  79  80  81  82  83  84  85  86
  Y   I   E   G   R   V   G   A   E
|<-------- F.Xa site--->|<-- mature III-----
|<------ 5 ----------------------->|
|tat|att|gaa|ggc|cgt|gtt|ggc|gcc|gaA|........
EcoO109I|                     | KasI  |
```

*Note.* Q is a segment that allows the A-SHO-compatible *Esp*I site to appear. These amino acids could be eliminated if a *Kpn*I site were engineered into gene III as shown above.

**TABLE 8**   Functional Regions of BPTI-Display Protein

| Category | Bases where site would be useful | Residues | Reason to want restriction site, to allow: |
|:---:|:---:|:---:|---|
| 1 | 1–66 | 1–22 | Insertion of new domains |
| 2 | 67–84 | 23–28 | Mutation of residues 29–40 |
| 3 | 121–147 | 41–49 | Mutation of 29–40 and 50–61 |
| 4 | 196–225 | 66–75 | Mutation of residues 50–61 |
| 5 | 226–258 | 76–86 | Insertion of new domains |
| 6 | 85–120 | 29–40 | Cut parental DNA in varied region |
| 7 | 148–195 | 50–65 | Cut parental DNA in varied region |

Next, make a list of the residues you plan to variegate. Table 8 shows the functional regions of the BPTI-display gene.

To alter the affinity of a Kunitz domain for serine proteases, variation of residues 10, 11, 13, 15, 16, 17, 18, 19, 20, 21, 31, 32, 34, 35, 36, 39, 40, 41, 42, and 46 may be desirable. Note that this group breaks into two segments: roughly 10–21 and 31–46. We will endeavor to have at least one restriction site in each of the regions 1–9, 10–21, 22–30, 31–46, 47–58. When introducing vgDNA into a display vector, residual uncleaved or once cleaved vector molecules can produce a significant undesirable background of parental DGPs. Having a restriction site in the region to be varied allows us to select against these vectors. One can also eliminate the unmodified background by having a parental vector having a frameshift in the region to be variegated. This is practical for protein VIII display or for protein III display in a phagemid where a functional pIII is not essential. One can also use as parental BD a variant that does not bind the target so that escapes pose no problem.

Compile a list of sites where restriction enzyme sites are compatible with amino acid sequence. (The outline of simple computer program to list all possible restriction sites is given in the web site.) Table 9 shows where restrictions sites (for enzymes that do not cut the vector A-SHO outside the display gene) can be built into a BPTI display gene consistent with the designed amino acid sequence. Following the possible location, the segment identifier (from Table 8) is given in brackets. Those sites that are not incorporated into the designed gene are shown ~~stricken~~. The choice of sites was as follows.

Category 5 allows only *Kas*I and *Eco*O109 I sites. The vector A-SHO contains a *Kas*I that aligns with bases 250–255 of the designed gene. Thus, we reject the *Kas*I sites at 49 and 226. Region 2 allows only an *Xho*I site with *Bst*1107I, *Eco*N I, and *Pfl*MI sites being possible on the 2–6 boundary. Thus, we elect the *Xho*I site. Region 3 allows *Tth*111I, *Stu*I, and *Sty*I sites, but only *Sty*I gives useful ends. We pick the *Stu*I and *Sty*I sites at 138 and 131 and reject the sites at 50 and 26. A-SHO has an *Bsp*E I site that aligns with the site at bases 60–65 of the designed gene; we pick this site and cancel the potential *Bsp* EI site at 46. We now pick other sites so that we have several sites in each region that allow multiple sites to arrive at the gene shown in Table 7. A restriction map of the designed fragment confirmed that the sites intended to be unique were so.

**TABLE 9**     Possible Restriction Enzyme Sites in BPTI-III Display and Those Incorporated in Final
Design

| Enzyme | No. possible | Where(base number)[Category] | | | |
|---|---|---|---|---|---|
| *Age*I | 1 | 88[6] | | | |
| *Apa*I | 2 | 90[6] | ~~91[6]~~ | | |
| *Bbs*I | 1 | 202[4] | | | |
| ~~*Bsg*I~~ | ~~3~~ | ~~50[1]~~ | ~~168[7]~~ | ~~197[4]~~ | |
| *Bsi*WI | 1 | 214[4] | | | |
| *Bsp*EI | 2 | ~~46[1]~~ | 60[1] | | |
| *Bss*HII | 1 | 103[6] | | | |
| *Bst*1107I | 2 | 84[2/6] | ~~157[7]~~ | | |
| ~~*Bst*XI~~ | ~~1~~ | ~~20[1]~~ | | | |
| *Eag*I | 2 | ~~58[1]~~ | 197[4] | | |
| ~~*Eco*NI~~ | ~~1~~ | ~~82[2/6]~~ | | | |
| *Eco*O109I | 4 | ~~90[6]~~ | ~~91[6]~~ | ~~214[4]~~ | 226[5] |
| *Esp*I | 2 | 52[1] | ~~175[7]~~ | | |
| *Kas*I | 3 | ~~49[1]~~ | ~~226[4]~~ | 250[5] | |
| *Kpn*I | 1 | 33[1] | | | |
| ~~*Nru*I~~ | ~~1~~ | ~~17[1]~~ | | | |
| ~~*Pfl*MI~~ | ~~1~~ | ~~82[2/6]~~ | | | |
| ~~*Pml*I~~ | ~~1~~ | ~~216[4]~~ | | | |
| ~~*Ppu*MI~~ | ~~2~~ | ~~90[6]~~ | ~~214[4]~~ | | |
| *Pst*I | 2 | 168[7] | ~~197[4]~~ | | |
| *Rsr*II | 1 | 90[6] | | | |
| ~~*Sac*II~~ | ~~1~~ | ~~197[4]~~ | | | |
| *Sap*I | 1 | 12[1] | | | |
| ~~*Spe*I~~ | ~~1~~ | ~~27[1]~~ | | | |
| *Sph*I | 1 | 209[4] | | | |
| *Stu*I | 2 | ~~58[1]~~ | 138[3] | | |
| *Sty*I | 2 | ~~26[1]~~ | 131[3] | | |
| ~~*Tth*111I~~ | ~~1~~ | ~~140[3]~~ | | | |
| ~~*Xcm*I~~ | ~~1~~ | ~~20[1]~~ | | | |
| *Xho*I | 1 | 73[2] | | | |

*Considered, but no site possible:*

| | | | | | | | |
|---|---|---|---|---|---|---|---|
| *Aat*II | *Afl*II | *Asc*I | *Avr*II | *Bcl*I | *Bsm*I | | *Bst*BI |
| *Bst*EII | | *Eco*RI | *Eco*RV | *Hpa*I | *Mlu*I | *Mun*I | *Nco*I |
| *Nhe*I | *Not*I | *Nsi*I | *Pme*I | | *Sac*I | *Sfi*I | *Sfr*I |
| *Sgr*AI | *Sse*8387I | *Xma*I | | | | | |

For display on pVIII, we need a gene fragment that encodes 50 amino acids of
mature pVIII. To avoid the possibility of genetic crossover, it is preferred that this
fragment be as different as possible from the wild-type gene which is also present in
the vector. The display gene or a new insert for an existing display gene may be
synthesized by PCR amplification of overlapping synthetic fragment (Jayaraman *et
al.,* 1991; Jayaraman and Puccini, 1992).

# VARIEGATION OF DISPLAYED PROTEINS

## Picking Amino Acid Positions to Vary

Which amino acid to vary may be the most important decision after choosing a parental protein. The major principle is that one seeks to vary surfaces that can contact the target. Thus it is preferred to pick several residues that all lie on a surface close enough together to touch the target at once. When such residues are varied together, the number of surfaces is the product of the number of types at each varied location. If the residues were unable to interact with the target simultaneously, the number of surfaces would be reduced to the sum of the number of substitutions at each varied location.

## Picking Amino Acid Types to Allow at Each Position

Having picked locations to vary, one then must decide how many amino acid types to allow at each position. There are two opposing considerations. If you let all 20 amino acids occur at very many locations, the number of variants gets out of hand very quickly. In many cases, there is not a unique amino acid sequence that gives high affinity, so you may not need all 20 at the varied locations to find out which positions are important. On the other hand, you are much more likely to select a protein that is allowed than one you did not program. (An interesting aspect of the method is that you can pick up unplanned mutants if they show high-affinity binding.) If Minority Sampling is deemed sufficient, then one can allow a large number of variants, e.g., $10^8$ or $10^9$. For example, in a Kunitz domain, there may be 10 residues that are likely to affect binding. If you vary all of these through all 20 types, you get $20^{10}$ or about $10^{13}$. Getting several times $10^{13}$ transformed cells is a huge amount of work and, fortunately, is unnecessary. It is highly unlikely that only 1 sequence out of $10^{13}$ will give high-affinity binding; several positions are likely to be relatively unimportant. Suppose that 3 of the positions are highly important (only 1 type will work) and 7 require only that one of 12 types appear (or that 1 of 8 types does not appear). If we use simple combinatorial DNA synthesis, there are $32^{10} = 1.13$ x $10^{15}$ DNA sequences in a library that has 10 fully varied positions. We will succeed if we have 1 of the $12^7 = 3.6$ x $10^7$ sequences that conform to the 3 required types and 7 permissive contexts. Thus the number of transformants we need to have 95% probability of success is 3. x $(1.13$ x $10^{15}/3.6$ x $10^7) = 9.4$ x $10^7$. Note that you do not need to know *a priori* which 3 positions are crucial and that the list of allowed types at the other 7 positions could be different at each position.

Using a Minority Sampling approach gives a high level of diversity. If one does not select a high-affinity binder, one cannot say that the designed scheme does not give high-affinity binders; perhaps the sampling was too small or just unlucky. To an extent, it is a question of taste. If the library contains $10^7$ or more transformants, with a DNA diversity of $32^5 = 3.3$ x $10^7$, and you get no good binders, the probability is that you did not vary the right residues. If you fail to obtain a good binder from a library of $10^7$ transformants taken from a diversity of $10^{10}$, however, it is very possible that a second library of $10^7$ transformants from the same DNA population will yield a good binder.

My preference is to allow a large number of possibilities at several positions, but not necessarily all 20 amino acids. The coding scheme NNT[3] encodes 15 amino acids with only 16 codons. You will miss it if Met, Trp, Lys, Glu, or Gln was essential, but this is not very likely. The scheme NNG encoded 13 different amino acids with 16 codons, as opposed to the NNS scheme NNS which utilizes 32 codons for 20 amino acids.

## Optimizing Number of Amino Acid Sequences/Number of DNA Sequences

If your DNA synthetic repertoire includes introducing specific triplets, then variegation is both more efficient and more demanding. One use of introducing codons is to allow all possible amino acids with only 20 codons. Nevertheless, using mixtures of nucleotides (as opposed to codons) can provide highly diverse populations in which the number of encoded sequences close to the number of DNA sequences. In particular, the NNt (equivalent to NNc), NNg, VNt, and several other codons give high efficiency.

## Introducing Variability into Displayed Proteins

Any suitable means of introducing variability may be used to generate a diverse pool of genes from which useful binding proteins may be selected. Means of introducing diversity into a genetic package can be characterized in two ways: (1) how the diversity is generated or captured, and (2) how the diversity is directed to a particular genetic locus.

Diversity can be obtained by: (1) DNA synthesis involving mixtures of substrates, (2) DNA synthesis involving mixtures of codons, (3) DNA synthesis involving separation of the growing chains into pools for differential treatment, (4) DNA synthesis involving base analogs, such as inosine, that can pair with more than one of the standard bases, (5) DNA synthesis using an error-prone enzyme, and (6) capture of diversity from nature. Each of these approaches has advantages in one or another circumstances. One can also mix these modes. For example, a natural DNA sequence could be the template for PCR using: (a) an error-prone enzyme, (b) inosine, and/or (c) partially mismatched primers. DNA synthesis with mixed bases works very well when a large number of types is to be allowed at 5 to 10 positions. It is regrettable that most DNA synthesizers allow only seven bottles for base substrates. As one usually needs A, C, G, and T bottles, only three mixtures are easily accommodated. Which three mixtures one should choose depends on the sequence being varied. DNA synthesis with mixed codons is very well suited if one wishes to scan a region with a particular substitution, such as (wild-type or Ala codons) or (wild-type or Cys codons).

[3]Ambiguous bases (N, B, D, H, V, K, M, R, S, W, and Y) are given in Table 1.

The diverse DNA can be directed to the *display binding-domain* gene either *in vitro* or *in vivo*. The mechanism can be one of: (1) ligation of diverse DNA to specifically cleaved receptor DNA (cassette mutagenesis), (2) annealing of partially mismatching ssDNA to a ssDNA template (with or without enzymatic completion of the annealed DNA), and (3) *in vivo* homologous recombination. For small proteins, the use of *in vivo* recombination is probably not practical as it is difficult to arrange for specific recombination sites within a coding region, particularly a very small one. Cassette mutagenesis is probably the best method.

Figure 2 (in Chapter 5) shows several ways in which synthetic DNA can be prepared for use in cassette mutagenesis. The restriction sites used to introduce the vgDNA are labeled ① and ②. Preferably, there are restriction sites in the parental vector that are not designed into the variable region; these are shown as dashed restriction sites in Fig. 2. In A, the vgDNA is synthesized and converted to dsDNA by enzymatic template-based synthesis. An alternative approach (Devlin *et al.*, 1991) is to make a second strand having inosine to match variable bases. B is quite similar to A, but there are two primers so that PCR can be used to generate more DNA (Scott and Smith, 1990). C was used by Cwirla *et al.* (1990). In D, the synthetic DNA is designed to fold back on itself to allow enzymatic completion followed by restriction cleavage (Christian *et al.*, 1992). E is useful when the regions to be variegated are separated in sequence, but the separation must be enough to allow annealing two synthetic fragments (Kay *et al.*, 1993). Methods C and E require at least 10 (very CG-rich) fixed bases (preferably, 15 or more) of DNA for annealing stubs or primers.

## Varying Sites Widely Separated in Amino Acid Sequence

When the amino acids of a protein are close in space but distant in sequence (a common occurrence), one has a few options available to improve the efficiency of selection.

Using the example of BPTI, one could first make a library that varies residues 10, 11, 13, 15, 16, 17, 18, and 19. This library is selected for binding, but not for many rounds. The first objective is to obtain a population of perhaps 1000 to 100,000 that have some affinity for the target. Using this population as parental DNA, we then vary positions 31, 32, 34, 35, 39, 40, 41, and 42 and select for several rounds. One could then go back and vary positions 10–19 and select again until a protein of sufficient affinity is found. If we used NNT or NNG codons at eight positions, we generate about $4.3 \times 10^9$ DNA sequences and $10^9$ transformants would give about 22% representation. Such a plan allows isolation of a collection of about $6 \times 10^{18}$ protein sequences while making only a few times $10^9$ transformants.

A second approach is to make one piece of DNA that varies the first region and a second that varies the second region and combine these in PCR to make a piece of DNA that varies both regions. Table 10 shows two pieces of synthetic DNA that overlap by 33 bases in the middle and which encode residues 25 to 67 of the designed BPTI-display gene. After converting the DNA, into a double-stranded molecule, it can be cleaved with *Xho*I and *Eag*I and cloned into a vector that carries the parental BPTI-display gene. Note that several restriction sites present in the parental gene have been eliminated.

**TABLE 10**    Variegation of BPTI in the Trypsin-Binding Site

```
                L    E    P    P9
                25   26   27   28
         5'-|cctcctccct|CTC|GAG|cct|ccG|-
                       |gag|ctc|gga|ggc|
                       |  XhoI  |


     S|C S|C      S|C            F|S F|S F|S F|S
     L|P L|P      L|T      L|S Y|C Y|C Y|C Y|C
     H|R H|R      H|R      W|P L|P L|P L|P L|P
     F|I F|I      F|I      Q|R H|R H|A H|A H|A
     T|N N|Y      N|Y      M|T I|T I|T T|R T|R
     V|A V|A      N|A      V|A N|V N|V N|V N|V
     D|G D|G      D|G      E|G D|G D|G D|G D|G
        Y    T    G    P    C    K    A    R    I    I19
        29   30   31   32   33   34   35   36   37   38
     |<---- 6 -------------- 6 ------------->|
     |NNT|NNt|ggT|NNT|tgt|NNG|NNt|NNt|NNt|NNt|-
      |NNa|NNa|cca|NNa|aca|NNc|NNa|NNa|NNa|NNa|


        R    Y    F    Y    N    A    K    A    G    L    C30
        39   40   41   42   43   44   45   46   47   48   49
     |<--- 3 -------------------- 3 -------------->|
     |cgc|tac|ttc|tac|aac|gcC|aaG|gcc|ggt|Ctc|tgc|-3'
  3'-|gcg|atg|aag|atg|ttg|cgg|ttc|cgg|cca|gag|acg|-
                      |  StyI  |


        F|S        F|S F|S            F|S F|S        F|S
     L|S Y|C      Y|C C|V      Y|C Y|C L|S Y|C
     W|P L|P      L|P L|P      L|P L|P W|P L|P
     R|I H|R      H|R H|R      N|I N|R Q|R N|I
     K|V I|N      I|N I|N      T|V T|V M|T T|V
     A|E V|A      A|D A|D      A|D D|G V|A A|D
     G|T D|G      G|T G|T      G|H G|H E|G G|H
        Q    T    F    V    Y    G    G    C    R    A    K    R42
        50   51   52   53   54   55   56   57   58   59   60   61
     |<----- 7 --------------------- 7 ------------>|<--- 4 --
      |NNa|NNt|ttc|NNt|NNc|ggc|ggt|tgc|NNt|NNt|NNG|NNt|
      |NNt|NNa|aag|NNa|NNg|ccg|cca|acg|NNa|NNa|NNc|NNa|-


        N    N    F    K    S    A48
        62   63   64   65   66   67
      |aac|aac|ttc|aaa|tcG|gcc|gaa|
      |ttg|ttg|aag|ttt|agc|cgg|cttacccaccc-5'
                      |  EagI  |
```

# SELECTION OF BINDING DOMAINS — PROCEDURES

## Presentation of Molecular Target on Matrix (MTM)

The target is attached to a matrix or to a capturable tag (*e.g.,* biotin) with one or another covalent chemistries or the target is adsorbed onto a plastic well. Having the target attached to two matrix materials is useful as it allows discrimination between binding to target and binding to matrix. For DGPs that display multiple copies of the PBD, the amount of target on the matrix is important. High target density encourages multiple binding. This can be useful if the IPBD has no affinity for the target. After a few rounds of selection and amplification, the amount of target is reduced to help select for the phage having the best single protein binder.

## Estimation of Completeness of Library

The completeness of a library can be estimated from two data: (1) the number of independently transformed cells, and (2) DNA sequencing of several unselected GPs. Using the values in Table 2, one can make an initial estimate of the completeness of the library for the designed DNA diversity. One needs to examine DNA sequences of several unselected GPs to determine how well the variegated codons conform to the planned distribution.

## Elution of DGP–MTM Complex

### General considerations on fractionation

$K_D(A,B) = k_{off} \div k_{on}$. Since $k_{on}$ is, for molecules like BPTI, limited to less than about $10^7/M$/sec, big improvements in $K_D$ must come by lowering $k_{off}$. If we assume $k_{on} = 10^7/M$/sec, then for $K_D = 1$ p$M$, $k_{off}$ is $10^{-5}$/sec and $\tau_{1/2} = 6.9 \times 10^4$sec. The "on" rate of a PBD tethered to a GP must be greatly reduced from that of the free protein because it must diffuse as a molecule having a molecular weight of millions of daltons. The "off" rates of the tethered molecule and the free molecule should be about the same.

The most common form of fractionation involves:

1. binding the population of GPs for some time to the immobilized target,
2. washing away the unbound phage, and
3. eluting the bound phage with pH 2.2 buffer.

It is assumed that pH 2.2 will cause the bound GPs to release without destroying their viability. Various scenarios are possible, including:

a. low pH causes protonation of an atom in or near the PBD-target interface so that the proteins release with each still in essentially native form,
   b. low pH denatures the target so that it releases the PBD,
   c. low pH denatures the PBD so that it releases the target,

d. low pH denatures both PBD and target so that they release,

e. low pH ruptures the linkage between target and matrix, and

f. low pH ruptures the linkage between PBD and GP.

In any large population of PBDs that contains at least some very strong binders, it is expected that there will also be a greater number of strong binders, an even greater number of good binders, and yet more still poor binders. In this case, the fractionation will be most successful if the process can separate very strong binders from good binders and eliminate the poor binders. It is expected that a strong interaction between PBD and target stabilizes both molecules with respect to denaturation by protons. It is reasonable that there will be some very strong binders that stabilize the PBD:target complex against protons more than do weaker binders. In Roberts *et al.,* (1992), we demonstrate that at stepwise pH elution can be used to recover binding phage from an hNE matrix. This procedure seems to have been very efficient in removing binders that are quite strong but not the best of the population.

To settle definitely when such a procedure improves the quality of binders isolated would require extensive experimentation. We have done only a few preliminary experiments on this question and find that there is apparent improvement in some cases and not in others. As the appropriateness of such a procedure depends on the distribution of binders of various strengths and on the mode of release (which might vary from one PBD to the next PBD in one library) it is difficult to say *a priori* whether to use a stepwise gradient or to "bump" the binders off with pH 2 buffer. Whether step gradients of pH, guanidinium HCl, urea, salt, and the like is useful is determined for the most part by what happens when a library is screened by a pH 2 bump.

Another method of fractionation that is not as well developed as low-pH elution is elution with a competitor. This method depends on first binding the library to the target, then washing away unbound GPs and then adding a compound (the "competitor") that will saturate the target. As GPs let go of the target (as all reversible binders will do), the competitor binds the target and makes rebinding of the PBD impossible. For this to work, one must consider the release times that are likely to be associated with any particular $K_D$. One might think that we could avoid selecting PBDs having very low $k_{on}$ by limiting the reaction time of library with target. Remember though that the $k_{on}$ of displayed PBDs is likely to be artificially lowered by their being tethered to a GP. After washing thoroughly and adding a competitor (such as a known ligand of the target), the time waited before harvesting phage may be a useful parameter to vary. As with pH-step gradients vs pH bumps, we have not pushed this to a conclusion, but preliminary experiments indicate that time-of-incubation-with-inhibitor is a meaningful variable.

A technique related to competitor elution is specificity elution. Dennis and Lazarus (1994a,b) screened APP-I derivatives (a Kunitz domain) for inhibitors of Factor VII$_a$ (F.VII$_a$) complexed to Tissue Factor (TF). They obtained inhibitors having $K_D$(F.VII$_a$:TF) of about 2 n*M,* about 150-fold better than APP-I. Their best F.VII$_a$ inhibitor surprisingly showed improved binding to Factor XI$_a$. To improve the specificity of their inhibitors, they selected again for binding to immobilized

F.VII$_a$:TF but with soluble F.XI$_a$ in the solution. This gave inhibitors that were of about the same affinity for F.VII$_a$:TF as before, but with greatly reduced affinity for F.XI$_a$. In this technique, the specificity selectant (or anti-target) could be mixed with the display GPs at any time and should be present throughout the binding because its job is to tie up those members of the population that have high affinity for both target and anti-target.

## Number of Rounds

It is not wise to decide beforehand how many rounds of fractionation to do. Let the results guide you. If the percentage of phage applied that is retained by the target is going up, you should keep fractionating until the percentage levels off, saving all the intermediate populations. Test several clones from each round of selection to be sure the binding is specific to the target; many people have isolated phage that bind plastic or the beads or phage that have the display segment deleted. In this regard, it is useful to have the target presented in at least two ways so that alternate rounds could be done on different matrices. This should eliminate phage that bind the matrix. Assume that the initial population contains $10^8$ or fewer variants. No variant that does not bind at least 10-fold better than the average of the library is likely to be worth much. Thus, eight rounds should be enough and three usually is enough. We have had cases in which very interesting variants appeared only after six rounds of selection. The rapidity with which a consensus emerges depends to some extent on the complexity of the initial library but also on the level of discriminations exerted by the fractionation and relative infectivity of the variants. When a library of 1000 BPTI variant was fractionated on hNE, there was still substantial diversity in the second round. When a library of 8,500,000 phage, displaying random peptides constrained by flanking cysteine residues, was fractionated against the monclonal antibody 3E-7 or against streptavidin, a very strong consensus emerged after three rounds. When libraries of about 500,000 phage, displaying CD4 variants, were fractionated against gp120, five rounds of fractionation were needed in some cases before a consensus was evident.

Note that each of the phage dispayed peptides reported in McLafferty *et al.* (1993) were present at 1 part in about 14,000,000 in the library. Thus, capture of a component present at less than 1 in $10^7$ has been demonstrated and capture of a component present at 1 in $10^8$ or $10^9$ should be possible. Other research groups have had similar efficiencies.

When the fraction of input phage that bind levels off, one should sequence a number of isolates. Some of the displayed proteins affect the infectivity of phage and cause small plaques. Sometimes, though, the "small-plaque" phenotype does not follow the phage faithfully. Pick some plaques of each type observed and sequence the DNA insert for at least 10, and, if possible, 30 to 50 clones.

Analysis of the DNA sequence and the deduced amino-acid sequence should show that one or more consensus sequences have been isolated, if the selection has worked as planned. Roberts *et al.* (1992) show that the MYMUT library of 1000 BPTI derivatives contained a much lower diversity after three rounds of selection for binding to hNE. Only eight sequences were found in 20 colonies. In this case, the sequences that appeared multiple times in the output represented proteins that actually bind tighter than variants whose sequences appeared only once.

## Expression and Testing of Selected AA Sequences

### Phage binding

We have found that relative binding of proteins on phage is a good indicator of the relative binding of the free proteins.

### Bacterial expression systems

*Modified genes and use of suppressors*     A number of groups have exploited the existence of suppressors strains that can translate a stop codon, at significant frequency, into an amino acid, usually glutamine. A suppressible stop codon is placed in the segment between the displayed domain and the anchor. When expressed in a suppressor strain, some fusion protein is made as is some secreted soluble display domain. If the same gene is expressed in a non-suppressor strain, then only the soluble protein is made.

*Fusion protein systems*     The gene fragment that encodes the BD can easily be transferred to a system such as the Mal-E fusion system (as described in the New England BioLab catalog). In Mal-E, the passenger protein is usually placed at the carboxy terminus of Mal-E. Because Mal-E is normally secreted, the likelihood that a MalE:X fusion will be secreted is fairly high. We have fused several proteins to MalE and have had success in most cases.

*Phage-born genes*     Some of the proteins we have selected turn out to be toxic to *E. coli*. For example, a gene comprising *LacUV5→PhoA-signal:epi-hne-1:stop* when placed on a plasmid gave very poor expression in *E. coli*; one gets either very sick cells or cultures in which deletion mutants have taken over. We placed the gene in the intergenic region of M13mp18. M13 susceptible cells were grown to high cell density and then infected with M13 carrying the expression gene. As no further cell growth was needed, there was no possibility for the cells to delete the gene. They got quite sick but they did produce some of the protein.

*Plasmid-born genes*     If the protein to be expressed is not toxic to bacteria, then you just cut out the selected gene and transfer it to a plasmid having a suitable promoter, Shine-Dalgarno sequence, signal sequence, transcription terminator, and whatever else you need.

### Yeast expression systems

*Saccharomyces cerevisiae*     We and others have expressed several Kunitz domains and other proteins in *S. cerevisiae*. Transformation of *S. cerevisiae* is a well developed art and obtaining a strain that makes 1 mg/liter, up to 20 mg/liter is possible. Cerevisiae does not seem well suited to producing higher concentrations.

***Pichia pastoris***   We and others have had greater success in *P. pastoris* where several Kunitz domains have been produced at levels of 100 mg/liter, up to several grams per liter (Vedvick *et al.,* 1991; Wagner *et al.,* 1992; Cregg *et al.,* 1993). *P. pastoris* is also very suitable for many other protein domains. Note that all yeast are able to glycosylate proteins having glycosylation signals (NX(T/S)) and that yeast glycosylated proteins have been found to be highly antigenic.

## Mammalian and insect expression

The proteins selected on phage can be produced in either mammalian or insect expression systems. This would seem to be indicated only if the bacterial and yeast systems are not acceptable. For example, one might need mammalian glycosylation.

## Chemical synthesis

For proteins of about 30 amino acids or less, it is possible to chemically synthesize the protein and variants. For example, conotoxins (Gray *et al.,* 1983, 1984; Becker *et al.,* 1989a, b) and ST-IA (Bhatnagar and Frantz, 1986) have been synthesized chemically. BPTI (Takahashi *et al.,* 1974) and proteins up to 100 amino acids have been synthesized, but these are not easy or highly efficient.

CMTI-I is a plant-derived protein of 29 amino acids and three disulfides which inhibits trypsin. Table 4 shows the amino acid sequences of CMTI and several related proteins. McWherter and colleagues have synthesized CMTI-I and several variants (McWherter *et al.,* 1989). We have screened libraries of CMTI derivatives and found proteins that inhibit other proteases. The 3D structure of CMTI has been determined (Bode *et al.,* 1994; Holak *et al.,* 1989a,b).

Suppose we select CMTI variants that have affinities between 30 and 1000 p$M$ for a particular target. Chemical synthesis of, for example, 10 mg of several of these is not a major undertaking. In addition, it is feasible to substitute non-genetically encodable amino acids to generate a extensive structure–activity relationship (SAR). This effort is greatly aided by having an initial SAR based on the selected sequences. Peptide chemists often substitute $D$ amino acids into peptides, and this is certainly possible with proteins of this size. Probably more important here is to augment the range of side groups in the $L$ amino acid series.

Making a series of discrete compounds to be tested *in vitro* is what DSPGP was invented to avoid—haven't we gone in a large circle? No! There are two important differences between modifying a tight-binding small protein and modifying weak-binding unstructured compounds such as peptides. First, high-affinity small proteins that have highly similar sequences are likely to bind the target in highly related ways, unlike peptides. Second, the structural information that one can obtain from such a set of proteins is much greater than one can obtain from peptides.

Consider hypothetical variants of an unstructured peptide that has, for example, 400 n$M$ affinity for the target. Suppose you get a new peptide that differs from the "parental" peptide by one amino acid and having 300 n$M$ affinity for the target. You do not know that the new peptide even binds in the <u>same site</u>, much less in a <u>related configuration</u>, as the parental peptide. Even if the peptides compete with each other for binding to the target, there is no certainty that the conformations and orientations of the peptides in the binding site are highly similar, except from structural analyses of the complex by X-ray diffraction or NMR.

Now consider hypothetical CMTI derivatives selected by DSPGP for binding the target. Table 11 shows a hypothetical set of selectants Hy-1 through Hy-9. The consensus $X_4Q_5L_6F_7D_8X_9$ is required to get binding better than 100 p$M$. A wide variety of amino acid types are tolerated at positions 4 and 9. It is highly likely that all these compounds are binding in the same site and in the same orientation; subnanomolar binding sites for molecules as rigid as CMTI are not common. The binding of the Hy-5 (having $Ala_6$) derivative is probably closely related to the others and we are seeing the effect of lost hydrophobic interactions. An X-ray or NMR structure of the Hy-1:Target complex would be very valuable in interpreting the binding data and in designing related compounds that are likely to have high affinity for the target. A 3D structure is not required for further progress, which is very important if one cannot determine the structure of the target (if, for example, the target is membrane bound, glycosylated, or not available in large amounts).

To extend the SAR that is embodied in proteins Hy-1 through Hy-9, we could synthesize mutants utilizing encodable amino acids which could be made by altering a gene. Alternatively, one could take advantage of the small size of CMTI by making derivative compounds *via* chemical synthesis and escaping the limitations of genetically encodable amino acids. Chemical synthesis allows the introduction of $D$ as well as $L$ amino acids, but I think that the more important possibility is to introduce different side groups and alternative main-chain linkages (pseudopeptides). One could make a synthetic library in which a variety of amino acids

**TABLE 11**    Hypothetical CMTI Derivatives Selected for Binding to Targetin

| Protein | Position | | | | | | $K_D$ (p$M$) |
| --- | --- | --- | --- | --- | --- | --- | --- |
| | 4 | 5 | 6 | 7 | 8 | 9 | |
| Wild type | P | R | I | L | M | E | >1,000,000 |
| Hy-1 | P | Q | L | F | D | Q | 30 |
| Hy-2 | P | Q | M | F | D | Q | 60 |
| Hy-3 | T | Q | L | F | D | E | 40 |
| Hy-4 | A | Q | L | F | D | D | 50 |
| Hy-5 | S | Q | A | F | D | R | 400 |
| Hy-6 | G | Q | V | F | D | S | 200 |
| Hy-7 | P | Q | M | F | D | A | 80 |
| Hy-8 | P | Q | L | L | D | Q | 1000 |
| Hy-9 | L | Q | L | F | E | G | 400 |
| Consensus | X | Q | L | F | D | X | — |

are allowed at one or more of the positions 5 through 8. The synthetic mix would then be fractionated for binding to the target and the best binder(s) discovered by sequencing.

Unlike many synthetic library applications, one starts with a very good binder and is looking for both better binders and more structural information. It may be valuable to limit the diversity of combinatorial libraries so that one has confidence that all combinations are present in detectable amounts. This would allow one to make inferences based on what is isolated and on what is not isolated. Since one starts from an excellent binder, the substitutions should be biased toward the best selected protein binder.

For example, consider residue 7 where the strong consensus of Hy-1 through Hy-9 is Phe. The Phe side group is bulky, hydrophobic, and planar. One could replace the -CH$_2$-phenyl side group with groups having methyl-, ethyl-, methoxy-, fluoro-, chloro-, bromo-, *etc.* substituents at each position. Since Tyr was not selected, one believes that the *para*-OH is deleterious to binding. Nevertheless, hydrogen-bonding groups (such as -OH, -CONH$_2$, -CO-NH-CH$_3$, and -NH-CO-H) may work in other positions and hydrogen-bonding groups, excluding -OH, may work at the *para* position. Different aromatic groups (*e.g.,* 1-naphthyl or 2-naphthyl) could be tried; one might also try non aromatic cyclic groups, such as cyclohexyl. The -CH$_2$- could be replaced with -CH$_2$-CH$_2$-. Each of Gln$_5$, Leu$_6$, Phe$_7$, and Asp$_8$ could be varied separately. At each position, groups related to the selected group would be used. If a 3D structure of the Target with one of the Hy proteins is available, it could suggest when substitutions would be expected to act independently, cooperatively, or anti-cooperatively. If the target is a macromolecule, such as a protein, one should not be surprised if some side groups rearrange when the binding partner is changed.

Taking the structural information embodied in CMTI derivatives having affinity of 1 to 1000 p$M$ into a 500-Da non-peptide compound will probably involve loss of binding of at least 100-fold relative to the best binders. Thus, one might obtain a compound having affinity between 10 and 1 n$M$. If one started with unstructured compounds having affinities in the range 1 n$M$ to 1 μ$M$ the non-peptidyl compound might have affinity of 1 μ$M$ or even 1 m$M$.

# SELECTION OF BINDING DOMAINS—EXAMPLES: BPTI DERIVATIVES BINDING hNE

Roberts *et al.* (1992) describe production of a small (1000 protein) library of BPTI derivatives and selection of eight sequences having very high affinity for human neutrophil elastase (hNE). One of these proteins, EPI-HNE-1, has affinity for hNE of about 1 p$M$, still the most potent reversible hNE inhibitor reported. This case illustrates several important aspects of DSPGP.

All of the indications of successful screening were present: (1) increased phage binding, (2) change in pH at which most phage eluted, (3) strong consensus binding, (4) loss of binding to other targets, and (5) very large increase in affinity ($K_D$(BPTI, hNE) > 3 μ$M$, $K_D$(EPI-HNE-1,hNE) = 1p$M$).

The appearance of Phe18 was quite surprising; nothing in the structures of BPTI or hNE or in the literature suggested that position 18 was of any importance, much less that Phe was the optimal amino acid to put there.

## TROUBLESHOOTING

Table 12 contains a partial list of the problems one might encounter in DSPGP with suggested questions and actions one might take.

**TABLE 12**   Some Possible Problems and Possible Solutions

| Problem | Possible reason | Questions and Actions |
|---|---|---|
| Lack of Display | I. PBD not secreted | Q1. Do polyclonal Abs (PBD) show any secretion? If yes, see II and V. |
| | | Q2. Which display? If D3, try D8 and *vice versa*. |
| | | Q3. Is signal:PBD junction as acceptable? Try changes at junction. |
| | | A1. Use different host cells. |
| | | A2. Use different signal sequence. |
| | | A3. Add leader before PBD |
| | | A4. Add linker between PBD and Anchor |
| | II. PBD cleaved from anchor | Q1. Evidence of free PBD in conditioned medium (CM)? |
| | | Q2. Does PBD or any linker have recognition site for known protease? |
| | | A1. Use protease deficient host strain. |
| | | A2. Add PMSF to medium. |
| | III. PBD toxic to host cells | Q1. Can free PBD be secreted from host cells? |
| | | Q2. Is PBD an enzyme? A protease?  Can you add an inhibitor? |
| | IV. PBD abrogates GP propagation | Q1. Is display on III? If yes, try "3+D3," "8+D8," "3mid," "3+3stump," or "8mid" display. |
| | | Q2. Is display on VIII? If yes, try a display on III. |
| | | A1. Move display in III to one of the high-glycine regions. |
| | V. PBD misfolded | Q1. Do polyclonal Abs bind GPs that should display PBD while mAbs or other AfM(PBD) specific for folded form do not?  If yes, try A1–A5. |
| | | A1. Use different host cells, particularly "folding enhanced" strains. |
| | | A2. Add a leader after signal and before PBD. |
| | | A3. Produce GPs at different temperatures. |
| | | A4. Switch from D3 to D8 and *vice versa*. |
| | | A5. Try refolding PBD. Expose to high urea and then dialyze away. |
| Display vector genetically unstable | VI. PBD interferes with GP propagation | Same as IV above. |
| | VII. PBD toxic to host cells | Same as III. |
| | VIII. pbd display gene contains recombination sites or restriction sites | Q1. Does the display gene have may repeats, palindromes, or DNA sequences known to favor rearrangement? If yes, change them. |
| | | Q2. Does host cell have restriction or methylation system? Change host strain. |
| Library has large component of parental BD | IX. Incomplete digestion when introducing vgDNA | A1. Design in restriction sites in *pbd* gene but not in library. Cleave library DNA with one or more enzymes that cut only the parent. |
| | | A2. Use PBD that does not bind target; parental component becomes minor nuisance. |

*continues*

**TABLE 12**   *Continued*

| Problem | Possible reason | Questions and Actions |
|---|---|---|
| | | A3. Use DNA that has frame shift to be repaired by vgDNA insert; problem cannot arise. |
| | | A4. Digest DNA very throughly. |
| No Binders in Library as evidenced by: no consensus or poor binding of free selected BDs | X. There are no proteins of the form you allowed that bind Target | A1. Vary different set of residues. |
| | | A2. Allow different set of amino acid types at postitions previously varied, particularly MORE amino acid types. |
| | | A3. Change parental BD. |
| | XI. Fractionation did not work | Q1. Does the GP itself bind Target? If yes, use different GP. |
| | | Q2. Does GP or PBD bind the matrix or the vessels (plastic ware)? If yes, change offending item. |
| | XII. Best binders stayed on matrix | A1. Wash very throughly. |
| | | A2. Culture the matrix with host cells and growth medium. |
| | | A3. Include cleavage site, such as F.Xa between PBD and Anchor or between Target and matrix. |
| | | A4. Wash matrix with proteases, urea, phenol &c and recover DNA; use PCR to amplify insert region and clone into vector. |
| | | A5. Use competitor to block rebinding of GPs once released. |
| | XIII. Level of variegation much less than planned | Q1. Do unselected clones show expected level of diversity? If no, may need to resynthesize vgDNA. |

## CONCLUSIONS

Proteins and microproteins having from about eight amino acids up to several hundred amino acids can be displayed on a variety of genetic packages. Phage M13 and phagemids derived frrom M13 are particularly useful. The phage proteins III and VIII have been used to anchor a variety of proteins to phage and phagemids. Libraries of $10^7$ and more have been made and fractionated for high-affinity binding to molecular targets. By using small proteins as parentals, one can obtain binding proteins that have the residues responsible for binding highly localized. Small proteins also offer potential advantages including higher stability, lower antigenicity, and easier production.

### References

Baneyx, F., and Georgiou, G. (1992). Degradation of secreted proteins in *Escherichia coli*. *Ann. N. Y. Acad. Sci.* **665,** 301–308.

Barbas III, C. F., Bain, J. D., Hoekstra, M., and Lerner, R. A. (1992). Semisynthetic combinatorial antibody libraries: A chemical solution to the diversity problem. *Proc. Natl. Acad. Sci. U.S.A.* **89,** 4457–4461.

Bass, S., Greene, R., and Wells, J. A. (1990). Hormone phage: An enrichment method for variant proteins with altered binding properties. *Proteins* **8,** 309–314.

Becker, S., Atherton, E., Michel, H., and Gordon, R. (1989). Synthesis and characterization of conotoxin IIIa. *J. Protein Chem.* **8,** 393–394.

Becker, S., Atherton, E., and Gordon, R. D. (1989). Synthesis and characterization of μ-conotoxin IIIa. *Eur. J. Biochem.* **185,** 79–84.

Bhatnagar, P. K., and Frantz, J. C. (1986). Synthesis and antigenic activity of *E. coli* ST and its analogues. *Dev. Biol. Standard* **63,** 79–87.

Bode, W., Greyling, H. J., Huber, R. Otlewski, J., and Wilusz, T. (1994). The refined 2.0 Å X-ray crystal structure of the complex formed between bovine beta-trypsin and CMTI-I, a trypsin inhibitor from squash seeds (*Cucurbita maxima*). Topological similarity of the squash seed inhibitors with the carboxypeptidase A inhibitor from potatoes. *FEBS Lett.* 242, 285–292.

Brown, S. (1992). Engineered iron oxide-adhesion mutants of the *Escherichia coli* phage lambda receptor. *Proc. Natl. Acad. Sci. U.S.A.* **89,** 8651–8655.

Charbit, A., Sobczak, E., Michel, M. L., Molla, A., Tiollais, P., and Hofnung, M. (1987). Presentation of two epitopes of the preS2 region of hepatitis B virus on live recombinant bacteria. *J. Immunol.* **139,** 1658–1664.

Charbit, A., Molla, A., Saurin, W., and Hofnung, M. (1988). Versatility of a vector for expressing foreign polypeptides at the surface of gram-negative bacteria. *Gene* **70,** 181–189.

Charbit, A., Van der Werf, S., Mimic, V., Boulain, J. C., Girard, M., and Hofnung, M. (1988). Expression of a poliovirus neutralization epitope at the surface of recombinant bacteria: First immunization results. *Ann. Inst. Pasteur Microbiol.* **139,** 45–58.

Choo, Y., and Klug, A. (1994). Toward a code for the interactions of zinc fingers with DNA: Selection of randomized fingers displayed on phage. *Proc. Natl. Acad. Sci. U.S.A.* **91,** 11163–11167.

Choo, Y., and Klug, A. (1994). Selection of DNA binding sites for zinc fingers using rationally randomized DNA reveals coded interactions. *Proc. Natl. Acad. Sci. U.S.A.* **91,** 11168–11172.

Chothia, C. (1974). Hydrophobic bonding and accessible surface area in proteins. *Nature* (1974). **248,** 338–339.

Chothia, C., and Janin, J. (1975). Principles of protein-protein recognition. *Nature* 25, 705–708.

Christian, R. B., Zuckermann, R. N., Kerr, J. M., Wang, L., and Malcolm, B. A. (1992). Simplified methods for construction, assessment, and rapid screening of peptide libraries in bacteriophage M13. *J. Mol. Biol.* **227,** 711–718.

Clackson, T., and Wells, J. A. (1995). A hot spot of binding energy in a hormone-receptor interface. *Science* **267,** 383–386.

Corey, D. R., Shiau, A. K., Yang, Q., Janowski, B. A., and Craik, C. S. (1993). Trypsin display on the surface of bacteriophage. *Gene* **128,** 129–134.

Cregg, J. M., Vedvick, T. S., and Raschke, W. C. (1993). Recent advances in the expression of foreign genes in *Pichia pastoris. Bio/Technol.* **11,** 905–910.

Creighton, T. E., and Charles, I. G. (1987). Biosynthesis, processing, and evolution of bovine pancreatic trypsin inhibitor. *Cold Spring Harb. Symp. Quant Biol.* **52,** 511–519.

Creighton, T.E. (1988). Disulphide bonds and protein stability. *BioEssays* **8,** 57–63.

Cull, M. G., Miller, J. F., and Schatz, P. J. (1992). Screening for receptor ligands using large libraries of peptides linked to the C terminus of the lac repressor. *Proc. Natl. Acad. Sci. U.S.A.* **89,** 1865–1869.

Cunningham, B. C., and Wells, J. A. (1993). Comparison of a structural and a functional epitope. *J. Mol. Biol.* **234,** 554–563.

Cwirla, S. E., Peters, E. A., Barrett, R. W., and Dower, W. J. (1990). Peptides on phage: A vast library of peptides for identifying ligands. *Proc. Natl. Acad. Sci. U.S.A.* **87,** 6378–6382.

Dallas, W. S. (1990). The heat-stable toxin I gene from *Escherichia coli* 18D. *J. Bacteriol.* **172,** 5490–5493.

DeGraaf, M. E., Miceli, R. M., Mott, J. E., and Fischer, H. D. (1993). Biochemical diversity in a phage display library of random decapeptides. *Gene* **128,** 13–17.

Dennis, M. S., and Lazarus, R. A. (1994). Kunitz domain inhibitors of tissue-factor factor VIIa. I. Potent inhibitors selected from libraries by phage display. *J. Biol. Chem.* **269,** 22129–22136.

Dennis, M. S., and Lazarus, R. A. (1994). Kunitz domain inhibitors of tissue factor-factor VIIa. II. Potent and specific inhibitors by competitive phage selection. *J. Biol. Chem.* **269,** 22137–22144.

Devlin, J. J., Panganiban, L. C., and Devlin, P. E. (1991). Random peptide libraries: A source of specific protein binding molecules. *Science* **249,** 404–406.

Duenas M., Vazquez, J., Ayala, M., Soderlind, E., Ohlin, M., Perez, L., Borrebaeck, C. A., and Gavilondo, J. V. (1994). Intra- and extracellular expression of an scFv antibody fragment in *E. coli*: Effect of bacterial strains and pathway engineering using GroES/L chaperonins. *Biotechniques* **16,** 476–483.

Dufton, M. J. (1985). Proteinase inhibitors and dendrotoxins. *Eur. J. Biochem.* **153,** 647–654.

Dwarakanath, P., Viswiswariah, S.S., Subrahmanyam, Y. V. B. K., Shanthi, G., Jagannatha, H. M., and Balganesh, T. S. (1989). Cloning and hyperexpression of a gene encoding the heat-stable toxin of *Escherichia coli. Gene* **81,** 219–226.

Eigenbrot, C., Randal, M., and Kossiakoff, A. A. (1990). Structural effects induced by removal of a disulfide-bridge: The X-ray structure of the C30A/C51A mutant of basic pancreatic trypsin inhibitor at 1.6 Å. *Prot. Eng.* **3,** 591–598.

Felici, F., Luzzago, A., Folgori, A., and Cortese, R. (1993). Mimicking of discontinuous epitopes by phage-displayed peptides, II: selection of clones recognized by a protective monoclonal antibody against the *Bordetella pertussis* toxin from phage peptide libraries. *Gene* **128,** 21–27.

Fioretti, E., Iacopino, G., Angeletti, M., Barra, D., Bossa, F., and Ascoli, F. (1985). Primary structure and antiproteolytic activity of a Kunitz-type inhibitor from bovine spleen. *J. Biol. Chem.* **260,** 11451–11455.

Francisco J. A., Earhart C. F., and Georgiou, G. (1992). Transport and anchoring of beta-lactamase to the external surface of *Escherichia coli. Proc. Natl. Acad. Sci. U.S.A.* **89,** 2713–2717.

Francisco J. A., Campbell R., Iverson B. L., and Georgiou, G. (1993). Production and fluorescence-activated cell sorting of *Escherichia coli* expressing a functional antibody fragment on the external surface. *Proc. Natl. Acad. Sci. U.S.A.* **90,** 10444–10448.

Goldenberg, D. P. (1988). Kinetic analysis of the folding and unfolding of a mutant form of bovine pancreatic trypsin inhibitor lacking the cysteine-14 and -38 thiols. *Biochemistry* **27,** 2481–2489.

Gray, W. R., Rivier, J. E., Galyean, R., Cruz, L. J., and Olivera, B. M. (1983). Conotoxin MI. Disulfide bonding and conformational states. *J. Biol. Chem.* **258,** 12247–12251.

Gray, W. R., Luque, F. A., Galyean, R., Atherton, E., Sheppard, R. C., Stone, B. L., Reyes, A., Alford, J., McIntosh, M., Olivera, B. M. (1984). Conotoxin GI: Disulfide bridges, synthesis, and preparation of iodinated derivatives. *Biochemistry* **23,** 2796–2802.

Haas S. J., and Smith, GP. (1993). Rapid sequencing of viral DNA from filamentous bacteriophage. *Biotechniques* **15,** 422–431.

Hillyard, D. R., Olivera, B. M., Woodward, S., Corpuz, G. P., Gray, W. R., Ramilo, C. A., and Cruz, L. J. (1989). A molluscivorus conus toxin: Conserved framework in conotoxins. *Biochemistry* **28,** 3583–3586.

Holak, T. A., Gondol, D., Otlewski, J., and Wilusz, T. (1989). Determination of the complete three-dimensional structure of the trypsin inhibitor from squash seeds in aqueous solution by nuclear magnetic resonance and a combination of distance geometry and dynamic simulated annealing. *J. Mol. Biol.* **210,** 635–648.

Holak, T. A., Bode, W., Huber, R., Otlewski, J., and Wilusz, T. (1989). Nuclear magnetic resonance solution and X-ray structures of squash trypsin inhibitor exhibit the same conformation of the proteinase binding loop. *J. Mol. Biol.* **210,** 649–654.

Hoogenboom, H., Griffiths, A., Johnson, K., Chiswell, D., Hudson, P., and Winter, G. (1991). Multi-subunit proteins on the surfaces of filamentous phage: Methodologies for displaying antibody (Fab) heavy and light chains. *Nucl. Acids Res.* **19,** 4133–4137.

Il'ichev, A. A., Minenkova, O. O., Tat'kov, S. I., Karpyshev, N. N., Eroshkin, A. M., Petrenko, V. A., and Sandakhchiev, L. S. (1989). Production of a viable variant of the M13 phage with a foreign peptide inserted into the basic coat protein. *Doklady Akademii Nauuk. S.S.S.R.* **307,** 481–484.

Jayaraman, K., Gingar, S. A., Shah, J., and Fyles, J. (1991). Polymerase chain reaction-mediated gene synthesis: Synthesis of a gene coding for isozyme c of horseradish peroxidase. *Proc. Natl. Acad. Sci. U.S.A.* **88,** 4084–4088.

Jayaraman, K., and Puccini, C. J. (1992). A PCR-mediated gene synthesis strategy involving the assembly of oligonucleotides representing only one of the strands. *BioTechiques* **12,** 392–398.

Kawasaki, G.H. (1991). Cell-free synthesis andisolation of novel genes and polypeptides. U.S. Patent Office WO91/05058.

Kay, B. K., Adey, N. B., He, Y-S., Manfredi, J. P., Mataragnon, A. H., and Fowlkes, D. M. (1993). An M13 phage library displaying random 38-amino-acid peptides as a source of novel sequences with affinity to selected targets. *Gene* **128,** 59–65.

Kido, H., Yokogoshi, Y., and Katunuma, N. (1988.) Kunitz-type protease inhibitor found in rat mast cells. *J. Biol. Chem.* **263,** 18104–18107.

Ladner, R. C., and Guterman, S. K. (1990). PCT Patent Application WO90/02809.

Ladner, R. C., Guterman, S. K., Roberts, B., Markland, W., Ley, A. C., and Kent, R. B. (1993). Directed evolution of novel binding proteins. U.S. Patent 5,223,409.

Lowman, H. B., Bass, S. H., Simpson, N., and Wells, J. A. (1991). Selecting high-affinity binding proteins by monovalent phage display. *Biochemistry* **30,** 10832–10838.

Lowman, H. B., and Wells, J. A. (1993). Affinity maturation of human growth hormone by monovalent phage display. *J. Mol. Biol.* **234,** 564–578.

Makowski, L. (1993). Structural constraints on the display of foreign peptides on filamentous bacteriophages. *Gene* **128,** 5–11.

Marks, C. B., Vasser, M., Ng, P., Henzel, W., and Anderson, S. (1986). Production of native, correctly folded bovine pancreatic trypsin inhibitor by *Escherichia coli. J. Biol. Chem.* **261,** 7115–7118.

Mattheakis, L. C., Bhatt, R. R., and Dower, W. J. (1994). An *in vitro* polysome display system for identifying ligands from very large peptide libraries. *Proc. Natl. Acad. Sci. U.S.A.* **91,** 9022–9026.

Matthews, D. J., and Wells, J. A. (1993). Substrate phage: Selection of protease substrates by monovalent phage display. *Science* **260,** 1113–1117.

McCafferty, J., Griffiths, A. D., Winter, G., and Chiswell, D. J. (1990). Phage antibodies: Filamentous phage displaying antibody variable domains. *Nature* **348,** 522–524.

McCafferty, J., Jackson, R. H., and Chiswell, D. (1991). Phage-enzymes: Expression and affinity chromatography of functional alkaline phosphatase on the surface of bacteriophage. *Prot. Eng.* **4,** 955–961.

McLafferty, M. A., Kent, R. B., Ladner, R. C., and Markland, W. (1993). M13 bacteriophage displaying disulfide-constrained microproteins. *Gene* **128,** 29–36.

McWherter, C. A., Walkenhorst, W. F., Campbell, E. J., and Glover, G. I. (1989). Novel inhibitors of human leukocyte elastase and cathepsin G. Sequence variants of squash seed protease inhibitor with altered protease selectivity. *Biochemistry* **28,** 5708–5714.

Olivera, B. M., Rivier, J., Scott, J. K., Hillyard, D. R., and Cruz, L. J. (1991). Conotoxins. *J. Biol. Chem.* **266,** 22067–22070.

Parmley, S. F., and Smith, G. P. (1988). Antibody-selectable filamentous fd phage vectors: Affinity purification of target genes. *Gene* **73,** 305–318.

Parmley, S. F., and Smith, G. P. (1989). Filamentous fusion phage cloning vectors for the study of epitopes and design of vaccines. *Adv Exp Med Biol* **251,** 215–218.

Roberts, B. L., Markland, W., Ley, A. C., Kent, R. B., White, D. W., Guterman, S. K., and Ladner, R. C. (1992). Directed evolution of a protein: Selection of potent neutrophil elastase inhibitors displayed on M13 fusion phage. *Proc. Natl. Acad. Sci. U.S.A.* **89,** 2429–2433.

Rossmann, M. G. (1987). Virus structure, function, and evolution. *Harvey Lectures* **83,** 107–120.

Sakaguchi, M., Ueguchi, C., Ito, K., and Omura, T. (1991). Yeast gene which suppresses the defect in protein export of a secY mutant of *E. coli. J. Biochem.* (Tokyo) **109,** 799–802.

Schweitz, H., Jheurteaux, C., Bois, P., Moinier, D., Romey, G., and Lazdnski, M. (1994). Calcicludine, a venom peptide of the Kunitz-type protease inhibitor family, is a potent blocker of high-threshold $Ca^{2+}$ channels with a high affinity for L-type channels in cerebellar granule neurons. *Proc. Natl. Acad. Sci. U.S.A.* **91,** 878–882.

Scott, J. K., and Smith, G. P. (1990). Searching for peptide ligands with an epitope library. *Science* **249,** 386–390.

Scott, J. K. (1992). Discovering peptide ligands using epitope libraries. *Trends Biochem. Sci.* **17,** 241–245.

Siekmann, J., Wenzel, H. R., Schroeder, W., and Tschesche, H. (1988). Characterization and sequence determination of six aprotinin homologues from bovine lungs. *Biol. Chem. Hoppe-Seyler* **369,** 157–163.

Smith, G. P. (1985). Filamentous fusion phage: Novel expression vectors that display cloned antigens on the surface of the virion. *Science* **228,** 1315–1317.

Smith, G. P., and Scott, J. K. (1993). Libraries of peptides and proteins displayed on filamentous phage. *Methods Enzymol.* **217,** 228–57.

Sprecher, C. A., Kisiel, W., Mathewes, S., and Foster, D. C. (1994). Molecular cloning, expression, and partial characterization of a second human tissue-factor-pathway inhibitor. *Proc. Natl. Acad. Sci. U.S.A.* **91,** 3353–3357.

Steidler, L., Remaut, E., and Fiers, W. (1993). LamB as a carrier molecule for the functional exposition of IgG-binding domains of the *Staphylococcus aureus* protein A at the surface of *Escherichia coli* K12. *Mol. Gen. Genet.* **236,** 187–192.

Sternberg, N., and Hoess, R. (1995). Display of peptides and proteins on the surface of bacteriophage lambda. *Proc. Natl. Acad. Sci. U.S.A.* **92,** 1609–1613.

Takahashi, H., Iwanage, S., Kitagawa, T., Hokama, Y., and Suzuki, T. (1974). Snake venom proteinase inhibitors. II. Chemical structure of inhibitor II isolated from the venom of Russell's viper *(Vipera russelli). J. Biochem.* **76,** 721–733.

Vaughan, T. J., Williams, A. J., Pritchard, K., Osborn, J. K., Pope, A. R., Earnshaw, J. C., McCafferty, J., Hodits, R. A., Wilton, J., and Johnson, K. S. (1996). Human antibodies with subnanomolar affinities isolated from a large non-immunized phage display library. *Nature Biotechnology I* **14,** 309–314.

Vedvick, T., Buckholz, R. G., Engel, M., Urcan, M., Kinney, J., Provow, S., Siegel, R. S., and Thill, G. P. (1991). High-level secretion of biologically active aprotinin from the yeast *Pichia pastoris. J. Industrial Microbiol.* **7,** 197–201.

Wagner, S. L., Siegel, R. S., Vedvick, T. S., Raschek, W. C., and Van Nostrand, W. E. (1992). High level expression, purification, and characterization of the Kunitz-type protease domain of protease nexin-2/amyloid beta-protein precursor. *Biochem. Biophys. Res. Commun.* **186,** 1138–1145.

Wells, J. A. (1990). Additivity of mutational effects in proteins. *Biochemistry* **29,** 8509–8517.

# 11

# A cDNA Cloning System Based on Filamentous Phage: Selection and Enrichment of Functional Gene Products by Protein/ Ligand Interactions Made Possible by Linkage of Recognition and Replication Functions

*Mark Suter, Maria Foti,
Mathias Ackermann, and Reto Crameri*

## INTRODUCTION AND RATIONALE

Established cDNA cloning systems make use of a variety of vectors and screening methods to isolate clones encoding genes of interest (Sambrook *et al.*, 1989). For the identification of the desired cDNA clones basically two different methods are used: (a) screening of the cDNA library with oligonucleotide probes and (b) screening of the library with ligands able to detect recombinant proteins from cDNA expression libraries.

An obvious limitation of oligonucleotide-based screening is the need for some sequence information related to the gene of interest. cDNA expression libraries overcome this crucial limitation but suffer from other technical drawbacks: first, in commercially available systems designed for screening cDNA expression libraries, the expressed proteins are bound to solid-phase membranes such as nitrocellulose or nylon membranes. The gene product is thus separated from its gene. This lack of physical linkage between a particular cDNA sequence and its product excludes the opportunity for enrichment or biological selection. To obtain the recombinant cDNA that corresponds to the selected protein multiple rounds of isolation are thus required (Sambrook *et al.,* 1989). Additionally, the absence of a physical linkage between the gene products and its genetic information necessitates that there is a threshold amount (nMol to pMol) of immobilized target for the probe to allow visualization by common methods. Second, adsorption of cDNA-encoded proteins to solid phase will induce alteration of their structure, ranging from discrete distortion of the three-dimensional structure to total denaturation as compared to the native proteins (for review, see Butler, 1992, 1995). In addition, the need to immobilize recombinant products from expression libraries to solid phase supports strongly limits the number of clones which can be screened with a reasonable effort.

Due to the inefficiencies listed above, a novel cDNA cloning system — the pJuFo system (Crameri and Suter, 1993)—has been developed (Crameri *et al.,* 1994; Gramatikoff *et al.,* 1994; Hottiger *et al.,* 1995). The rationale for the development of the pJuFo system (Crameri and Suter, 1993) was:

1. To link the gene product physically with the genetic information required for its production and expose the product on the surface of filamentous phage to allow direct enrichment of genes by gene product–ligand interactions.

2. To direct secretion of translated proteins encoded by cDNA into the periplasmic space of *Escherichia coli* to achieve folding of the gene products.

3. To allow efficient access to very large libraries employing a powerful enrichment technique and thus to make rare mRNA products available for analysis. Additionally, pJuFo allows one to isolate cDNA products and thus their encoding genes using p*M*ol amounts or less of ligand (see below).

In this chapter we show that in the pJuFo phage system both the cDNA and its product can be selected by a specific ligand and the phage amplified in *E. coli.* Therefore, cDNA cloning is achieved by *direct selection and enrichment* of a single unit containing cDNA and its product.

# THE pJuFo VECTOR

## Design of the pJuFo Vector

Specific prokaryotic peptide leader/signal sequences such as the *ompA* (Henning *et al.,* 1979) or the *pelB* leader/signal sequence (Diolez *et al.,* 1986) direct the translated gene products into the periplasmic space of *E. coli.* The oxidizing milieu of the periplasmic microenvironment allows disulfide bond formation and folding of gene products (Skerra and Plückthun, 1988). This cellular transport and folding system is also utilized by filamentous phage for the production of its capsid proteins as well as displayed proteins (Kang *et al.,* 1991, Barbas *et al.,* 1991, Winter and Milstein, 1991).

For the construction of pJuFo (Crameri and Suter, 1993), the phagemid pComb3 (Barbas *et al.,* 1991) was used as a starting vector. The pComb3 vector contains two *lac*-promoter/operators. Each promoter is followed by the *pelB* leader/signal sequence. This vector or similar vectors (Winter and Milstein, 1991) has been used to display heterodimeric proteins linked to the surface of the filamentous phage (Kang *et al.,* 1991; Winter and Milstein, 1991).

As described in other parts of this book, phage will display sequences at the N-terminus of either mature proteins III or VIII. The pIII fusion proteins consist of either an N-terminal fusion to the full-length pIII, or a fusion in which the N-terminal domain of pIII is replaced by the sequence to be displayed (Crissman and Smith, 1984; Cwirla *et al.,* 1990). The N-terminal domain of pIII (aa 1–197) binds to the F′ pili, allowing infection of *E. coli.;* thus, when it is replaced, complementation with a wild-type pIII is essential. The C-terminal domain (aa 198–406) is required for anchoring pIII to the phage surface and capping the trailing end of the filamentous virus as it is extruded from infected bacteria cells (Chang *et al.,* 1979; Crissman and Smith, 1984).

Unmodified pComb3 is not suitable for cDNA cloning because the stop codons from full-length cDNAs would terminate translation before the pIII fusion protein could be synthesized. [Note, however, that one could display randomly primed libraries as a pIII fusion. In this case cDNAs lacking stop codons would be expressed in place of the N terminal domain of pIII; wild-type pIII (for infectivity) would be supplied by helper phage. However, the cDNA in an unmodified pComb vector would have to be in-frame with both the 5′ leader/signal sequence <u>and</u> the 3′ domain 2 of pIII.] On the other hand, a cDNA-product fused to the C terminus of pIII would hamper incorporation of the coat protein into the virion. To display full-length cDNA-encoded proteins on the phage surface, a strategy was developed to covalently link the N-terminus of cDNA encoded proteins to the phage pIII (Crameri and Suter, 1993; Fig. 1). The strategy chosen made use of the strong association (*Kd* approx. 100 nM; Pernelle *et al.,* 1993) of the Jun and Fos leucine zipper domains (Gentz *et al.,* 1989). Utilizing this strategy, the cDNA is cloned as a C terminal fusion to Fos and the Fos/cDNA fusion protein is secreted into the periplasm. Jun, the binding partner of Fos, replaces the N terminal domain of pIII, and thus is displayed both on the inner membrane during morphogenesis and ultimately on the mature phage. While in the periplasm, the anchored Jun is bound by the soluble Fos fusion protein. Thus, the cDNA-encoded protein is indirectly bound to pIII. To stabilize the zipper domains formed, we introduced flanking cysteine residues at both ends of the zippers which formed covalent intermolecular bonds. This is a requirement to avoid interphage product exchange and thus dissociation of the linkage between genetic information and gene product (Crameri and Suter, 1993).

In summary, the Jun zipper is expressed from the first promoter of pComb3 as a fusion protein that is fused to domain 2 (the C terminal domain) of pIII (Fig. 1). The

**198**

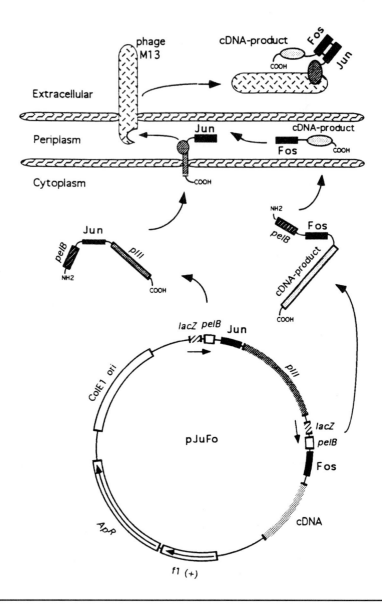

**FIGURE 1**   The essential elements for cloning cDNA into pJuFo phagemid and the expression of recombinant proteins on filamentous phage are shown. The nt sequence of the pelB leaders, the GlyGlyGlySer flexible linker, and the C-terminal domain of pIII are based on the phagemid pComb3 (Barbas *et al.,* 1991). The gene encoding the Jun Leucine zipper is inserted as *Hind*III–*Bam*HI fragment. The gene encoding the Fos Leucine zipper is inserted as *Sac*I–*Xba*I fragment. The *Bgl*II, *Xba*I, and *Kpn*I sites allow construction of cDNA expression libraries whose gene products are displayed on the phage surface. The Jun and Fos Leucine zippers are each flanked by the amino acids (aa) CysGlyGly and GlyGlyCys at the N- and C-termini of the zippers, respectively. This allows covalent linking of the Jun and Fos Leu zippers by disulfide bond formation as described (Crameri and Suter, 1993). The diagram illustrates the composition of the pJuFo vector and the proposed pathway for the capture of cDNA products on the phage surface. Reprinted from *Gene* **137**, Crameri, R., and Suter, M., Display of biologically active proteins on the surface of filamentous phages, 69–75 (1993) with kind permission from Elsevier Science-NL, Sara Burgerhartstraat 25, 1055 KV Amsterdam, The Netherlands.

second lac promoter of pComb 3 drives the expression of the Fos zipper peptide which is fused to the N-terminus of the cDNA products. As shown, Fos–protein fusions are covalently bound, via Fos binding the Jun/pIII fusion, to the outside of the filamentous phage and are thus accessible for molecular interactions. The C terminus of the cDNA product is free, and thus stop codons do not interfere with expression.

# CONSIDERATIONS FOR CDNA CLONING INTO PJUFO AND PREPARATION OF PHAGE PARTICLES

As for any cDNA cloning procedure, a representative library with large cDNA inserts is desirable. This is because for biological functions (i.e., enzymatic activity) or molecular interactions multiple domains or complete molecules may be required (Crameri and Suter, 1993; Hottiger *et al.*, 1995). The cloning sites chosen for the construction of libraries in pJuFo are those restriction enzymes which infrequently cleave eukaryotic genes. Furthermore, the restriction sites *Bgl*II/*Xba*I/*Kpn*I were selected to facilitate the transfer of cDNA libraries from λZap vectors to pJuFo (Short *et al.*, 1988).

---

**PROTOCOLS**

## Generation of cDNA libraries displayed on phage surface from premade λ-Zap libraries

### Preparation of pJuFo vector

1. Spread *E. coli* XL-1 Blue cells harboring pJuFo or an equivalent phagemid cDNA display vector on LB-agar [Ampicillin (Amp) 100, Tetracycline (Tet) 10 μg/ml] to obtain single colonies.
2. Pick bacteria from a single colony with a sterile loop and transfer to a sterile tube containing 10 ml LB-medium (Amp 100, Tet 10 μg/ml). Grow the culture until $OD_{600} = 0.8$ is reached.
3. Transfer bacteria to 500 ml SOB medium (Amp 100, Tet 10 μg/ml) and grow overnight at 37°C, 220 rpm.
4. Collect bacteria by centrifugation, 10 min, 4000 rpm, 4°C.
5. Lyse the cells and prepare plasmid DNA according to the Qiagen MAXI-prep. protocols.
6. Set up restriction digests as follows:

| | |
|---|---|
| Purified pJuFo DNA | 5 μg |
| 10x restriction buffer | 5 μl |
| *Xba*I | 10 U |
| *Kpn*I | 10 U |
| Water to | 50 μl |

Digest at 37°C for 4 hr

7. Purify restriction mixtures on 0.8% agarose gels. Excise band on long-wave UV box, electroelute DNA in an appropriate device or in a dialysis bag for 2 hr at

100 V using 0.5x TBE-buffer, reverse current for 2 min before collecting the DNA if dialysis bags are used.

**Note:** The restricted pJuFo vector should have a size of about 4.3 kb.

8. Collect fluid and clear by 2 min centrifugation in a microcentrifuge. Transfer supernatant to a new vial and ethanol precipitate DNA for 30 min at -20°C.
9. Spin tubes for 20 min at 14,000 rpm, air dry the pellets, and resuspend DNA in 40 µl TE. Store at -20°C until use.
10. Test the background of the vector by ligation of 1 µl vector DNA and electroporation into competent *E. coli* XL-1 Blue cells.

**Note:** To reduce background, pJuFo can be treated with Calf Intestine alkaline Phosphatase (CIP) (Boehringer) after step 6. However, this might reduce ligation efficiency. Alternatively, the vector can be redigested (step 6) and repurified (step 7). Make sure the background is low (< 1%) when compared to uncut vector.

## Preparation of inserts from premade λ-Zap libraries: mass excision

11. Infect 20 ml of *E. coli* XL-1 Blue cells ($OD_{600}$ = 0.8 ) with $10^9$ recombinant λ-phage particles (representing 10–100x the primary library size) together with $10^{11}$ pfu of R408 helper phage. Incubate 15 min at 37°C.
12. Add 200 ml 2x YT medium (100 µg/ml Amp) and continue incubation at 37°C, 220 rpm for 6 hr.
13. Heat the culture at 70°C for 20 min, then centrifuge 10 min at 8000*g*. Decant supernatant into a new tube and store the excised pBluescript phagemid at 4°C. Titer the colony forming units.
14. To prepare DNA, infect 20 ml *E. coli* XL-1 Blue (OD = 1.0) grown in the presence of tetracycline (10 µg/ml) with $10^{10}$ cfu excised phagemid, add 250 ml LB-medium (Amp 100, Tet 10 µg/ml), and further incubate at 37°C, 220 rpm overnight.
15. Isolate DNA according to steps 4 and 5, determine the concentration and store at -20°C until use.
16. Set up a restriction digest as in step 6 and purify DNA as in step 7. There are three ways in which to purify insert from λ-Zap in order to transfer it to the pJuFo vector:

(a) Cut out the cDNA inserts with >500 bp length, avoiding the contamination with the λ-Zap generated plasmid DNA, and proceed with steps 8 and 9. Check the background by self-ligating the insert preparation and electroporating it to be sure that it is free of λ-Zap-generated pBluescript DNA. Note that some transcripts have long (>800 bp) nontranslated tails and collecting small inserts may bias an oligo dT generated library away from coding regions.

(b) In theory, the digested cDNA sample can be treated with CIP and cloned into a nonphosphatased vector. This avoids loss of cDNAs that are the same size as Bluescript. Although the pBluescript DNA remains in the population, it should not be able to self-ligate or accept inserts; its origin of replication should make it an unlikely competitor for ligation into pJuFo. As above, check the insert population by self-ligating and electroporating it to be sure that the CIP reaction was successful. This approach may require more optimization than the other two approaches (a and c) which have been performed routinely in several labs.

(c) Using biotinylated oligos homologous to Bluescript, PCR directly from the λ-Zap phage. [It is best to use an error-free polymerase for this procedure, as PCR-generated errors or bias are the risks one takes in utilizing this less labor-intensive approach].

PCR primers: the commercially available "reverse primer":

5'GGAAACAGCTATGACCATG3' is the 5' primer.

The commercially available "M13 -20 primer" 5'GTAAAAC-GACGGCCAGT 3'" is the 3' primer.

PCR program: 10 (100-μl) reactions using 1μl each of λ-Zap phage of $10^6$ pfu/μl and standard PCR reagents (Boehringer Mannheim (BM) *Taq* DNA polymerase and buffer, nucleotide concentrations as recommended by BM and Pfu DNA polymerase from Stratagene used at the same concentration as the BM Taq polymerase), were performed utilizing the following program:

95°C, 1 min
50°C, 30 sec
72°C, 1 min
25 cycles of the above program, followed by 1 extension cycle of
    10 min at 72°C.

Digest the PCR products with the appropriate pJuFo cloning enzymes and remove the biotinylated ends with streptavidin agarose beads (Gibco). The beads can be used in batch or in a column. The batch procedure works best if the equilibrated beads are added to the PCR product DNA in an Eppendorf tube, incubated at 4°C for 1 hr, and then spun. The supernatant is transferred to another tube and fresh beads are added, the incubation is repeated, and the beads spun out again. The supernatant may have to be spun several times to remove all of the beads. A long tipped-pipette facilitates supernatant removal from the beads. To verify that the ends have been removed, compare aliquots from the cut and untreated sample to the cut and agarose streptavidin bead-treated sample on a 20% acrylamide gel. The inserts are then phenol-chloroform extracted, ethanol precipitated, and cloned directly into the cut and CIP-treated pJuFo vector.

### Generation of a cDNA phage surface display library

17. Set up test ligations to determine the best vector:cDNA insert ratio. It is useful to do twofold dilutions of insert to a constant amount of vector. This

emperical method is simple and avoids errors due to the difficulty of accurately quantifying insert concentrations.

| | |
|---|---|
| Linearized vector DNA | 1 µg |
| 10x ligase buffer | 1 µl |
| T4 DNA ligase (>1 Weiss U) | 1 µl |
| Insert DNA | x µl |
| Water to an end volume of | 10 µl |

**Note:** Optimal conditions are normally obtained with a vector:cDNA insert ratio of 1:2.

18. Determine electroporation efficiency (using electrocompetent male cells such as XL1-Blue) of a test ligation and then scale-up the ligation using the optimal vector:insert ratio in order to obtain the desired library complexity ($10^7$–$10^9$ cfu). Note that your efficiency must be 100-fold over the vector to self-ligation in order to maintain the insert population through amplification.
19. Precipitate the large-scale ligation mixture by addition of 0.1 vol 3 M sodium acetate (pH 5.2) and 2.5 vol of cold ethanol for 2 hr at -20°C. Recover DNA as in step 9, air dry the pellet, and resuspend in 30 µl of 0.1x TE. Store at -20°C.
20. Transform electrocompetent *E. coli* XL-1 Blue cells (1 µg per electroporation $\times$ 10 tubes to generate $10^8$ cfu per primary library), add 1 ml SOC medium to each transformation and pool the samples.
21. Immediately titer the cfu to determine the primary library size (typically $10^7$–$10^9$ cfu). PCR from these titer plates with oligos that flank the insertion site to evaluate the size range and frequency of insertion.
22. Grow the culture for 1 hr at 37°C, 330 rpm, then add SOB medium (Amp 100, Tet 10 µg/ml) to 100 ml and shake at 330 rpm, 37°C, for an additional hour.
23. Add VCS M13 or an equivalent helper phage ($10^{10}$ pfu/ml of culture) and continue shaking for an additional 2 hr. After 2 hr add 70 µg/ml kanamycin and further incubate for 10–12 hr.

**Note:** Longer growth time might result in an increase of phage carrying smaller inserts. Growth at 30°C might increase the proportion of phage carrying larger or properly folded inserts.

24. Clear the culture by centrifugation (4000 rpm, 15 min, 4°C). Freeze the cell containing pellets in 20% glycerol. Precipitate phage from the supernatant by addition of 4% (w/v) polyethylene glycol 8000 and 3% (w/v) NaCl, keep the culture for 1 hr on ice, and recover phage by centrifugation (8000g, 20 min 4°C).
25. Resuspend phage pellets in 2 ml 50 mM phosphate buffer, pH 7.2, 150 mM NaCl, and clear by centrifugation for 2 min. Transfer phage to a fresh tube and titer the cfu. Store aliquots at -20°C directly or at -80°C in 20% glycerol, or at 4°C with a cocktail of protease inhibitors. Use the phagemids in a panning experiment as soon as possible. Some loss in titer may result from prolonged freezer storage.

26. Due to the risk of proteolysis, it is best to plan experiments such that phagemids are panned soon after they are isolated. If this is not possible there are several ways to store libraries:

(a) Do a large ligation and store the frozen ligation. Use 1 to 10 µg per library electroporation and per screening experiment. Thus, one returns to the original ligation and reelectroporates for each new experiment. The risk one takes using this storage method is that of having a slightly different diversity population per library; however, this should be compared to the risk of starting with a library already once biased by amplification. Most of the other storage methods recommended below incorporate the risk of reamplifying a primary amplification.

(b) Use the cell pellet from the primary library to isolate supercoiled DNA, and electroporate this into cells to generate a fresh library. Supercoiled DNA gives a high electroporation efficiency and thus the plasmid DNA serves as a good library reservoir.

(c) Use the stored phagemids directly in a panning. We use frozen peptide libraries directly in pannings; however, the monovalent display of the phagemid and the greater risk of proteolysis make this approach riskier than in the M13 phage display protocol.

(d) Use the stored phagemids to reinfect cells to generate a new library, or grow up the frozen cell pellet under appropriate selection to generate a new library.

### Testing the complexity of the library

Insert size and frequencies should be evaluated by PCR from the cfu's obtained from amplified library titers. Compare the ratio of insert to no insert before and after the overnight amplification. Less than 1% background (no insert phagemid) is desirable, though not essential.

### Media and buffers

- LB medium (Luria Bertani)
  To 950 ml of deionized $H_2O$ add:

  | | |
  |---|---|
  | Bacto - tryptone (Difco) | 10 g |
  | Bacto yeast extract (Difco) | 5 g |
  | NaCl | 5 g |

  Shake until the solutes have dissolved. Adjust the pH to 7.0 with 5 N NaOH (approx. 0.2 ml) . Adjust the volume of the solution to 1 liter deionized $H_2O$. Sterilize by autoclaving for 20 min at 15 lb/square inch on liquid cycle.

- 2x YT medium

  To 900 ml of deionized $H_2O$ add:

  | | |
  |---|---|
  | Bacto - tryptone (Difco) | 16 g |
  | Bacto yeast extract (Difco) | 10 g |
  | NaCl | 5 g |

  Adjust pH and sterilize as for LB medium.

- SOB medium

  To 900 ml of deionized $H_2O$ add:

  | | |
  |---|---|
  | Bacto - tryptone (Difco) | 20 g |
  | Bacto yeast extract (Difco) | 5 g |
  | NaCl | 0.5 g |

  Adjust pH and sterilize as for LB medium.

- SOC medium

  Per 100 ml:

  5 ml 20% glucose

  1 ml 1 *M* MgCl2

  1 ml 1 *M* MgSO4

  Add SOB to 100 ml; sterilize by membrane filtration.

- 3 *M* sodium acetate for ethanol precipitation

  27.22 g sodium acetate·3 $H_2O$

  5.75 ml glacial acetic acid

  75.4 ml $H_2O$

### Material and equipment obtained from commercial sources

Below is a selective list of material and equipment we obtained from commercial sources. Other commercial sources may be as adequate or better.

- Handling of DNA: For DNA purification the silicon based columns from Qiagen were used. The protocols given by the manufacturer are easy to follow without modification.
- Libraries and competent cells: The competent XL-1 Blue cells and the λ library systems including the *in vivo* excision system were from Stratagene. The company provides protocols or reagents which are summarized above and can be used step by step as described in the protocols given by the company.
- Electroporation using the Bio-Rad Gene Pulser: The instructions of the manufacturer are precise and can be followed without modification. For the construction of libraries new electroporation cuvettes were always used to avoid any contamination with previous libraries.

**NOTE:** (1) For the electroporation the cuvette holder is precooled at -20°C and the cuvettes kept at -4°C. (2) The final cell suspension for the electroporation must be salt-free. Make certain that cuvettes are free from moisture before placing them into the cuvette holder.

• Handling phage: Working with phage—as with any virus—contamination can easily happen. When pipetting phage, tips (10 to 1000 μl) with plugs (Molecular Bio-products Inc., San Diego, CA) were used to avoid contamination.

## SCREENING OF PHAGE

### Make Sure Your Ligand is Functional!

The most important consideration of phage screening using the pJuFo system is that the ligand used to screen the phage is tagged or immobilized in such a way that the ligand is still biologically functional. It is therefore mandatory to test the bioactivity of the ligand after immobilization or tagging. We draw attention to the fact that many proteins, when adsorbed directly to solid phase by hydrophobic interaction, are altered in their three-dimensional structure and lose (for review, see Butler, 1992, 1995). For instance, it has been shown that some monoclonal antibodies (mAbs), when adsorbed directly to plastic, can lose their antigen-binding capacity (Suter and Butler, 1986). Alternative strategies for immobilizing mAbs or other ligands must be developed (Suter *et al.,* 1989), such as modification of ε amino groups of lysine with biotin and using immobilized avidin or streptavidin (Suter *et al.,* 1989). However, even "mild" chemical modifications and immobilization of a ligand have to be evaluated with care. It is possible that: (a) the chemical modifications *per* se may alter the function of the ligand (Dudler *et al.,* 1994); (b) the immobilization procedure may distort the three-dimensional structure of the protein when bound to solid phase (Suter *et al.,* 1989); or (c) the solid-phase bound ligand is inaccessible to the phage due to steric hindrance. Useful alternatives to chemical modifications of proteins are fusions to glutathione S-transferase (Hottiger *et al.,* 1995) or tagging with [His]$_6$ (Hochuli *et al.,* 1987, Stüber *et al.,* 1990). However, even these modifications need to be tested for their influence on the ligand.

### Screening for Binding to a Desired Ligand—Theoretical Considerations

In general, the known or expected characteristics of the ligand will dictate the screening procedure used. Irrespective of a particular ligand chosen, some considerations are warranted to predetermine the molar ratios of ligand to receptor. Mass law considerations related to this issue have been discussed in detail elsewhere (Peterman and Butler, 1989). We consider a screening system published by Crameri *et al.,* (1994) as an example for ligand–receptor interaction using phage. The goal was to find allergens recognized by serum IgE antibodies (Ab) from allergic individuals. A monoclonal antibody (mAb; TN-142) specific to the Cε 2 domain of human IgE was used as capture antibody (cAb) and coated to 96-well microtiter plates (Fig. 2). After washing the plates, the bound cAb was used to bind serum IgE from allergic individuals containing allergen-specific IgE (i.e., ligand) for the

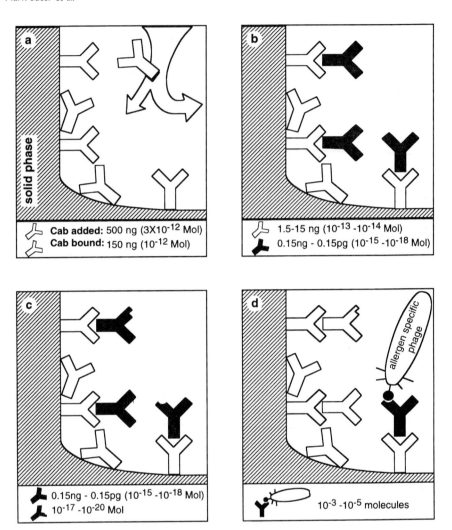

**FIGURE 2**    The example (Crameri *et al.,* 1994) described in detail in the text is illustrated; a, b, c, d symbolize four steps in the assay to select filamentous phage molecules displaying allergens. (a) Solid phase is coated with capture antibody (cAb). The amount of cAb added (mAb specific to human IgE) and the amount of cAb bound after washing the plates is indicated. (b) Only 1–10% of the cAb bound are functional and can bind serum IgE. (c) A proportion of serum IgE will be allergen specific, indicated by differently shaped Ab symbols. (d) In the final step of the assay the number of polyclonal IgE molecules specific for one single particular allergen are approximately equal to the number of phage carrying this single particular allergen. After washing noncognate phage, IgE-bound phage can be released, amplified, and reselected by several rounds repeating the same procedure.

purpose of affinity-selecting allergens expressed by filamentous phage (Crameri and Suter, 1993) using the pJuFo system (Fig. 2).

The successful isolation of allergens by this system may be surprising at first glance. This is because more than 90% of the cAb is denatured when adsorbed directly to solid phase (Suter and Butler, 1986; Butler *et al.*, 1993) and most of the poly-clonal serum IgE from allergic individuals are not specific to the particular class of allergens screened (i.e., *Aspergillus fumigatus*). Considerations of the molar amount of allergen-specific IgE and phage used and the power of phage selection (in this case wild-type pJuFo; Crameri and Suter, 1993) will explain the data and highlight the sensitivity of this system for selection of cDNA clones.

## Example:

If 500 ng of anti IgE (5 µg/ml, 100 µl volume used) is added to polystyrene wells, approx. 150 ng of Ig protein will be bound to the solid phase (Fig. 2), with more than 90% of it denatured. We thus assume that the anti-IgE cAb bound to the mi-crotiter plates may be able to capture 1–15 ng of IgE (Butler *et al.*, 1993). Of the serum IgE added, some 0.01–1% may be specific against a certain group of aller-gens (i.e., *A. fumigatus*-specific allergens).

### Antibody capture system

150 ng cAb anti IgE bound to solid phase $= 10^{-12}$ mol of Ig
1.5–15 ng serum IgE captured $= 10^{-13}–10^{-14}$ mol of IgE
0.01–1% functional allergen specific IgE $= 10^{-15}–10^{-18}$ mol of IgE Ab
$= 10^3 — 10^6$ molecules

If we assume in a patient there are antibodies recognizing 100 different allergens (Moser *et al.*, 1992; Crameri *et al.*, 1994) and they are represented equally in the serum pool, for one single allergen there is an average of $10^{-17}–10^{-20}$ mol of spe-cific IgE Ab bound to the solid phase ready to capture allergen bound to filamentous phage. This means that a polyclonal population of $10^3$ to $10^6$ molecules of IgE Ab molecules specific against one particular single allergen is bound to solid phase ready to bind its allergen displayed on phage particles.

### Antigen displayed on filamentous phage

$10^{23}$ phage represent 1 mol of phage carrying 1 mol of antigen.

For library screening, approx. $10^{13}$ phage can be used.

This represents $10^{-10}$ mol of phage carrying $10^{-10}$ mol of antigen.

The diversity of the original library is assumed to be $10^7$ different phage particles. Therefore, $10^{13}$ phage particles contain $10^6$ of each phage or approx. $10^{-17}$ mol of each potential antigen.

Based on this calculation, the screening system employed by Crameri *et al.*, (1994) can be considered roughly equimolar related to the IgE Ab/allergen (phage) ratio used. Depending on the results the system could be made more affinity/avidity dependent by lowering the antigen concentration used for screening.

Mass law considerations predict some minimal requirement for the affinity needed for affinity selection of a ligand by a specific receptor. There have been several

**TABLE I**   Alignment of Three Phage Clones Isolated by Selection with SPA

| | | | Fc part human IgG1 | | |
|---|---|---|---|---|---|
| | CH1 | Hinge region | | CH3 | |
| aa position | 208 | 225 | 233/242 | 469 | 478 |
| Fc IgG1 aa sequence | CNVNHKPSNTKVDKKVEPKSCDKTHTCPPCP ........... | | | QKSLSLSPGK* | |
| 1 | | DKTHTCPPCP ........... | | QKSLSLSPGK* | |
| 2 | | HTCPPCP ........... | | QKSLSLSPGK* | |
| 3 | | DKTHTCPPCP ........... | | QKSLSLSPGK* | |

*Note.* Clones 1–3 are compared to the aa sequence of human IgG1 Fc. aa 225 marks the start of the hinge region. aa 233 and 242 are used for interdomain disulfide bond formation pairing CH2 and CH3 of IgG1. aa 469 to 478 mark the last 10 aa of CH3.
   *, End of the polypeptide chain.

experiments addressing this issue with pJuFo (Hottiger *et al.,* 1995; K. Gramatikoff, unpublished data). Random-primed cDNAs, prepared from human peripheral blood lymphocytes immortalized with Epstein Barr virus, were introduced in pJuFo and screened with Staphylococcal protein A (SpA) immobilized on agarose beads (Pharmacia, Uppsala, Sweden). After three rounds of affinity selection, five independent clones were chosen and analyzed. The nucleotide sequence from three of the five clones was determined and found to be identical with each other. The amino acid (aa) sequences translated from the inserts are shown in Table 1 and correspond to aa 231 to 478 and 228 to 478 of human IgG1 hinge region CH2 and CH3 (Kabat *et al.,* 1991).

The IgG–SpA complex has been studied in some detail (Deisenhofer *et al.,* 1978; Deisenhofer, 1981). Positions 252, 253, 254, 309, 310, 311, 434, 435, and 436 of CH2 and CH3 (Deisenhofer *et al.,* 1978; Deisenhofer, 1981) interact with SpA. Interestingly, isolated CH3 did not permit crystal formation with SpA. The clones isolated by Gramatikoff (Table 1) confirm these results. All the clones isolated started at the hinge region, including the interdomain disulfide bonds (aa 233 and 242), and terminated at the correct aa position 478.

We conclude that the cDNA clones isolated by Gramatikoff must allow homodimer formation of the Fc part of IgG1. This is based on the presence of the necessary cysteine residues aa 233 and 242 of the hinge region naturally used for interdomain disulfide bond formation and the complete aa sequence including CH2 to CH3. Furthermore, structural data obtained by Deisenhofer *et al.,* (1978; Deisenhofer, 1981) predict that an intact Fc dimer is required for binding to SpA.

Therefore, the affinity of $10^{-8}$ *M* between SpA and the Fc part of immunoglobulin (Hottiger *et al.,* 1995) was sufficient to isolate the respective clones. No clones were found extending toward CH1 of IgG1 which is not involved in binding to SpA. In contrast, CH2 and CH3, which are involved in binding SpA, were present in all clones (Table 1), even though a random-primed library was used. The data indicate that a specific ligand may select functional proteins exposed on filamentous phage with the minimal length but adequate affinity for isolation.

## PRACTICAL ASPECTS OF SCREENING PHAGE EXPRESSED BY pJuFo SYSTEM

Basically all screening methods used with other filamentous phage systems apply to pJuFo without restriction. The reader is referred to the appropriate articles in this book. In this chapter we focus only on developments related to pJuFo. When panning this library, the first two rounds should be washed gently. Recall that only 10% of the phagemids may be displaying and only one copy of the cDNA product may be displayed per phagemid. The first round will sustain great losses since the selected phagemids may be present at low numbers in the original library. The second round has been amplified only once and should still be handled with caution. Titering inputs and outputs to each round should indicate enrichment by the fourth round if the selection is proceeding well. Once the selection has been amplified it may be useful to use shorter amplification times than used initially.

Polyclonal or monoclonal antibodies adsorbed directly to solid phase (Crameri and Suter, 1993) or captured with solid-phase-adsorbed isotype-specific antibodies have been used to select phage encoding allergenic proteins from an *A. fumigatus* cDNA library displayed as recombinant proteins on phage surface (Crameri *et al.,* 1994). The power for the selection of a desired phage from a population was previously demonstrated by sorting phage displaying *A. fumigatus* allergen I/a (Moser *et al.,* 1992) from a mixture of noncognate phage at a ratio as low as 1 in $10^7$ phage (Crameri and Suter, 1993).

## APPLICATIONS OF THE pJuFo VECTOR

In other experiments we have demonstrated that phage surface display using pJuFo is not limited to naturally secreted proteins. In one example, mouse dihydrofolate reductase (DHFR) was subcloned as a *Bam*HI/*Xba*I fragment from plasmid pDS 76 (Stüber *et al.,* 1990) into the *Bgl*II/*Xba*I restriction sites of pJuFo. Phage particles were generated by superinfection with helper phage as usual (Crameri and Suter, 1993) and shown to display biologically active DHFR as demonstrated by the measurement of enzymatic activity (R. Crameri and D. Stüber, unpublished). In a second example, an array of strictly intracellular cyclin-dependent kinases from man and hamster were demonstrated to be displayed on phage surface by Western blot analysis (R. Jaussi and R. Crameri, unpublished); functional analyses of these proteins displayed on phage surface are in progress. Therefore, the experiments show that intracellular proteins which are not secreted in the original biologic system can be forced into the periplasmic space of *E. coli,* incorporated into phage particles, and secreted into the growth medium attached to phage particles. These experiments indicate that proteins displayed on phage using pJuFo may refold to a conformation able to confer to the protein the ability to exert their original biologic function as shown for the displayed *E. coli* alkaline phosphatase (Crameri and Suter, 1993), the *A. fumigatus* allergen I/a (M. Moser, R. Crameri, and M. Suter, unpublished), and DHFR (see above and Table 2).

**TABLE 2**   Proteins Expressed in pJuFo and Found to be Biologically Functional

| Name | Characteristics and ligand interaction | Reference |
|---|---|---|
| *Asp f* I | Allergen from *A. fumigatus*, ribotoxin, isolated by mAb | Moser *et al.,* (1992), Crameri and Suter (1993) |
| phoA | *E. coli* alkaline phosphatase, cloned into pJuFo | Crameri and Suter (1993) |
| *Asp f* X | Allergens (up to 70 kDa) from *A. fumigatus,* isolated by polyclonal human IgE | Crameri *et al.* (1994) |
| β-Actin | Interaction with large β subunit of HIV reverse transcriptase | Hottiger *et al.* (1995) |
| IgG1 | Human IgG1 Fc, dimer, interacting with Staphylococcal protein A | Hottiger *et al.* (1995), Table 1 |
| DHFR | dihydrofolate reductase, cytoplasmic enzyme, cloned into pJuFo | Crameri and Stüber, unpublished |
| cdk | Intracellular cycline-dependent kinases from man and hamster, cloned into pJuFo | Jaussi *et al.,* unpublished |
| λ phage repressor | DNA binding protein, dimer, binds to λ phage DNA, biologically active, cloned into pJuFo | Crameri, unpublished |

Theoretical considerations implied that it should be possible to isolate DNA-binding proteins using cDNA libraries constructed into the pJuFo system. A problem may be the fact that DNA-binding proteins are often homo- or heterodimers (Johnson *et al.,* 1979; Abate *et al.,* 1990). To test the feasibility of isolating DNA-binding proteins displayed on a phage surface with DNA as ligand, the well known temperature sensitive λ-phage repressor (Johnson *et al.,* 1979) was used as a model; λ-phage repressor is a homodimer with a unit size of 236 amino acids. Natural λ-phage repressor binds as dimer to the operator elements of the phage λ genome blocking transcription of genes located 3′ of the promoters. Thus *E. coli* cells containing repressor molecules at concentrations higher than the $K_d$ of the homodimer formation ($2 \times 10^{-8}$ $M$) became resistant to λ-phage super infection (Szybalski and Szybalski, 1979; Ptashne *et al.,* 1980).

To clone the gene encoding the λ-phage repressor into pJuFo, the corresponding gene fragment of λ was amplified by PCR and directionally inserted as a *Bgl*II–*Xba*I fragment 5′ of the Fos leucine zipper (5′ primer: 5′ GA AGA TCT ATG AGC ACA AAA AAG AAA CCA TTA3′; 3′ primer: 5′ GC TCT AGA TCA GCC AAA CGT CTC TTC AGG CC5′). After cloning, the function of the phagemid containing the λ-repressor was tested by the ability to repress λ-plaque formation. For this, *E. coli* XL-1 Blue cells were transformed with pJuFo:: λ-repressor and the lac-promoter was induced with IPTG (1 m$M$). Transformed cells expressing the λ-repressor were infected with λ-phage at permissive and nonpermissive temperatures. The pJuFo:: λ-transfected *E. coli* XL-1 Blue cells did not produce λ-phage plaques at permissive temperature demonstrating the biological activity of Fos λ-phage repressor fusion protein. The intriguing fact that Fos::λ-repressor fusion, which is preceded by the *pelB* leader and thus expected to be transported into the periplasmic space, is able to suppress λ-plaque formation at permissive temperature can be explained by the fast production of recombinant protein after induction of the lac-promoter, compared to translocation into the periplasm (Perez-Perez *et al.,* 1994). Thus, recombinant proteins destined for the periplasm accumulate in the cytoplasm (Wilkinson and Harrison, 1991) when expressed from a strong promoter. We conclude that the accumulation of λ-phage repressor in the cytoplasm allows interaction with λ-phage DNA promoter regions and suppress of plaque formation at permissive temperatures.

## EXAMPLE OF A POSITIVE CONTROL SYSTEM

To test the possibility to enrich for DNA-binding proteins, the pJuFo:: λ-repressor phagemid was converted to phage by superinfection of bacterial cells with helper phage (Barbas and Lerner, 1991). Subsequently, pJuFo:: λ-repressor phage was mixed at different ratios with nonrecombinant phage (Crameri and Suter, 1993). This phage mixture was enriched for pJuFo:: λ-repressor phage using solid-phase immobilized λ-phage DNA by five rounds of selection. To perform the pJuFo:: λ-repressor phage enrichment 2 μg of λ-phage DNA was first adsorbed to a Qiagen Type 5 column (Qiagen, Hilden, Germany), subsequently washed with 1 ml of QA buffer (Qiagen) and equilibrated with 1 ml of SOB medium, pH 7.4 (Barbas *et al.,* 1991). The equilibrated column carrying immobilized λ-phage DNA was loaded with 0.3 ml (titer: $10^{11}$ cfu/ml) of phage in SOB medium mixed in a ratio of 1 in $10^4$ to $10^5$ of pJuFo:: λ-repressor phage to parental pJuFo phage. The phage mixture was left in the column for 30 min at room temperature. Thereafter, the column was washed with 10 ml of SOB medium, pH 7.4, at a flow rate of 0.1 ml min$^{-1}$. Phage adhering to the column was eluted with 0.7 ml of F - buffer (Qiagen), titrated, and used to reinfect *E. coli* XL-1 Blue cells (Barbas and Lerner, 1991; Crameri and Suter, 1993). Amplified phage was used for a further round of selection on λ-phage DNA. After five rounds of selection, one in three phage (33%) contained the expected λ-repressor insert starting from a ratio of one recombinant phage mixed in $10^5$ nonrecombinant phage (R. Crameri, unpublished).

This example shows that the pJuFo system allows expression of homodimers using one copy of the λ-phage repressor gene. The dimeric proteins expressed have the capacity to exert their expected function in *E. coli* as shown by suppressing λ-phage plaque formation at the permissive temperature. In addition, the desired phage can be isolated by a simple affinity isolation of solid phase bound DNA. Application of such a simple approach may lead to the isolation of other DNA-binding proteins from different sources. Using similar approaches it appears feasible to identify the DNA region encoding the DNA binding site or even to determine the precise nucleotide sequence(s) required for protein/DNA interaction of newly isolated DNA binding proteins.

pJuFo is presently being used in many different laboratories and the goals are as diverse as the laboratories. The original aim of pJuFo as a tool for improved cDNA cloning (Crameri and Suter, 1993; Crameri *et al.,* 1994) has been extended to very basic questions of biology: the interaction of molecules, such as protein/protein or protein/ DNA interactions. We are therefore confident that the pJuFo system described will help to speed up not only cloning of cDNA but also analyzing gene products.

## POTENTIAL AND LIMITATIONS OF THE pJuFo SYSTEM

The most attractive feature of the pJuFo cDNA cloning system is without doubt the fact that the recombinant cDNA product and the corresponding cDNA are physically linked and packed as one entity into a filamentous phage particle. Like other phage display systems this opens up the possibility of a direct selection and enrichment of this entity. In established cDNA cloning systems some $10^{-10}$ mol of a specific product has to be present for a successful detection with a ligand (Sambrook *et al.*, 1989). In a phage system a homogeneous population of as few as $10^3$ to $10^4$ copies of a given recombinant phage may be sufficient for a successful cloning (Crameri *et al.*, 1994). However, $10^3$ to $10^4$ copies of a given phage translates into some $10^{-19}$ to $10^{-20}$ mol of a product. So, the pJuFo system appears to be approximately $10^{10}$ times more sensitive than established cDNA cloning systems.

Even though *E. coli* allow folding of recombinant proteins and assembly of homo- or heterodimers (Winter and Milstein, 1991; Kang *et al.*, 1991; Crameri and Suter, 1993; this chapter), there are biologic limitations related to the host. It is well known that *E. coli* cannot glycosylate proteins or perform post-translational modifications like phosphorylations, farnesylations, geranylations, or prenylations which often occur in eukaryotic systems. Furthermore, one should be aware that proteins may be handled differently depending on the compartment within a eukaryotic and that post-translational modifications of proteins may depend on the activation status of the eukaryotic cell per se. However, these limitations are common in all prokaryotic expression systems.

### Acknowledgments

We thank Drs. Jaussi and Stüber for providing information on the use of pJuFo prior to publication. Drs. Gramatikoff and Hottiger, University of Zürich, allowed us to use preprints and helped us to sharpen our ideas with stimulating discussions. In addition, Dr. Gramatikoff has been very generous in providing unpublished data and in discussing them with us in detail. M. S., M. F., and M. A. were supported by grants of the Kanton of Zürich, the National Funds (GRD), and the Bundesamt für Bildung und Wissenschaft (BBW No. 94.0154) which is part of Eu Grant BMH1CT941184). The work at SIAF was supported in part by a Swiss National Science Foundation Grant (31-30185.90 to R. C.).

### References

Abate, C., Luk, D., Gentz, R., Rauscher, F. J., III, and Curran, T. (1990). Expression and purification of the leucine zipper and DNA-binding domains of Fos and Jun: Both Fos and Jun contact DNA directly. *Proc. Natl. Acad. Sci. U.S.A.* **87,** 1032–1036.

Barbas, C. F. III, and Lerner, R. A. (1991). Combinatorial immunoglobulin libraries on the surface of phage (Phabs): Rapid selection of antigen-specific Fabs. *Methods: Companion Methods Enzymol.* **2,** 119–124.

Barbas, C. F., III, Kang, A. S., Lerner, R. A., and Benkovic, S. J. (1991). Assembly of combinatorial antibody libraries on phage surfaces: The gene III site. *Proc. Natl. Acad. Sci. U.S.A.* **88,** 7978–7982.

Butler, J. E. (1992).The behavior of antigens and antibodies immobilized on a solid phase. *In* "Structure of Antigens" Vol. 1, pp. 208–259. (M. H. V. vanRegenmortel, ed.), CRC Press, Boca Raton, FL.

Butler, J. E. (1995). Solid-phases in Immunoassay. *In* (E. P. Diamandis and T. K. Christopoulos, eds.) Chapter 9. "Textbook of Immunological Assays" Am. Assoc. Clin. Chem. Press (in press).

Butler, J. E., Li Ni, Brown, W. R., Joshi, K. S., Chang, J., Rosenberg, B., and Voss, E. W. (1993). The immunochemistry of sandwich ELISAs. VI. Greater than 90% of monoclonal and 75% of polyclonal antifluorescyl capture antibodies (cAbs) are denatured by passive adsorption. *Mol. Immunol.* **30,** 1165–1175.

Chang, C. N., Model, P., and Blobel, G. (1979). Membrane biogenesis: Cotranslational integration of the bacteriophage f1 coat protein into an *Escherichia coli* membrane fraction. *Proc. Natl. Acad. Sci. U.S.A.* **76,** 1251–1255.

Crameri, R., and Suter, M. (1993). Display of biologically active proteins on the surface of filamentous phages: A cDNA cloning system for selection of functional gene products linked to the genetic information responsible for their production. *Gene* **137,** 69–75.

Crameri, R., Jaussi, R., Menz, G., and Blaser, K. (1994). Display of expression products of cDNA libraries on phage surfaces—a versatile screening system for selective isolation of genes by specific gene-product/ligand interaction. *Eur. J. Biochem.* **226,** 53–58.

Crissman, J. W., and Smith, G. P. (1984). Gene-III protein of filamentous phages: Evidence for a carboxyl-terminal domain with a role in morphogenesis. *Virology* **132,** 445–455.

Cwirla, S. E., Peters, E. H., Barrett, R. W., and Dower, W. J. (1990). Peptides on phage: A vast library of peptides for identifying ligands. *Proc. Natl. Acad. Sci. U.S.A.* **87,** 6378–6382.

Deisenhofer, J. (1981). Crystallographic refinement and atomic models of a human Fc fragment and its complex with fragment B of protein A from *Staphylococcus aureus* at 2.9- and 2.8- Å resolution. *Biochemistry* **20,** 2361–2370.

Deisenhofer, J., Jones, T. A., Huber, R., Sjödahl, J., and Sjöquist, J. (1978). Crystallization, crystal structure analysis and atomic model of the complex formed by a human Fc fragment and fragment B of Protein A from *Staphylococcus aureus*. *Hoppe-Seyler's Z. Physiol. Chem.* **359,** 975–980.

Diolez, A., Richaud, F., and Coleno, A. (1986). Pectate lyase gene regulatory mutants of *Erwinia chrysanthemi*. *J. Bacteriol.* **167,** 400–403.

Dudler, T., Schneider, T., Annand, B., Gelb, M., and Suter, M. (1994). Antigenic surface of bee venom allergen phopholipase A2: Structural and functional analysis of human IgG4 antibodies reveals potential role in protection. *J. Immunol.* **152,** 5514–5522.

Gentz, R., Rauscher, F. J., III, Abate, C., and Curran, T. (1989). Parallel association of Fos and Jun leucine zippers Juxtaposes DNA binding domains. *Science* **245,** 1695–1699.

Gramatikoff, K., Georgiev, O., and Schaffner, W. (1994). Direct interaction rescue, a novel filamentous phage technique to study protein-protein interactions. *Nucleic Acids Res.* **22,** 5761–5762.

Henning, U., Royer, H. D., Teather, R. M., Hindennach, I., and Hollenberg, C. P. (1979). Cloning of the structural gene (*omp*A) for an integral outer membrane protein of *Escherichia coli* K-12. *Proc. Natl. Acad. Sci. U.S.A.* **76,** 4360–4364.

Hochuli, E., Döbeli, H., and Schacher, A. (1987). New metal chelate adsorbent selective for proteins and peptides containing neighbouring histidine residues. *J. Chromatogr.* **411,** 177–184.

Hottiger, M., Gramatikoff, K., Georgiev, O., Chaponnier, C., Schaffner, W., and Hübscher, U. (1995). The large subunit of HIV-1 reverse transcriptase interacts with β-actin. *Nucleic Acids Res.* **23,** 736–741.

Johnson, A. D., Meyer, B. J., and Ptashne, M. (1979). Interaction between DNA-bound repressor govern regulation by the λ phage repressor. *Proc. Natl. Acad. Sci. U.S.A.* **76,** 5061–5065.

Kabat, H., Wu, T. T., Reid-Miller, M., Perry, H. M., and Gottesman, K.S. (1991). "Sequences of Proteins of Immunologic Interest," 5th ed. US Department of Health and Human Services, Public Health Service, National Institutes of Health, Washington, DC.

Kang, A. S., Barbas, C. F., III, Janda, K. D., Benkovic, S. J., and Lerner, R. A. (1991). Linkage of recognition and replication functions by assembling combinatorial antibody Fab libraries along phage surfaces. *Proc. Natl. Acad. Sci. U.S.A.* **88,** 4363–4366.

Moser, M., Crameri, R., Menz, G., Schneider, T., Dudler, T., Virchow, C., Gmachl, M., Blaser, K., and Suter, M. (1992). Cloning and expression of recombinant *Aspergillus fumigatus* allergen I/a (rAsp f I/a) with IgE binding and type I skin test activity. *J. Immunol.* **149,** 454–460.

Pérez-Pérez, J., Márquez, G., Barbero, J., and Gutiérrez, J. (1994). Increasing the efficiency of protein export in *Escherichia coli*. *Bio/Technology* **12,** 178–180.

Pernelle, C., Clerc, F. F., Dureuil, C., Bracco, L., and Tocque, B. (1993). An efficient screening assay for the rapid and precise determination of affinities between leucine zipper domains. *Biochemistry* **32,** 11682–11687.

Peterman, J. M., and Butler, J. E. (1989). Application of theoretical considerations to the analysis of ELISA data. *Biotechniques* **7,** 608–615.

Ptashne, M., Backman, K., Humayun, M. Z., Jeffery, A., Maurer, R., Meyer, B. J., Pabo, C. O., Roberts, T. M., and Sauer, R. T. (1980). How the Lambda repressor and cro work. *Cell (Cambridge, Mass.)* **19,** 1–11.

Sambrook, J., Fritsch, E. F., and Maniatis, T. (1989). "Molecular Cloning: A Laboratory Manual," 2nd ed. Cold Spring Harbor Lab., Press, Cold Spring Harbor, NY.

Short, J. M., Fernandez, J. M., Sorge, J. A., and Huse, W. D (1988). λ Zap: A bacteriophage λ expression vector with *in vivo* excision properties. *Nucleic Acids Res.* **16,** 7583–7600.

Skerra, A., and Plückthun, A. (1988). Assembly of a functional immunoglobulin Fv fragment in *Escherichia coli. Science* **240,** 1038–1041.

Stüber, D., Matile, H., and Garotta, G. (1990). System for high level production in *E. coli* and rapid purification of recombinant proteins: Application to epitope mapping, preparation of antibodies and structure-function analysis. *In* I., Lefkovits, and B. Pernis, eds., "Immunological Methods" pp. 121–152. Academic Press, San Diego.

Suter, M., and Butler, J. E. (1986). The immunochemistry of sandwich ELISAs. II. A novel system prevents the denaturation of capture antibodies. *Immunol. Lett.* **13,** 313–316.

Suter, M., Butler, J. E., and Peterman, J. H. (1989). The immunochemistry of sandwich ELISAs. III. The stoichiometry and efficacy of the protein-avidin-biotin capture (PABC) system. *Mol. Immunol.* **26,** 221–230.

Szybalski, E. H., and Szybalski, W. (1979). A comprehensive molecular map of bacteriophage Lambda. *Gene* **7,** 217–270.

Wilkinson, D. L., and Harrison, R. G. (1991). Predicting the solubility of recombinant proteins in *Escherichia coli. Bio/Technology* **9,** 178–180.

Winter, G., and Milstein, C. (1991). Man-made antibodies. *Nature (London)* **349,** 293–299.

# *12*

# Multicombinatorial Libraries and Combinatorial Infection: Practical Considerations for Vector Design

*Régis Sodoyer, Luc Aujame, Frédérique Geoffroy, Corinne Pion, Isabelle Peubez, Bernard Montègue, Paul Jacquemot, and Jocelyne Dubayle*

## INTRODUCTION AND RATIONALE

In 1985, Smith developed a powerful phage display system for peptide library presentation and identification of epitopes (see also Parmley and Smith, 1988). A few years later, several groups (McCafferty *et al.*, 1990; Barbas *et al.*, 1991; Breitling *et al.*, 1991; Garrard *et al.*, 1991) took advantage of this methodology to express antibody fragments on the phage surface. Both Fab and sFV combinatorial libraries have now been successfully displayed on M13 phage or M13-derived cloning vectors.

Thus, this process may be considered as a valuable complement to the classical hybridoma technology of Milstein and Köhler (1975). Moreover, antibody phage display can provide direct and easy access to human antibodies and the possibility of generating anti-self antibodies (Griffiths *et al.*, 1993).

Even though many interesting results have been obtained with phage-displayed peptides and antibody libraries, there is still a need to improve this technology as the size of libraries obtained reaches, in the best of cases, $10^8$ clones. This represents only a very small fraction of the natural antibody repertoire (about $10^{12}$). The construction of very large repertoires is an indispensable step to obtain "high-affinity binders," antibodies directed against nonclassical antigens such as small organic molecules or antibodies with peculiar properties such as catalytic activity. For peptide combinatorial libraries, the problem is exactly the same, as $10^8$ represents also a very small portion of all possible combinations of random peptides, even if the peptide length is only 8 amino acids (i.e., $20^8 = 2.56$ x $10^{10}$). In fact the problem is even worse if one uses 64 codons to encode 20 different amino acids (see Appendix). Because of the large number of combinations, libraries need to be as large (complex) as possible. Currently, the bottleneck is the efficiency of the transformation (electroporation) step. We and others have proposed circumventing this limitation through combinatorial infection (C.I.), as infection frequencies are much more efficient than other means of transfecting genetic material. In this method, *Escherichia coli* can be transformed with a repertoire of random peptides or light-chain genes (i.e., $10^6$ clones) encoded on a plasmid and then infected with another repertoire of random peptides or heavy-chain sequences (i.e., $10^6$ clones) encoded on a phage or phagemid. In each bacterial cell, fusion proteins are expressed simultaneously from each vector and assembled at the surface of the phage particles. Two similar approaches have been described (Collet *et al.,* 1992; Hogrefe and Shopes, 1994).

A physical association between the two vectors, or part of them, is absolutely necessary for the recovery of genetic information. Such an association can be obtained by a process of site-specific recombination. The site-specific recombination systems that could be adapted to this approach include: the *lox*–Cre system of bacteriophage P1 (Abremski *et al.,* 1983), the FLP recombinase of the yeast 2 μ plasmid (Andrews *et al.,* 1985), the resolvases of the γδ or Tn3-like transposons (Grindley and Reed, 1985), and the *att* attachment sites of phage λ (Campbell, 1962).

Important parameters to be considered when choosing between these systems are:

- the availability and inducibility of the recombinase activity,
- the irreversible or leaky character of the vector association,
- the rapid identification and easy recovery of "true" recombinants.

The *lox*–Cre system of bacteriophage P1 has already been tested for the assembly of two different vectors (Waterhouse *et al.,* 1993). Using this approach, antibody libraries with a size more representative of the natural immune repertoire have been obtained (Griffiths *et al.,* 1994; K. S. Johnson, personal communication, 1994).

The final phage preparation consists of a mixture of combined and nonrecombined phage. To date, the system does not allow for selection of recombinants.

# THE *ATT* SYSTEM

This system takes advantage of the site-specific integration properties of bacteriophage λ (Craig, 1988; Miller, 1988). Utilization of the *att* system leads to the association of heavy and light antibody chains, encoded by two different vectors, within an infected bacterium. The recombination event is irreversible in the model we have designed and there is a direct selection for recombinants.

## Lysogeny and Phage Integration

A circular form of bacteriophage λ integrates into the bacterial chromosome during the establishment of lysogeny. Two sequences are involved in intermolecular recombination, a 240-bp phage-encoded *attP*-specific site and a 23-bp *attB* site in the bacterial chromosome. The phage-encoded protein Int and the *E. coli*-encoded protein Integration Host Factor (IHF) are both required for this process. Two different *att* sites, *attL* (99-bp) and *attR* (164-bp), are created upon recombination. The presence of a phage-encoded excision factor (Xis) is absolutely essential to reverse the integration mechanism (Fig. 1).

To make use of the *att* recombination system, we have designed two families of vectors (plasmid and phagemid) and modified an *E. coli* strain.

## Vector Design

pM834 is the prototype of the *phagemid* family, carrying *attP*, the *cat* promoter, two replication origins ColE1, and f1 and a resistance gene to ampicillin. pM827 is the prototype of the *plasmid* family, carrying *attB*, the *cat* gene, a distinct (to ColE1) bacterial replication origin p15A, and a resistance gene to kanamycin. The construction of these two vectors has been described elsewhere in full detail (Geoffroy *et al.,* 1994).

Upon recombination, one obtains a new phagemid (pM835) carrying both antibody chains or encoded peptides and characterized by the functional association of the *cat* resistance gene with its promoter (Fig. 2). The absence of *xis*, the phage-encoded excision factor, prevents any reverse recombination.

## Strain Modification

The recombination strain was obtained by transformation of *E. coli* D1210HP (Hasan and Szybalski, 1987) with the F′ episome derived from XL-1 Blue (Stratagene, La Jolla, CA). Strain D1210HP carries the heat-inducible Int recombinase. The presence of the F′ episome, maintained by its resistance to tetracycline, allows infection by filamentous phages (Geoffroy *et al.,* 1994). The genotype of the new strain (D1210HP-F′) is HB101, *lacI*q, *lacY*+, λ, *c*Its857 *xis*-, *kil*- [F′, *proAB*+, *lacI*qZΔM15, Tn*10*(*tet*R)]. The overall recombination protocol is shown in Fig. 3.

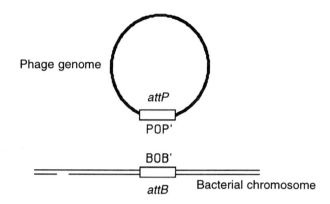

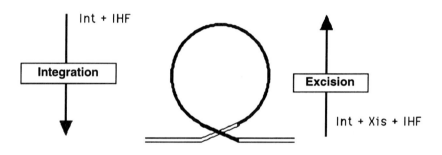

**FIGURE 1**     Integration of bacteriophage λ. *attP* and *attB* are, respectively, the attachment sites of λ phage and bacteria. A common core sequence called O is flanked by the regions P,P′, B, and B′, each of them being distinct in sequence. Integration requires recognition between *attP* (POP′) and *attB* (BOB′), while excision requires recognition between *attL* (BOP′) and *attR* (POB′). Int (phage encoded integrase) and IHF (bacterial encoded Integration Host Factor) are involved in both integration and excision. Xis (phage encoded excision factor) controls the excision and inhibits integration.

Our system thus has three major features:

1. The recombinase activity is perfectly controlled, as Int expression is heat-shock inducible;

2. The recombination process is irreversible;

3. An easy selection of recombinants and a calculation of library complexity are made possible by the creation of a new genetic marker on the recombined phagemid.

## Application of the System

We have tested the functionality of our system on the recombination of anti-HIV gp160 antibody light- and heavy-chain sequences (Geoffroy *et al.*, 1994) in which the light-chain gene was inserted into pM827 and the heavy-chain into pM834. As shown in Fig. 2, the cloning sites for the light-chain and heavy-chain sequences are, respectively, *Sac*I–*Xba*I and *Xho*I–*Spe*I. These sites are available for the insertion of other coding sequences (antibody variable regions, random peptides, etc.).

Different aspects of the recombination process were successively analyzed, and the functionality was assessed in terms of recombination efficiency, infectivity, and ability to express soluble molecules (Fab in the example described).

## Recombination Efficiency and Stability

This first point was evaluated through the cotransfection of both pM827 and pM834 into D1210HP-F' bacteria, using the standard experimental conditions defined by Hasan and Szybalski (1987). After a 30 min exposure to heat-shock at 42°C, followed by 2 hr at 37°C, cells were recovered for plasmid DNA extraction by standard techniques. The presence and the correct size of the recombined vector (pM835) were shown by gel electrophoresis after digestion at a unique restriction site (Geoffroy *et al.*, 1994). Furthermore, as confirmed by sequencing, the junctions (*attL* and *attR*) obtained after recombination between the parental vectors were in perfect accordance with the model described by Craig (1988) for λ phage. Efficient, but not optimal (see Assembly of a New Marker), expression of chloramphenicol resistance was demonstrated after transformation of an *E. coli* strain by the recombined vector pM835, isolated by gel electrophoresis. After plating, only colonies containing the full-size (unmodified) recombined phagemid were able to grow on a selective medium.

## Infectivity

In order to test infectivity, a phage culture of the recombined vector was prepared on XL1-Blue by coinfection with VCSM13 helper phage (Stratagene) (m.o.i. = 10–100). A titer of $10^{11}$ to $10^{12}$ cfu/ml was obtained by plating on Cm, thus demonstrating the capacity of pM835 to be correctly packaged into infectious particles.

## Production and Characterization of Soluble Fab

The designed vector allows either phage or soluble gene III-associated protein (Fab in our case) production. Translation of the fusion protein stopped at the TAG codon was obtained from pM835, by making use of the amber mutation located 5' to the gene III coding sequence. A non-suppressor bacterial strain TOPP2 (Stratagene) was transformed by the recombined phagemid and grown at 37°C until an

**FIGURE 2**     Construction of the recombination vectors (Geoffroy *et al.,* 1994). Phagemid pM834: The parental vector pBluescript SKII+ (Stratagene, La Jolla, CA) was modified by insertion of a synthetic linker including *Sac*II, *Xho*I, *Spe*I, *Nhe*I, and *Not*I between *Sac*I and *Kpn*I sites (phagemid pM828). The following DNA fragments were sequentially cloned in the modified parental vector: (i) the synthetic leader sequence derived from *pelB,* (ii) part of the M13 phage gene *III* encoding sequence (Pro 216–Ser 424) amplified by PCR, (iii) a fragment of the heavy-chain sequence of an anti-HIV gp160 Ab (684-bp amplified by PCR from a cDNA library) (C. Pion and I. Peubez, unpublished data), (iv) the *attP* recombination sequence (240-bp amplified by PCR from λ phage), (v) the amber codon TAG introduced by PCR at the 3′-end of the heavy-chain, (vi) the promoter for the *cat* gene (170 bp amplified by PCR from pBR328, nt positions 3270 to 3440). Plasmid pM827: A 2050-bp fragment, including the Km^R gene, the P15A *ori,* and a unique *Eco*RI site, was amplified from pACYC177 (NE Biolabs, Beverly, MA) after insertion of a synthetic linker comprising *Sac*I, *Kpn*I, *Xba*I, and *Sph*I into the *Eco*RI site. Sequential insertion was as follows: (i) the synthetic *attB* recombination sequence (23 bp), (ii) a 960-bp *Nhe*I-*Xba*I fragment, including the *lac* promoter, the leader sequence derived from *pelB,* and the light-chain sequence of the same anti-HIV gp160 Ab (derived from pM452; C. Pion and I. Peubez, unpublished data), (iii) the *cat* gene devoid of promoter (770 bp amplified by PCR from pBR328, nt positions 3441 to 4210). Plasmid pM 840 was obtained by substituing the *cat* gene by the *acC1* gene also devoid of promoter (nt positions 1235 to 1783 according to nomenclature adopted for plasmid pRAACC1 — Genebank Accession No. X15852). **A,** *Asc*I; **B,** *Bam*HI; **E,** *Eco*RI; **K,** *Kpn*I; **M,** *Mlu*I; **N,** *Nhe*I; **S,** *Sac*I; **S2,** *Sac*II; **Sh,** *Sph*I; **Sp,** *Spe*I; **X,** *Xba*I; **Xh,** *Xho*I. Arrows indicate the direction of transcription.

**FIGURE 3**     Recombination protocol.

**FIGURE 4**     Phage preparation. Upon recombination the plasmid (low copy) is almost completely lost. The two remaining species, large-size recombined phagemid and starting phagemid, can be packaged and secreted. VCS M13 helper phage (multiplicity of infection = 10–100) is used for phage preparation. Phage titers obtained are typically $10^{10}$–$10^{11}$ and only 1–10% result from the packaging of the large-size recombined vector. Consequently a strong reduction or complete elimination of the starting phagemid is necessary before assembly of large repertoires (see text).

**FIGURE 5**     Permutation of the origins of replication and cloning of the antiterminator. Permutation of the origins of replication (1) PCR was used to amplify, from pM839, a fragment containing ColE1 flanked by two unique restriction sites (*Mlu*I and *Bam*HI). A linear *Mlu*I–*Bam*HI DNA fragment containing plasmid pM840, devoid of its origin of replication, was also obtained by amplification. Ligation of these two fragments gave rise to pM842, a high-copy number plasmid. (2) The low-copy number phagemid pM841 was obtained by the same kind of strategy, the restriction sites used to assemble the new vector being *Sph*I and *Xba*I. Stipled box = antibody fragment, random peptide, or two different proteins that heterodimerize. Cloning of the nutL/N antiterminator: Phagemid pM839 was obtained after insertion of a 112-bp double-stranded synthetic DNA, containing the nutL sequence, into the unique *Kpn*I site 3′ to the *cat* promoter. The pN coding sequence was amplified by PCR from phage λ DNA and linked to the 5′ end of the gentamycin resistance gene of plasmid pM842; the plasmid obtained is named pM843.

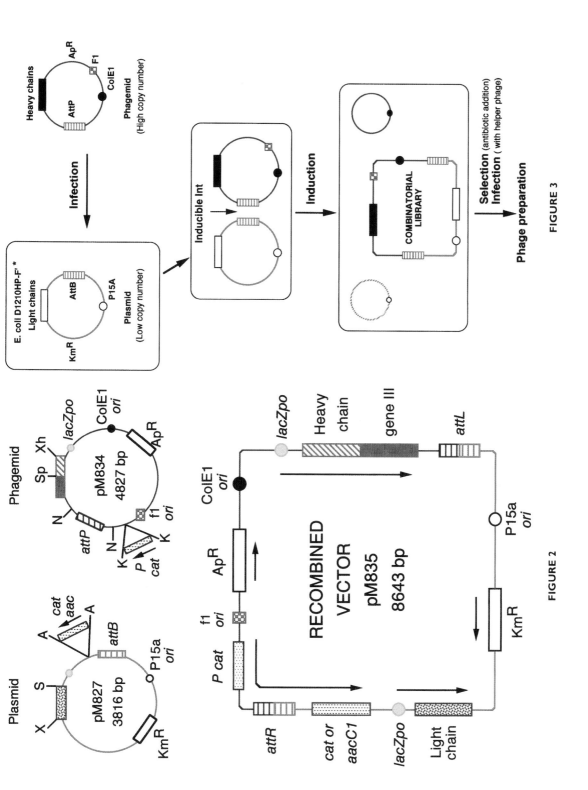

FIGURE 3

FIGURE 2

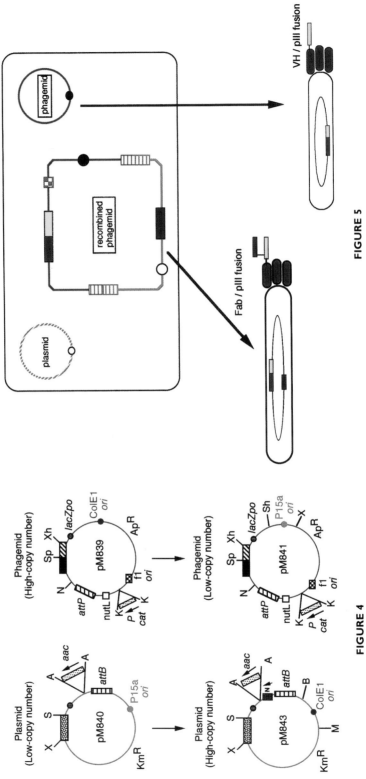

**FIGURE 5**

**FIGURE 4**

OD$_{600}$ of 0.6 was achieved. A final concentration of 1 m$M$ IPTG was used for induction. Soluble Fab was recovered from an overnight culture by cell lysis through sonication at 4°C (10–12 10-sec bursts at maximum power, allowing the sample to cool for 30 sec between each burst). Unpurified supernatant was used directly for ELISA testing against HIV gp160. The soluble Fab was revealed by an anti-human F(ab')$_2$-alkaline phosphatase Ab (1/500 in PBS), 50 μl per well.

Fab production and periplasmic localization were confirmed in a Western blot experiment (data not shown).

## Preliminary Conclusion

We have developed a system, based on λ phage *att* recombination sites, that allows one to control the irreversible association and the expression of sequences encoded by separate vectors. The most exciting applications of the system would undoubtedly be the construction of very large antibody or peptide repertoires.

# FULL-SCALE LIBRARY CONSTRUCTION—OPTIMIZATION OF THE SYSTEM

The *att* system has proven to be a very efficient means of inducing the irreversible association between two vector partners. However, some improvements or optimizations have to be carried out for the construction of very large repertoires.

## Assembly of a New Genetic Marker

In the example described previously, a functional *cat* gene was created. Chloramphenicol acetyl transferase activity was our first choice because of the availability of promoter and gene from commercial plasmids. Chloramphenicol has "bacteriostatic" activity, but, for our purpose, a real "bactericidal" agent would probably be better to ensure complete removal of nonrecombined phagemid-containing cells. To that effect, the *cat* gene of plasmid pM827 was replaced by the gentamycin resistance gene (*aacC1*), amplified by PCR from plasmid pSS1129 (Stibitz, 1994), to give plasmid pM840. The functionality of the new construct was then tested upon recombination by plating the cells on different concentrations of gentamycin. Selection was only efficient for very low antibiotic concentrations, indicating a weak expression of the resistance gene. This may be due to the lack of proximity between the gentamycin resistance gene and cat promoter. To overcome this problem, two different strategies have been conducted in parallel or in association:

1. Insertion of an antitermination sequence between the resistance gene and the promoter (Hasan and Szybalski, 1987),
2. Use of a different or/and stronger promoter.

The synthetic *nutL* antiterminator recognition site, derived from λ DNA (Daniels *et al.,* 1983), was inserted into phagemid pM834 (Fig. 4), creating phagemid pM839.

The pN encoding sequence, responsible for antitermination activity, was inserted 5′ to the gentamycin resistance gene of plasmid (see below), the resulting plasmid being pM843. Recent data suggest that these modifications have improved the expression of the gentamycin resistance gene and that recombinants can indeed be selected with this antibiotic at 10 to 15 µg/ml.

## Phagemid Persistence After Recombination

In our system, plasmid and phagemid have distinct replicons, p15A and ColE1, respectively. Since ColE1, is a high copy number replicon, the phagemid is present in large excess as compared to the plasmid with its p15A replicon. After induction of the Int recombinase, all the plasmid copies available are probably completely lost; the two remaining species (phagemid and large-size recombined phagemid) act as competitors to maintain themselves within the bacteria (Fig. 5). In such a situation, the recombined phagemid is certainly at a disavantage in terms of replication rate (larger size) or packaging time. A way to minimize phagemid persistence would be to take advantage of the phagemid/plasmid balance before recombination. For example, if the plasmid is overrepresented as compared to the phagemid, one can assume that the majority of the phagemid copies will be involved in recombination and therefore will not interfere further. This situation can be obtained by a simple reciprocal exchange of the origins of replication. Such a strategy has recently been applied in our laboratory, using PCR to create unique restriction sites and amplify the ori sequences, as shown in Fig 4. The new recombination partners are named phagemid pM841 and plasmid pM843, respectively.

Preliminary experiments have shown that phagemid persistence has been extremely reduced, and that ori permutation is an undeniable improvement.

**TABLE I**   Infection Efficiency

| $A_{600nm}$ | Temperature of infection | Time of infection | % of infection |
|---|---|---|---|
| 0,5 | 37°C | 60 min | 5% |
| 0,5 | 37°C | 90 min | 10% |
| 0,5 | 37°C | 120 min | 10–15% |
| 0,2 | 37°C | 90 min | 10–15% |
| 0,2 | 37°C | 120 min | 20% |
| 0,1 | 37°C | 90 min | 95% |
| 0,1 | 37°C | 120 min | 95% |
| 0,5 | 30°C | 90 min | 10% |
| 0,5 | 30°C | 120 min | 15% |
| 0,2 | 30°C | 90 min | 10–15% |
| 0,2 | 30°C | 120 min | 10–15% |
| 0,1 | 30°C | 90 min | 30% |
| 0,1 | 30°C | 120 min | 50% |

**TABLE 2**    Analysis of Bacterial Survival

| Temperature of induction | Time of induction | % survival |
|---|---|---|
| 42°C | 60 min | 5–10% |
| 42°C | 30 min | 20–25% |
| 42°C | 15 min | 40–45% |
| 40°C | 30 min | 70–75% |
| 40°C | 15 min | 80–85% |

Among the problems which still need to be solved, some are intrinsically related to the very large size of the library. To ensure the expected results, obtaining $10^{11}$ or $10^{12}$ different clones, all the components or steps for the assembly of the library have to be optimized.

## Infection Efficiency

We have designed a series of experiments to evaluate infection efficiency in terms of absorbance, temperature, and time of infection. The percentage of infection is directly accessible, after plating, as a ratio of ampicillin-resistant colonies (phagemid-infected) over kanamycin-resistant ones (plasmid-containing). The raw data, presented in Table 1, indicate that infection efficiency seems to be dependent on temperature and is extremely sensitive to the density of the bacterial culture. In our hands, the best results have been obtained at an $OD_{600}$ of 0.1 ($10^7$–$10^8$ cells/ml).

More recently, transformation assays were performed after resuspension of cells grown at 37°C in 10 m$M$ $MgSO_4$ at an $OD_{600}$ of 1 or 2 (>$10^8$ cells/ml) with transformation efficiencies of close to 100%. This approach makes it easier to handle large size library constructions as it considerably reduces the volumes employed. We now use this procedure routinely and have been able to associate VL and VH libraries of $10^5$ and $10^6$ respectively with high efficiencies.

## Recombination Efficiency and Bacteria Survival

As a source of recombinase activity, our system makes use of *E. coli* strain D1210HP, containing a heat-shock-inducible integrase. A stress due to sustained heat shock is often deleterious to bacteria, so it was essential to analyze that point precisely. As a first step, bacteria survival was evaluated for two different heat-shock temperatures and three different time periods. The results are summarized in Table 2.

In further experiments, efficient expression of recombinase activity was confirmed for the best survival conditions. A 15-min 40°C heat shock provides a reasonable compromise between Int induction and cell viability. According to these data, the following recombination protocol have been defined:

1. Strain D1210HP-F′ transformed with plasmid pM843 ($OD_{600}$ of 0.01—obtained from an overnight preculture at 37°C in the presence of 25 µg/ml Km and 10 µg/ml Tc) is grown at 37°C in SB medium with 25µg/ml Km and 10 µg/ml Tc to an $OD_{600}$ of 0.6. This is centrifuged at 1500$g$ and resuspended in the appropriate volume of 10 m$M$ $MgSO_4$ to give an $OD_{600}$ of 1 or 2.

2. In parallel, strain XL1-Blue or any compatible strain is transformed with phagemid pM841 and infected with VCS M13 helper phage (m.o.i. 10–100) to obtain a phage culture ($10^{10}$ or more pfu/ml) which is used to infect the D1210HP-F′ (pM843) culture (m.o.i. 2–10).

3. After a 60 min infection at 37°C, the culture is heat-shocked at 40°C, while shaking, for 15 min, to induce Int-dependent recombination.

4. The culture is then returned to 37°C and diluted 10- to 20-fold in SB supplemented with Km and Tc.

5. Gentamycin (10 to 15 µg/ml) is added 3 to 4 hr later to allow for the selection of recombinants, and left overnight at 37°C while shaking.

This culture now constitutes the stock recombinant library from which phage recombinants may be obtained following infection with VCS M13 helper phage. This last step is preferentially done in small-scale (21) fermenters to increase phage yield.

## Library Preparation, Phage Production, and Panning

The construction of a multicombinatorial library cannot be considered as a simple benchtop experiment: the access to small-scale (2–10 liters) fermentation technology is essential. Phage particle production techniques are extremely well documented in Chapter 6 of this book; nevertheless, it is important to emphasize an aspect inherent to very large antibody or peptide repertoires. In a standard panning procedure, one would expect to use, at least, a 100-fold overrepresentation of each single clone; this would mean $10^{13-14}$ phage particules in the case of a $10^{11-12}$ combinatorial repertoire. Consequently, under standard conditions, 10–100 ml of a high-titer phage preparation would be required for each round of panning (polyethyleneglycol (PEG) precipitation can be used to increase the phage titer and reduce the volume). In addition, a large amount of antigen may also be necessary for the panning of large libraries and, in some cases, might be a serious limitation. Consequently, the most suitable panning procedure is likely the one making use of biotin-linked antigen, or a closely related technique. The limitation of panning such large libraries may foster the developement of more efficient selection procedures. For example, the interesting SAP (Selection and Amplification of Phage) procedure, which has been described recently by Duenas and Borrebaeck (1994), could be of some help in the case of very large repertoires. In this technique, antigen recognition and phage replication are associated, as the antigen and gene III product are covalently linked. More recently, a quantitative comparison of selection procedures for phage-displayed antibodies has been published by Kretzschmar et al., (1995). These authors indicate that the optimization of microselection procedures using "immunotubes" can improve the efficiency of selection by a factor of 30 or more.

## New Promises, New Problems

Multicombinatorial libraries and related approaches can provide a breakthrough in biotechnology research. Many biological problems can benefit from the phage display system which may be considered as an ideal screening tool for new drug discovery. The association of variable polypeptides encoded on two different vectors gives access to unexpected experiments such as the improvement of binding affinity of heterodimeric receptors, and many other experiments involving the coexpression of two polymorphic molecules.

Nevertheless, new problems, directly related to the size of the libraries obtained by multicombinatorial approaches, are emerging. For example, new technologies have to be set up to produce, handle, and conserve such large repertoires.

The system we describe provides a potential access to very large "coliclonal" (Chiswell and McCafferty, 1992) antibody or peptide repertoires. However, many improvements have to be carried out before multicombinatorial library construction becomes routine.

### Acknowledgments

The authors thank Dr. Pierre Meulien for encouragement. This work was supported by the French "Ministère de l'Enseignement Supérieur et de la Recherche."

### References

Abremski, K., Hoess, R., and Sternberg, N. (1983). Studies on the properties of P1 site-specific recombination: Evidence for topologically unlinked products following recombination. *Cell (Cambridge, Mass.)* **32,** 1301–1311.

Andrews, B. J., Proteau, G. A., Beatty, L. G., and Sadowski, P. D. (1985). The FLP recombinase of the 2 m circle DNA of yeast: Interaction with its target sequences. *Cell (Cambridge, Mass.)* **40,** 795–803.

Barbas, C. F., III, Kang, A. S., Lerner, R. A., and Benkovic, S.J. (1991). Assembly of combinatorial antibody on phage surfaces: The gene III site. *Proc. Natl. Acad. Sci. U.S.A.* **88,** 7978–7982.

Breitling, F., Dübel, S., Seehaus, T., Klewinghaus, I., and Little, M. (1991). A surface expression vector for antibody screening. *Gene* **104,** 147–153.

Campbell, A. M. (1962). Episomes. *Adv. Genet.* **11,** 101–145.

Chiswell, D. J., and McCafferty, J. (1992). Phage antibodies: Will new "coliclonal" antibodies replace monoclonal antibodies? *Trends Biotechnol.* **10,** 80–84.

Collet, T. A., Roben, P., O'Kennedy, R., Barbas, C. F., III, Burton, D. R., and Lerner, R. (1992). A binary plasmid system for shuffling combinatorial antibody libraries. *Proc. Natl. Acad. Sci. U.S.A.* **89,** 10026–10030.

Craig, N. L. (1988). The mechanism of conservative site–specific recombination. *Annu. Rev. Genet.* **22,** 77–105.

Daniels, D. L., Schroeder, J. L., Szybalski, W., Sanger, F., Coulson, A. R., Hong, G. F., Hill, D. F., Petersen, G. B., and Blattner, F. R. (1983). Complete annotated lambda sequence. In "Lambda II" (R. W. Hendrix, J. W. Roberts, F. W. Stahl, and R. A. Weisberg, eds.) pp. 519–676. Cold Spring Harbor Lab., Cold Spring Harbor, NY.

Duenas, M., and Borrebaeck, C. A. K. (1994). Clonal selection and amplification of phage displayed antibodies by linking antigen recognition and phage replication. *Bio/Technology* **12,** 999–1002.

Garrard, L. J., Yang, M., O'Connell, M. P., Kelley, R. F., and Henner, D.J. (1991). Fab assembly and enrichment in a monovalent phage display system. *Bio/Technology* **9,** 1373–1377.

Geoffroy, F., Sodoyer, R., and Aujame, L. (1994). Multicombinatorial libraries: Access to very large antibody repertoires. *Gene* **151,** 109–113.

Griffiths, A. D., Malmqvist, M., Marks, J. D., Bye, J. M., Embleton, M. J., McCafferty, J., Baier, M., Holliger, K. P., Gorick, B. D., Hughes-Jones, N. C., Hoogenboom, H. R., and Winter, G. (1993). Human anti-self antibodies with high specificity from phage display libraries. *EMBO J.* **12,** 725–734.

Griffiths, A. D., Williams, S. C., Hartley, O., Tomlinson, I. M., Waterhouse, P., Crosby, W. L., Kontermann, E., Jones, P. T., Low, N. M., Allison, T. J., Prospero, T. D., Hoogenboom, H. R., Nissim, A., Cox, J. P. L., Harrison, J. L., Zaccolo, M., Gherardi, E., and Winter, G. (1994). Isolation of high affinity human antibodies directly from large synthetic repertoires. *EMBO J.* **13,** 3245–3260.

Grindley, N. D. F., and Reed, R. R. (1985). Transpositional recombination in prokaryotes. *Annu. Rev. Biochem.* **54,** 863–896.

Hasan, N., and Szybalski, W. (1987). Control of cloned gene expression by promoter inversion in vivo: Construction of improved vectors with a multiple cloning site and the p $_{tac}$ promoter. *Gene* **56,** 145–151.

Hogrefe, H. H., and Shopes, B. (1994). Construction of phagemid display libraries with PCR-amplified immunoglobulin sequences. In *PCR Methods appl.,* **4,** S109–S122.

Kretzschmar, T., Zimmermann, C., and Geiser, M. (1995). Selection procedures for nonmatured phage antibodies: A quantitative comparison and optimization strategies. *Anal. Biochem.* **224,** 413–419.

McCafferty, J., Griffiths, A. D., Winter, G., and Chiswell, D. J. (1990). Phage antibodies: Filamentous phage displaying antibody variable domains. *Nature (London)* **348,** 552–554.

Miller, H. I. (1988). Viral and cellular control of site-specific recombination. In K. B., Low, (ed.), "The Recombination of Genetic Material" pp. 361–384. Academic Press, San Diego, CA.

Milstein, C., and Köhler, G. (1975). Continuous culture of fused cells secreting antibody of predefined specificity. *Nature (London)* **256,** 52–53.

Parmley, S. F., and Smith, G. P. (1988). Antibody-selectable filamentous fd phage vectors: Affinity purification of target genes. *Gene* **73,** 305–318.

Smith, G. P. (1985). Filamentous fusion phage: Novel expression vectors that display cloned antigens on the virion surface. *Science* **228,** 1315–1317.

Stibitz, S. (1994). Use of conditionally counterselectable suicide vectors for allelic exchange. *In* "Methods in Enzymology (V. L. Clark and P. M. Baroil, eds.), vol. **235,** pp. 458–465. Academic Press, San Diego, CA.

Waterhouse, P., Griffiths, A. D., Johnson, K. S., and Winter, G. (1993). Combinatorial infection and in vivo recombination: A strategy for making large phage antibody repertoires. *Nucleic Acids Res.* **21,** 2265–2266.

# Screening Phage-Displayed Random Peptide Libraries

*Andrew B. Sparks, Nils B. Adey,*
*Steve Cwirla, and Brian K. Kay*

## INTRODUCTION

Screening phage-displayed random peptide libraries represents a powerful means of identifying peptide ligands for targets of interest (Smith, 1985). Typically, M13 bacteriophage expressing binding peptides are selected by affinity purification with a target of interest; peptide sequences are subsequently deduced from DNA sequence data. Ligands identified in this manner frequently interact with natural binding site(s) on the target molecule and often resemble the target's natural ligand(s). These characteristics have allowed the use of phage-displayed random peptide libraries to study protein–protein interactions in a variety of contexts. For example, phage-displayed random peptide libraries have been used to map the epitopes of monoclonal (Scott and Smith, 1990) and polyclonal (Kay *et al.,* 1993) antibodies, to identify critical residues in calmodulin-binding sequences (Dedman, 1993), to determine the sequence preferences of the ER chaperonin BiP (Blond-Elguindi *et al.,* 1993), and to characterize the ligand specificities of several SH3 domains (Sparks *et al.,* 1994, 1996).

Because phage-displayed random peptide libraries afford functional access to the peptides and provide a physical link between phenotype (the displayed peptide) and genotype (the encoding DNA), these libraries lend themselves to a screening process

*Phage Display of Peptides and Proteins*
Copyright © 1996 by Academic Press, Inc. All rights of reproduction in any form reserved.

in which binding clones are separated from nonbinding clones by affinity purification. This approach has several advantages over traditional screening methods, including the facility with which highly complex libraries may be screened, the ability to screen using solution-phase binding, and the ability to exert almost unlimited control over binding and elution conditions. Here, we discuss several techniques which take advantage of the power of affinity purification for screening phage-displayed expression libraries.

Fundamentally, screening phage-displayed random peptide libraries for binding peptides involves the following steps (see Fig. 1): obtaining a suitable library and source of target; incubating the library with target and capturing binding phage:target complexes onto solid phase; washing away the nonbinding phage remaining in solution; eluting the binding phage from immobilized target; propagating the eluted phage; and iterating this process until the binding phage are purified to homogeneity. Binding phage are subsequently isolated as individual clones and may be characterized using a variety of methods. This chapter discusses methods relating to each of these processes. Protocols 1–3 describe three different approaches to affinity purification of binding phage and Protocols 4–8 describe techniques used to isolate, propagate, and characterize individual binding clones.

In addition to being used as a vector for accessible expression of peptide libraries, M13 has been used to display mutant libraries of entire proteins, including antibodies (Clackson *et al.*, 1991) and growth hormone (Lowman and Wells, 1993). Whereas this chapter emphasizes the screening and analysis of peptide libraries, these methods are readily adaptable to analysis of libraries of proteins.

## GENERAL CONSIDERATIONS

### Quality of Phage Library

To maximize representation of sequence space, it is important to screen a highly complex library. Assuming a Poisson distribution of sequences within the population, approximately $5 \times 10^9$ unique clones are required to represent all possible hexamer peptide sequences at a 99% confidence level (Clackson and Wells, 1994). Whereas libraries of this complexity are readily attainable, biological selection against sequences incompatible with phage propagation, as well as the nonuniform distribution of amino acids within the genetic code, impose additional constraints upon representation. It is therefore important to screen as complex a library as possible. Alternatively, the diversity of amino acid sequences represented by a given number of clones may be increased by constructing libraries of long peptides, each of which contains several different short peptide sequences (Kay *et al.*, 1993).

To minimize problems arising from instability of phage-displayed peptides, phage should be protected from proteolytic degradation. Displayed peptides are susceptible to degradation during phage propagation in bacterial culture; the degree of susceptibility varies among recombinants. Therefore, phage used in screening experiments should be from a freshly amplified stock.

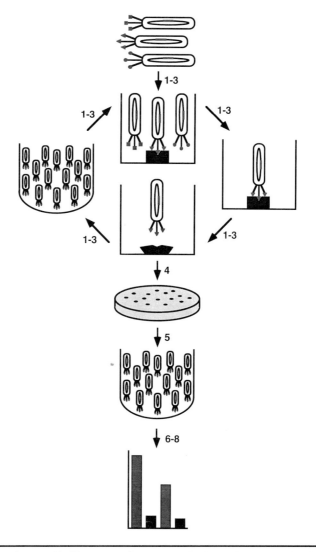

**FIGURE I**   Schematic of random peptide library screening process. Numbers refer to Protocols within which each step is described in detail.

## Quality of Target

The quality of target is important in several respects. The target must retain activity during affinity purification of binding phage. To this end, it is useful to have a functional test for the target of interest, such as ligand binding or an enzyme assay. Furthermore, the target should be relatively pure or consideration must be given to means whereby clones that bind the target may be discriminated from those that interact with contaminants in the binding reaction.

## Affinity Purification of Binding Phage

Affinity selection of binding phage from phage-displayed random peptide libraries has been performed using two fundamentally different approaches. One approach entails incubating library phage with preimmobilized target such that binding phage are temporarily immobilized onto solid-phase while nonbinding phage are washed away. The alternative approach entails incubating library phage with target in solution and subsequently capturing binding phage:target complexes onto solid phase and washing away nonbinding phage. Which approach is more appropriate will depend upon the particular application. For example, because most phage-displayed random peptide libraries express multiple peptides per phage, affinity purification with immobilized target may afford multivalent attachment. The avidity provided by such interactions may be critical for the isolation of peptides with weak, albeit biologically relevant, affinity. Alternatively, solution binding with monovalent target avoids multivalent attachment, thereby enhancing discrimination of high affinity binding clones during affinity purification.

## Recovery of Binding Phage by Elution

Elution of phage from the target:phage complex is an important aspect of screening phage-displayed random peptide libraries. Generally, bound phage are eluted with drastic pH changes (i.e., glycine-HCl, pH 2; ethanolamine, pH 12) which result in target or peptide denaturation without loss of phage infectivity. Phage may also be eluted by treatment with 6 $M$ guanidine-HCl, 2 $M$ urea, or 10 m$M$ DTT, although these reagents must be removed prior to phage amplification. Similarly, direct elution of phage ssDNA by treatment with 1% SDS or phenol requires DNA transformation to regenerate infectious phage particles. It is also possible to use an elution protocol based on the binding characteristics of the target. For example, if the target requires metal ion for binding activity, phage may be eluted with EDTA or EGTA. Other elution reagents include natural or synthetic ligands for competitive elution from receptors, or interacting proteins for competitive displacement by formation of multimeric complexes.

## Amplification of Recovered Phage

One round of affinity purification typically effects the recovery of approximately 1% of PFU representing any given binding clone. To ensure binding clones are not lost to incomplete recovery, we screen 1000 library equivalents ($10^{11}$–$10^{12}$ PFU) in the first round of affinity purification. However, the number of phage particles representing any given binding clone that are recovered from the first round is reduced to the point that binding clones may be lost if subjected to a second round of purification without intervening amplification. Amplification should result in a $10^{5}$- to $10^{7}$-fold increase in the titer of any given clone from the previous round of affinity purification. This amplified titer is generally sufficient to allow two subsequent rounds of affinity purification without intervening amplification.

Phage amplification may be performed on petri plates or in liquid culture. A recent publication suggests that there is little difference in the growth of phage by either method (McConnell *et al.*, 1995). Liquid culture can accommodate a higher amount of input phage ($10^6$ PFU/ml culture) than amplification on plates ($10^4$ PFU/100-mm plate). Amplification on plates, however, has the advantage of minimizing effects of growth or infectivity biases on the representation of different phage clones. Moreover, the relative number of recovered viral particles can be assessed on petri plates.

## Iterative Rounds of Purification

One round of affinity purification with a target immobilized on ELISA plates typically results in an ~$10^3$-fold enrichment of binding over nonbinding phage. Thus, three rounds of affinity purification are generally sufficient for the isolation of binding phage from libraries with complexities of <$10^9$. However, rates of enrichment will vary from target to target and must be taken into account when deciding on the number of rounds of affinity purification to perform. For example, a mere 10-fold enrichment was reported for MHC-binding phage (Hammer *et al.*, 1992).

Enrichment of binding phage may be assessed by doping a library aliquot with an equal number of nonbinding phage which are distinguishable from library phage. For example, libraries encoded by α-complementing phage vectors may be doped with nonbinding, noncomplementing phage. After affinity purification, an increase in the ratio of blue versus white plaques indicates enrichment of phage from the library (see Fig. 2).

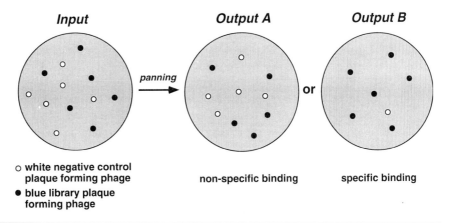

○ white negative control
plaque forming phage

● blue library plaque
forming phage

non-specific binding          specific binding

**FIGURE 2**   Use of an internal control to monitor library screening process. *lac* Z- nonbinding phage are mixed with *lac* Z+ library phage. If binding phage clones are not enriched during affinity purification, the ratio of white to blue plaques will not change appreciably (Output A). However, if library members are enriched during affinity purification, the proportion of blue to white plaques will increase in the output (Output B).

---

**PROTOCOL I**   **Affinity purification with target immobilized on ELISA plates**

### Introduction

Several different methods for target immobilization and subsequent affinity purification of target-binding phage have been reported in the literature. The most common strategy entails adsorption of target protein onto ELISA-treated microtiter plates. However, the target may be immobilized onto any number of solid supports, such as Sepharose (Dedman *et al.*, 1993) or paramagnetic beads (see Protocol 2). Although the ideal target immobilization strategy will vary from application to application, the methods described in Protocols 1–3 are adequate for most purposes.

Some proteins may become denatured and lose binding activity when bound to plastic or other solid supports. These problems may be circumvented by using an intermediate capture protein to immobilize the target. For example, immobilized protein A may be used to capture an antibody via its Fc while still allowing access to its Fab. Similarly, immobilized anti-target antibodies may be used to capture target. The methods described in Protocols 1–3 are readily adapted to this strategy.

Immobilization of target onto wells of ELISA-treated microtiter plates facilitates simultaneous affinity purification of multiple libraries with different targets. Passive adsorption is typically sufficient to immobilize the modest quantities of target required for successful isolation of binding phage. Using this format, a single well is typically required per round of purification per target/library combination.

### Procedure

### Immobilization of target onto ELISA-treated microtiter plates

1. Add 1–10 µg of target protein in 100 µl of 100 m$M$ $NaHCO_3$ (pH 8.5) or PBS to each microtiter well. Seal the wells (e.g., with Scotch tape) to avoid evaporation, and incubate the plate at 25°C for 1–3 hr or at 4°C overnight.
2. Block the wells by adding 150 µl 1.0% BSA or 1.0% lowfat milk in 100 m$M$ $NaHCO_3$ (pH 8.5) or PBS to each well. Seal the wells and incubate the plate at 25°C for 1–3 hr or at 4°C overnight.
3. Discard the solution by inverting the plate and shaking out its contents. Remove residual liquid by slapping the inverted plate on a mat of paper towels several times.
4. Remove unbound protein by washing the wells four times with 200 µl PBS–Tween 20. Discard the solution as described in step 3.

### Affinity purification of binding phage: round I

5. Add an appropriate titer (typically 1000 times the library complexity) of library phage in 150 µl PBS–Tween 20 to each well containing target protein. Seal the wells and incubate the plate at 25°C for 1–3 hr. Remove nonbinding phage by washing the wells as described in step 4.
6. Elute bound phage by adding 150 µl 50 m$M$ glycine-HCl (pH 2.0) to each well and incubating the plate at 25°C for 15 min. Neutralize the solution by transferring the eluted phage to a new well containing 150 µl 200 m$M$ $NaPO_4$ buffer (pH 7.5).

## Amplification of recovered binding phage

7A. Dilute 50 µl of an overnight culture of F′ *Escherichia coli* (e.g., DH5αF′) in 5 ml sterile 2x YT. Add 290 µl Round I output phage and incubate the culture at 37°C, 220 rpm for 6–8 hr. To minimize proteolytic degradation of displayed peptides, do not incubate longer than 8 hr. Pellet cells by centrifugation at 4°C, 4000g (~6000 rpm using a J20 rotor in a Beckman J2-21 centrifuge) for 10 min and transfer the phage supernatant to a new tube.

Alternatively:

7B. Overlay a 2x YT agar plate with 290 µl recovered binding phage + 3 ml lique-fied 0.8% top agar + 200 µl DH5αF′ overnight. Allow the agar to harden and incubate the plate inverted at 37°C for 8 hr. Cover the plate with 5 ml PBS–Tween 20 and rock at 4°C for 2 hr to promote diffusion of the phage particles from the top agar. Decant the phage supernatant into a new tube. Centrifuge at 4°C 4000g (~6000 rpm using a J20 rotor in a Beckman J2-21 cen-trifuge) for 10 min to pellet any cell or agar debris or use 0.45-µm filtration.

8. It is sometimes desirable to concentrate the phage or conduct the binding reaction in a solution other than bacterial media. If this is deemed important, precipitate 2 ml of the phage supernatant by adding 500 µl filter-sterilized 30% PEG-8000/1.6 $M$ NaCl, incubating at 4°C for 1 hr, and centrifuging at >10,000g (~8000 rpm using a J20 rotor in a Beckman J2-21 centrifuge or 14,000 rpm using an Eppendorf microcentrifuge) for 30 min. The phage pellet is visible as a light film along the side of the tube. Resuspend the pellet in 200 µl PBS–Tween 20.

## Affinity purification of binding phage: rounds 2 and 3

9. Immobilize the target protein on microtiter wells as described in steps 1–4. Add 150 µl amplified phage ($10^{11}$–$10^{12}$ PFU) to each well. Seal the wells and in-cubate the plate at 25°C for 1–3 hr. Remove nonbinding phage by washing the wells as described in step 4.

10. Elute bound phage by adding 100 µl 50 mM glycine-HCl (pH 2.0) to each well and incubating the plate at 25°C for 15 min. Transfer the eluted phage to a new well containing 100 µl 200 mM NaPO$_4$ buffer (pH 7.5).

11. Immobilize the target protein on microtiter wells as described in steps 1–4. Add 100 µl PBS–Tween 20 and 190 µl Round 2 output phage to each well. Seal the wells and incubate the plate at 25°C for 1–3 hr. Remove nonbinding phage by washing the wells as described in step 4.

12. Elute bound phage by adding 150 µl 50 mM glycine-HCl (pH 2.0) to each well and incubating the plate at 25°C for 15 min. Transfer the eluted phage to a microcentrifuge tube containing 150 µl 200 mM NaPO$_4$ buffer (pH 7.5). This represents the output phage population from the affinity purification. Any-where from 0 to $10^5$ PFU may be present in the output. Refer to Protocols 4–8 for instructions regarding isolation, propagation, and analysis of individ-ual phage clones.

## Solutions

1. 100 mM NaHCO$_3$ (pH 8.5)

   1 M stock solution:
     83 g NaHCO$_3$
     1 liter ddH$_2$O

   Autoclave
   Dilute 1:10 with sterile ddH$_2$O.

2. 1.0% BSA or milk

   10x stock solution:
     10 g BSA Fraction V or Carnation Instant Nonfat Dried Milk
     100 ml sterile ddH$_2$O

   Store at 4°C short-term or at -20°C long-term.

3. 1X PBS: 137 mM NaCl, 3 mM KCl, 8 mM Na$_2$HPO$_4$, 1.5 mM KH$_2$PO$_4$

   10x stock solution:
     80.0 g NaCl
     2.0 g KCl
     11.5 g Na$_2$HPO$_4$ •7H$_2$O
     2.0 g KH$_2$PO$_4$
     1 liter ddH$_2$O

   Confirm pH 7.5
   Autoclave
   Dilute 1:10 with sterile ddH$_2$O.

4. PBS–Tween 20

   1 liter 1X PBS
   1 ml Tween 20

5. 50 mM Glycine-HCl (pH 2.0)

   1 M stock solution:
     111.6 g Glycine
     1 liter ddH$_2$O

   Adjust pH to 2.0 with HCl
   Autoclave
   Dilute to 50 mM with sterile ddH$_2$O.

6. 200 mM Na PO$_4$ buffer (pH 7.5)

   1 M stock solution:
     44.16 g NaH$_2$PO$_4$•H$_2$O
     450.66 g Na$_2$HPO$_4$•7H$_2$0
     1 liter ddH$_2$O

Autoclave

Dilute to 200 mM with sterile ddH$_2$O.

7. 2x YT media, 2x YT top agar, and 2x YT bottom agar

10 g tryptone
10 g yeast extract
5 g NaCl
1 liter ddH$_2$O

+15 g Bacto-agar for 2x YT bottom agar
+8 g Bacto-agar for 2x YT top agar
Autoclave.

8. 30% PEG-8000 / 1.6 M NaCl

300.0 g polyethylene glycol 8000
92.8 g NaCl
ddH$_2$O to 1 liter
Filter-sterilize using 0.22 μm filter.

---

## PROTOCOL 2   Affinity purification with target immobilized on paramagnetic beads

### Introduction

Target immobilization onto paramagnetic beads offers the potential advantage of increased surface area (and therefore target concentration) per unit volume relative to target immobilization onto microtiter plates. However, a larger amount of protein is typically required for immobilization.

### Procedure

### Immobilization of target onto BioMag 8-4100 paramagnetic beads

1. Add the desired amount of BioMag beads (10 mg beads per 1 mg target protein) to a 100-ml tissue culture flask.
2. Wash the beads by adding 50 ml PBS–Tween 20 and gently swirling (50 rpm) at 25°C for 10 min. Use a strong magnetic source to draw the beads to one side of the flask. Remove the liquid by aspiration. Wash the beads with PBS–Tween 20 three additional times.
3. Activate the beads by adding 20 ml 5% glutaraldehyde and swirling (50 rpm) at 25°C for 3 hr. Wash the beads with PBS–Tween 20 three times as described in step 2.
4. Couple the target protein to the beads by adding the appropriate amount of protein in 50 ml coupling buffer. If the target is not limiting, add 25–75 mg to 50 ml of beads; if it is limiting, add 1 mg target + 2 mg BSA to 50 ml beads. Incubate overnight at 250 rpm, 25°C. Remove the liquid by aspiration.

5. Block the beads by adding 50 ml 50 mM glycine (pH 7.5) and agitating at 25°C for 10 min. Wash the beads with PBS–Tween 20 three times as described in step 2. Store the beads at 4°C in PBS–Tween 20 + 0.1% sodium azide.

### Affinity purification of binding phage: round 1

6. Aliquot 1 mg target-coupled beads in 500 μl PBS–Tween 20 per microcentrifuge tube. Add the appropriate titer of library phage in 750 μl PBS–Tween 20 to each tube. Tumble the tubes (50 rpm) at 4°C for 1–3 hr. Remove nonbinding phage by washing the beads four times with 1 ml PBS–Tween 20 as described in step 2.

7. Elute binding phage by resuspending the beads in 250 μl 50 mM glycine-HCl (pH 2.0). Tumble the tubes at 25°C for 10 min. Pellet the beads by microcentrifugation at 12,000 rpm, 25°C, for 5 min. Neutralize the eluted phage solution by transferring the solution to a new tube containing 250 μl 200 mM NaPO$_4$ buffer (pH 7.5).

### Amplification of recovered binding phage

8A. Dilute 50 μl of an overnight culture of DH5αF$'$ in 5 ml sterile 2x YT. Add 490 μl Round 1 output phage and incubate the culture at 37°C, 220 rpm, for 6–8 hr. To minimize proteolytic degradation of phage-displayed random peptides, do not incubate longer than 8 hr. Pellet cells by centrifugation at 4000$g$ (~6,000 rpm using a J20 rotor in a Beckman J2-21 centrifuge) for 10 min and transfer the phage supernatant to a new tube.

Alternatively:

8B. Overlay a 2x YT agar plate with 490 μl recovered binding phage + 3 ml liquefied 0.8% top agar + 200 μl DH5αF$'$ overnight. Allow the agar to harden and incubate the plate inverted at 37°C for 8 hr. Cover the plate with 5 ml PBS–Tween 20 and rock at 4°C for 2 hr to promote diffusion of the phage particles from the top agar. Decant the phage supernatant into a new tube. Centrifuge at 4000$g$ (~6000 rpm using a J20 rotor in a Beckman J2-21 centrifuge) for 10 min to pellet any cell or agar debris.

9. It is sometimes desirable to concentrate the phage or conduct the binding reaction in a solution other than bacterial media. If this is deemed important, precipitate 2 ml of the phage supernatant by adding 500 ml filter-sterilized 30% PEG-8000/1.6 $M$ NaCl, incubating at 4°C for 1 hr, and centrifuging at >10000$g$ (~8000 rpm using a J20 rotor in a Beckman J2-21 centrifuge or 14,000 rpm using an Eppendorf microcentrifuge) for 30 min. The phage pellet is visible as a light film along the side of the tube. Resuspend the pellet in 200 μl PBS–Tween 20.

### Affinity purification of binding phage: rounds 2 and 3

10. Aliquot 1 mg target-coupled beads in 500 μl PBS–Tween 20 per microcentrifuge tube. Add 750 μl amplified phage ($10^{11}$–$10^{12}$ PFU) to each tube. Tumble the tubes at 4°C for 1–3 hr. Remove nonbinding phage by washing the beads four times with 1 ml PBS–Tween 20 as described in step 2.

11. Elute binding phage by resuspending the beads in 250 µl 50 mM glycine-HCl (pH 2.0). Tumble the tubes at 25°C, for 10 min. Pellet the beads by microcentrifugation at 12,000 rpm, 25°C, for 5 min. Neutralize the eluted phage solution by transferring the solution to a new tube containing 250 µl 200 mM $NaPO_4$ buffer (pH 7.5).

12. Aliquot 1 mg target-coupled beads in 500 µl PBS–Tween 20 per microcentrifuge tube. Add 500 µl Round 2 output phage to each tube. Tumble the tubes at 4°C for 1–3 hr. Remove nonbinding phage by washing the beads four times with 1 ml PBS–Tween 20 as described in step 2.

13. Elute binding phage by resuspending the beads in 250 µl 50 mM glycine-HCl (pH 2.0). Tumble the tubes at 25°C for 10 min. Pellet the beads by microcentrifugation at 12,000 rpm, 25°C for 5 min. Neutralize the eluted phage solution by transferring the solution to a new tube containing 250 µl 200 mM $NaPO_4$ buffer (pH 7.5). This represents the output phage population from the affinity purification. Anywhere from 0 to $10^5$ PFU may be present in the output. Refer to Protocols 4–8 for instructions regarding isolation, propagation, and analysis of individual phage clones.

## Solutions

1. 5% Glutaraldehyde

2. 50 mM Glycine (pH 7.5)

   5.58 g glycine
     1 liter ddH2O

   Adjust pH with NaOH
   Autoclave.

3. 1X PBS: 137 mM NaCl, 3 mM KCl, 8 mM $Na_2HPO_4$, 1.5 mM $KH_2PO_4$

   10x stock solution:
     80.0 g NaCl
     2.0 g KCl
     11.5 g $Na_2HPO_4$ •$7H_2O$
     2.0 g $KH_2PO_4$
     1 liter dd$H_2O$

   Confirm pH 7.5
   Autoclave
   Dilute 1:10 with sterile dd$H_2O$.

4. PBS–Tween 20

   1 liter 1X PBS
   1 ml Tween 20

5. 50 mM Glycine-HCl (pH 2.0)

1 *M* stock solution:
111.6 g Glycine
1 liter ddH$_2$O

Adjust pH to 2.0 with HCl
Autoclave
Dilute to 50 m*M* with sterile ddH$_2$O.

6. 200 m*M* Na PO$_4$ buffer (pH 7.5)

1 *M* stock solution:
44.16 g NaH$_2$PO$_4$·H$_2$O
450.66 g Na$_2$HPO$_4$·7H$_2$0
1 liter ddH$_2$O

Autoclave
Dilute to 200 m*M* with sterile ddH$_2$O.

7. 2x YT media, 2x YT top agar, and 2x YT bottom agar

10 g tryptone
10 g yeast extract
5 g NaCl
1 liter ddH$_2$O

+15 g Bacto-agar for 2x YT bottom agar
+8 g Bacto-agar for 2x YT top agar
Autoclave.

8. 30% PEG-8000 / 1.6 *M* NaCl

300.0 g polyethylene glycol 8000
92.8 g NaCl
ddH$_2$O to 1 liter

Filter sterilize using 0.22-µm filter.

---

**PROTOCOL 3    Affinity purification by capture of biotinylated target:phage complexes**

### Introduction

Several methods have been devised for capture of target:phage complexes during affinity purification of binding phage. In one of the first implementations of phage-displayed random peptide libraries, Scott and Smith (1990) screened a hexamer library with biotinylated protein in solution and captured phage:target complexes with immobilized streptavidin. Alternatively, Sepharose-conjugated antibodies may be used to immunoprecipitate antigen:phage complexes, even when the antigen is not known. As mentioned above, when using a capture protein to recover target:phage complexes, it is critical to minimize recovery of phage which bind the capture protein. For example, phage expressing the peptide sequence HPQ have been shown to bind to streptavidin. These phage can be avoided by blocking unbound streptavidin with excess biotin after phage:target capture.

## Procedure

### Biotinylation of target protein

1. Ensure the target protein is in a Tris-free buffer, as Tris amines will quench the biotinylation reagent. Add the appropriate amount of target to 1 ml NLS-biotinylation reagent in 200 m$M$ Borate buffer. Incubate at 25°C for 4 hr.

2. Quench the biotinylation reaction by addition of 100 µl 1 $M$ Tris-HCl (pH 8.0). Dialyze the biotinylated protein against 1000X volume of PBS at 4°C overnight. Successful biotinylation may be assessed by Western blotting with streptavidin or target immobilization followed by detection with streptavidin-alkaline phosphatase.

### Immobilization of streptavidin onto ELISA-treated microtiter plates

3. Add 10–25 µg streptavidin in 100 µl of 100 m$M$ NaHCO$_3$ (pH 8.5) or PBS to each microtiter well. Seal the wells to avoid evaporation, and incubate the plate at 25°C for 1–3 hr or at 4°C overnight.

4. Block the wells by adding 150 µl 1.0% BSA or 1.0% milk in 100 m$M$ NaHCO$_3$ (pH 8.5) or PBS to each well. Seal the wells and incubate the plate at 25°C for 1–3 hr or at 4°C overnight.

5. Discard the solution by inverting the plate and shaking out its contents. Remove residual liquid by slapping the inverted plate on a mat of paper towels several times.

6. Remove unbound protein by washing the wells four times with 200 µl PBS–Tween 20. Discard the solution as described in step 5.

### Affinity purification of binding phage: round 1

7. Mix the appropriate titer of library phage with ~1 µg biotinylated target in a total of 250 µl PBS–Tween 20. Incubate the tube at 25°C for 1–3 hr. Capture target:phage complexes by transferring the solution to a streptavidin-coated microtiter well. Incubate at 25°C for 1–3 hr.

8. Block unbound streptavidin by addition of 25 µl 10 m$M$ biotin, pH 8.0. Incubate at 25°C for 15 min. Remove nonbinding phage by washing the wells 4 times with PBS–Tween 20 as described above.

9. Elute bound phage by adding 150 µl 50 m$M$ glycine-HCl (pH 2.0) to each well and incubating the plate at 25°C for 15 min. Neutralize the solution by transferring the eluted phage to a new well containing 150 µl 200 m$M$ NaPO$_4$ buffer (pH 7.5).

## Amplification of recovered binding phage

10A. Dilute 50 μl of an overnight culture of DH5αF′ in 5 ml sterile 2×YT. Add 290 μl Round 1 output phage and incubate the culture at 37°C, 220 rpm, for 6–8 hr. To minimize proteolytic degradation of displayed peptides, do not incubate longer than 8 hr. Pellet cells by centrifugation at 4°C, 4000g (~6000 rpm using a J20 rotor in a Beckman J2-21 centrifuge) for 10 min and transfer the phage supernatant to a new tube.

Alternatively:

10B. Overlay a 2×YT agar plate with 290 μl recovered binding phage + 3 ml liquefied 0.8% top agar + 200 μl DH5αF′ overnight. Allow the agar to harden and incubate the plate inverted at 37°C for 8 hr. Cover the plate with 5 ml PBS–Tween 20 and rock at 4°C for 2 hr to promote diffusion of the phage particles from the top agar. Decant the phage supernatant into a new tube. Centrifuge at 4°C, 4000g (~6000 rpm using a J20 rotor in a Beckman J2-21 centrifuge) for 10 min to pellet any cell or agar debris.

11. It is sometimes desirable to concentrate the phage or conduct the binding reaction in a solution other than bacterial media. If this is deemed important, precipitate 2 ml of the phage supernatant by adding 500 μl filter-sterilized 30% PEG-8000/1.6 M NaCl, incubating at 4°C for 1 hr, and centrifuging at >10,000g (~8000 rpm using a J20 rotor in a Beckman J2-21 centrifuge or 14,000 rpm using an Eppendorf microcentrifuge) for 30 min. The phage pellet is visible as a light film along the side of the tube. Resuspend the pellet in 200 μl PBS–Tween 20.

## Affinity purification of binding phage: rounds 2 and 3

12. Mix 200 μl amplified phage with ~1 μg biotinylated target in a total of 250 μl PBS–Tween 20. Incubate the tube at 25°C for 1–3 hr. Capture target:phage complexes by transferring the solution to a streptavidin-coated microtiter well. Incubate at 25°C for 1–3 hr. Remove nonbinding phage by washing the wells four times with PBS–Tween 20 as described above.

13. Elute bound phage by adding 100 μl 50 mM glycine-HCl (pH 2.0) to each well and incubating the plate at 25°C for 15 min. Neutralize the solution by transferring the eluted phage to a new well containing 100 μl 200 mM NaPO$_4$ buffer (pH 7.5).

14. Mix 190 μl Round 2 output phage with ~1 μg biotinylated target in a total of 250 μl PBS–Tween 20. Incubate the tube at 25°C for 1–3 hr. Capture target:phage complexes by transferring the solution to a streptavidin-coated microtiter wells. Incubate at 25°C for 1–3 hr.

15. Block unbound streptavidin by addition of 25 μl 10 mM biotin, pH 8.0. Incubate at 25°C for 15 min. Remove nonbinding phage by washing the wells four times with PBS–Tween 20 as described above.

16. Elute bound phage by adding 150 μl 50 mM glycine-HCl (pH 2.0) to each well and incubating the plate at 25°C for 15 min. Neutralize the solution by trans-

ferring the eluted phage to a new well containing 150 μl 200 mM $NaPO_4$ buffer (pH 7.5). This represents the output phage population from the affinity purification. Anywhere from 0 to $10^5$ PFU may be present in the output. Refer to Protocols 4–8 for instructions regarding isolation, propagation, and analysis of individual phage clones.

## Solutions

1. 200 mM Borate buffer

   1.24 g boric acid
   100 ml sterile $ddH_2O$

   Adjust pH to 10 with 10 M NaOH.

2. 5 mM Biotin-LC-Hydrazide solution (Pierce Chemical Company)

   0.019 g Biotin-LC-Hydrazide
   10 ml labeling solution

   Store at 4°C and use within 2 months.

3. 1M Tris-HCl

   157.6 g Tris-HCl
   1 liter $ddH_2O$

   Autoclave.

4. 100 mM $NaHCO_3$ (pH 8.5)

   1 M stock solution:
     83 g $NaHCO_3$
     1 liter $ddH_2O$

   Autoclave
   Dilute 1:10 with sterile $ddH_2O$.

5. 1.0% BSA or milk

   10% stock solution:

   10 g BSA Fraction V or Carnation Instant Nonfat Dried Milk
   100 ml sterile $ddH_2O$

6. 1X PBS: 137 mM NaCl, 3 mM KCl, 8 mM $Na_2HPO_4$, 1.5 mM $KH_2PO_4$

   10x stock solution:
     80.0 g NaCl
     2.0 g KCl
     11.5 g $Na_2HPO_4$ •$7H_2O$
     2.0 g $KH_2PO_4$
     1 liter $ddH_2O$

Confirm pH 7.5
Autoclave
Dilute 1:10 with sterile ddH$_2$O.

7. PBS–Tween 20

1 liter 1X PBS
1 ml Tween 20

8. 10 m$M$ Biotin

2.04 mg d-Biotin
1 m liter sterile ddH$_2$O

Store -20°C.

9. 50 m$M$ Glycine-HCl (pH 2.0)

1 $M$ stock solution:
111.6 g Glycine
1 liter ddH$_2$O

Adjust pH to 2.0 with HCl
Autoclave
Dilute to 50 m$M$ with sterile ddH$_2$O.

10. 200 m$M$ Na PO$_4$ buffer (pH 7.5)

1 $M$ stock solution:
44.16 g NaH$_2$PO$_4$·H$_2$O
450.66 g Na$_2$HPO$_4$·7H$_2$0
1 liter ddH$_2$O

Autoclave
Dilute to 200 m$M$ with sterile ddH$_2$O.

11. 2x YT media, 2x YT top agar, and 2x YT bottom agar

10 g tryptone
10 g yeast extract
5 g NaCl
1 liter ddH$_2$O

+15 g Bacto-agar for 2xY T bottom agar
+8 g Bacto-agar for 2x YT top agar
Autoclave.

12. 30% PEG-8000/1.6 $M$ NaCl

300.0 g polyethylene glycol 8000
92.8 g NaCl
ddH$_2$O to 1 liter
Filter-sterilize using 0.22-μm filter.

**PROTOCOL 4**   **Isolation of affinity-purified phage clones**

### Introduction

In this protocol, affinity-purified phage stocks are plated to determine titers and yield isolated plaques from which clonal cultures are produced for further analysis. We use methods designed for rapid dilution of multiple phage samples; 12 different samples may be accommodated per microtiter plate.

### Procedure

1. Using a 12-channel pipetter, fill the wells of a non-ELISA 96-well microtiter plate with 180 µl PBS. Add 20 µl of each phage stock to be titered in row A. Mix by pipetting 180 µl up and down ~5 times.

2. Using a 12-channel pipetter set to draw 180 µl and loaded with tips marked to indicate 20 µl volume, transfer 20 µl from row A to row B. Mix by pipetting 180 µl up and down ~5 times. Wick any residual liquid from the tips with a paper towel.

3. Repeat to continue the dilution series through the remaining rows. In this way, one may perform 12 separate 10-fold dilution series over 8 orders of magnitude in 10 min using only 24 pipette tips.

4. Overlay a 2x YT agar plate with 3 ml liquefied 1.2% top agar + 200 µl DH5αF' overnight (+ 30 µl 2% IPTG + 30 µl 2% X-Gal for detection of β-galactosidase activity, if desired). Harden the agar by incubating the plate at 4°C for 15 min. The plates can be used immediately or stored at 4°C for 1–2 weeks prior to use.

5. Flame-sterilize a 48-pronged replica-plater, allow it to cool, and place it in the dilution series microtiter plate. Carefully rest the replica-plater prongs onto the plate prepared above; this procedure will transfer ~2 µl per prong onto the plate surface.

6. Incubate the inverted plate at 37°C. Plaques should be visible after 8 hr (see Fig. 3) and may be used to estimate the titer of the original phage stock. Isolated plaques may be used to generate clonal phage stocks.

7. If isolated plaques are not obtained, use an appropriate well from the serial dilution plate to serve as a source of phage in the following procedure. Overlay a 2x YT agar plate with 30–300 PFU diluted phage + 3 ml liquefied 0.8% top agar + 200 µl DH5αF' overnight (+ 30 µl 2% IPTG + 30 µl 2% X-Gal for detection of β-galactosidase activity, if desired). Incubate the inverted plate at 37°C; isolated plaques should be visible after 8 hr and may be used to generate clonal phage stocks.

### Solutions

1. 1X PBS: 137 mM NaCl, 3 mM KCl, 8 mM $Na_2HPO_4$ 1.5 mM $KH_2PO_4$

   10x stock solution:
   80.0 g NaCl
   2.0 g KCl
   11.5 g $Na_2HPO_4$ •$7H_2O$

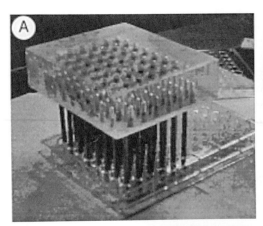

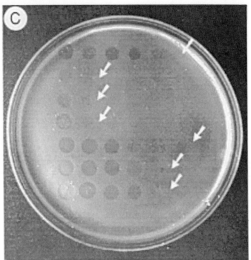

**FIGURE 3**  Isolation of individual binding clones using microtiter plate dilutions and replica-plater. A 48-prong replica-plater is shown transfering phage particles from 48 wells of a microtiter plate (A) onto a growing bacterial lawn (B). (C) The result of transferring liquid from 48 separate microtiter wells onto the lawn. Arrows denote individual plaques. More details are given in Chapter 1.

> 2.0 g KH₂PO₄
> I liter ddH₂O
> Confirm pH 7.5
> Autoclave
> Dilute 1:10 with sterile ddH₂O.

2. 2x YT media, 2x YT top agar, and 2x YT bottom agar

10 g tryptone
10 g yeast extract
5 g NaCl
1 liter ddH$_2$O

+15 g Bacto-agar for 2x YT bottom agar
+ 8 g Bacto-agar for 2x YT top agar
+12 g Bacto-agar for 2x YT pronging agar
Autoclave.

3. 2% IPTG

0.2 g isopropyl-β-*D*-thiogalactopyranoside
10.0 ml sterile ddH$_2$O

Filter-sterilize
Store at -20°C.

4. 2% X-Gal

0.2 g 5-bromo-4-chloro-3-indoyl-β-*D*-galactoside
10.0 ml DMSO or DMF

Store at -20°C; limit exposure to light.

---

**PROTOCOL 5**    **Propagation of individual phage clones**

**Procedure**

1. Dilute 200 μl of a DH5αF′ overnight 1:100 in 20 ml sterile 2x YT. Using a 12-channel pipetter, aliquot 200 μl into each well of a sterile round-bottom 96-well microtiter plate.
2. Using sterile toothpicks, carefully touch isolated plaques into individual wells of the microtiter plate, and incubate the plate at 37°C, 220 rpm for 6–8 hr. A covered sequencing tip box provides a convenient reservoir for the plate. To minimize evaporation, place a damp paper towel at the bottom of the box.
3. Pellet the cells by centrifugation with a microtiter plate rotor at 4000*g* for 10 min. The supernatant may serve as a source of phage (~10$^{12}$ PFU/ml) for binding experiments or for single-stranded DNA sequencing. The cell pellets may be frozen in sterile 20% glycerol as a source of infected bacteria for future phage propagation.
4. For production of larger amounts of phage, pick an isolated plaque, inoculate 3 ml of a 1:100 diluted DH5αF′ overnight culture, and incubate the culture at 37°C, 220 rpm, for 6–8 hr.
5. Pellet the cells by centrifugation at 4000*g* (6000 rpm using a J20 rotor in a Beckman J2-21 centrifuge) for 10 min. The cell pellet may serve as a source of double-stranded DNA for sequencing, while the supernatant may serve as a source of phage (~10$^{12}$ PFU/ml) for binding experiments.

## SOLUTIONS

1. 2x YT media

   10 g tryptone
   10 g yeast extract
   5 g NaCl
   1 liter ddH$_2$O

   Autoclave.

---

**PROTOCOL 6**   **Rapid confirmation of binding activity of affinity-purified phage clones by ELISA**

### Introduction

   Although affinity purification serves to enrich for binding clones, not all selected phage represent binding clones. For example, we have isolated plastic-binding phage from a variety of screening experiments. An anti-phage ELISA detection system facilitates the simultaneous characterization of multiple phage clones. Using this protocol, 48 clones may be screened in a single microtiter plate in as little as 4 hr (see Fig. 4).

### Procedure

1. For each clone to be tested, immobilize ~1 µg target protein in an ELISA microtiter well (see Protocol 1). As a negative control, block a separate well with 150 µl 1.0% BSA or 1.0 % lowfat milk in 100 m$M$ NaHCO$_3$ (pH 8.5) or PBS.
2. Add 100 µl PBS–Tween 20 to each well containing immobilized protein. Add 50 µl phage stock representing each clone to a separate pair (positive/negative) of wells. Seal the wells and incubate the plate at 25°C for 1–3 hr. Remove nonbinding phage by washing the wells four times with PBS–Tween 20 (see Protocol 1).
3. Dilute horseradish peroxidase-conjugated anti-phage antibody (Pharmacia, Catalog No. 27-9402-01) 1:5000 in PBS–Tween 20. Add 100 µl of the diluted conjugate to each well. Seal the wells and incubate the plate at 25°C for 1 hr. Wash the wells as described above.
4. Add 100 µl ABTS reagent containing 0.05% H$_2$O$_2$ to each well. Incubate the plate at 25°C until the color reaction develops (10–30 min). Quantify the reaction by measuring the absorbance at 405 nm with a microtiter plate reader. Positive interactions produce absorbencies in the range of 0.5 to 3.0, while negative signals typically range between 0.05 and 0.3.

### Solutions

1. 100 m$M$ NaHCO$_3$ (pH 8.5)

   1 $M$ stock solution:
      83 g NaHCO$_3$
      1 liter ddH$_2$O

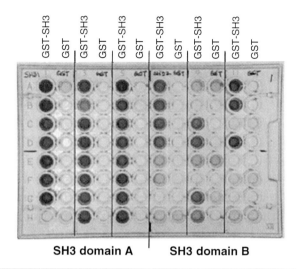

SH3 domain A          SH3 domain B

**FIGURE 4**      Confirmation of binding activity of phage clones using anti-phage ELISA. 48 GST–SH3 affinity-purified
phage clones were screened by anti-phage ELISA for binding to 1 µg GST–SH3 or GST alone. Most
clones bind GST–SH3 but not GST. An example of a clone with weak binding activity against the GST
moiety in both target constructs is found in wells E9 and E10.

Autoclave
Dilute 1:10 with sterile ddH$_2$O.

2. 1.0% BSA or milk

10% stock solution:
10 g BSA fraction V or Carnation Instant Nonfat Dried Milk
100 ml sterile ddH$_2$O

3. 1X PBS: 137 m$M$ NaCl, 3 m$M$ KCl, 8 m$M$ Na$_2$HPO$_4$, 1.5 m$M$ KH$_2$PO$_4$

10x stock solution:
80.0 g NaCl
2.0 g KCl
11.5 g Na$_2$HPO$_4$ •7H$_2$O
2.0 g KH$_2$PO$_4$
1 liter ddH$_2$O

Confirm pH 7.5
Autoclave
Dilute 1:10 with sterile ddH$_2$O.

4. PBS–Tween 20

1 liter 1X PBS
1 ml Tween 20

5. ABTS (2′,2′-azino-bis 3-ethylbenzthiazoline-6-sulfonic acid) solution

50 mM citric acid:
  10.5 g citrate monhydrate
  1 liter sterile ddH$_2$O

Adjust pH to 4 with approximately 6 ml of 10 M NaOH
Add 220 mg ABTS.
Filter-sterilize, store at 4°C.
Immediately before use, add 30% H$_2$O$_2$ to 0.05%.

---

**PROTOCOL 7**    **Quantitative characterization of binding activity of phage clones by titering**

### Introduction

Phage clones displaying specific peptides represent convenient reagents for comparing the binding characteristics of different peptide sequences. Whereas the ELISA assay described above allows for qualitative comparison of different phage clones, we use a PFU-based assay for quantitative analysis of phage binding (see Fig. 5). This assay measures the percentage of input PFU retained by immobilized target protein and has a signal to noise ratio of $10^3$–$10^4$. It should be emphasized that this assay measures phage (rather than peptide) binding. As each phage displays multiple peptides, high target protein concentrations may produce binding results which reflect phage avidity rather than peptide affinity. We have found that limiting target protein concentrations yield the greatest linear response.

### Procedure

1. For each clone/target protein combination to be tested, immobilize ~1 µg target protein in an ELISA microtiter well (see Protocol 1). If multiple clones and/or target proteins are to be tested, care must be taken to ensure that equimolar amounts of target protein are immobilized in each well.
2. Using a 12-channel pipetter, add 100 µl PBS–Tween 20 to each well in the microtiter plate. Add an equal titer ($10^9$–$10^{11}$ PFU) of each phage stock to the appropriate set of microtiter wells. Seal the wells and incubate the plate at 25°C for 1–3 hr. Remove nonbinding phage by washing the wells (see Protocol 1).
3. Elute bound phage by adding 150 µl 200 mM glycine-HCl (pH 2.0) to each well and incubating the plate at 25°C for 15 min. Transfer the eluted phage to a new well containing 150 µl 1 M NaPO$_4$ buffer (pH 7.5). This represents the output from the binding reaction.
4. Using a 12-channel pipetter and a non-ELISA microtiter plate, perform serial dilutions upon the binding reaction output (see Protocol 4). Use the replica-plater method to determine output titer. Calculate the percentage of input PFU recovered from the binding reaction.

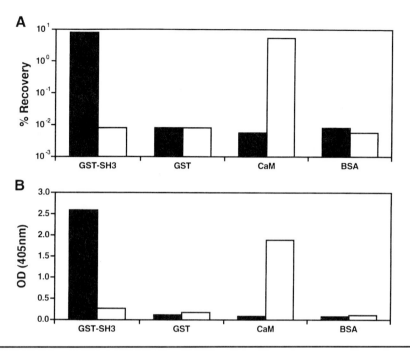

**FIGURE 5**     Comparison of results obtained from ELISA and PFU assays.  Phage representing a Src SH3-binding clone (solid) and a CaM-binding clone (open) were assayed for binding to 1μg immobilized GST-Src SH3, CaM, GST, or BSA. (A) Results from PFU assay. Values are % recovery of input PFU. (B) Results from anti-phage ELISA assay. Each point was performed in triplicate; values are average absorbence at 405 nm wavelength.

## Solutions

1. 100 m$M$ NaHCO$_3$ (pH 8.5)

   1 $M$ stock solution:
      83 g NaHCO$_3$
      1 liter ddH$_2$O

   Autoclave
   Dilute 1:10 with sterile ddH$_2$O.

2. 1.0% BSA or milk

   10% stock solution:
      10 g BSA Fraction V or Carnation Instant Nonfat Dried Milk
      100 ml sterile ddH$_2$O

   Store at 4°C short-term or at -20°C long-term.

3. 1X PBS: 137 m$M$ NaCl, 3 m$M$ KCl, 8 m$M$ Na$_2$HPO$_4$, 1.5 m$M$ KH$_2$PO$_4$

   10x stock solution:
      80.0 g NaCl

2.0 g KCl
11.5 g $Na_2HPO_4 \cdot 7H_2O$
2.0 g $KH_2PO_4$
1 liter $ddH_2O$

Confirm pH 7.5
Autoclave
Dilute 1:10 with sterile $ddH_2O$.

4. PBS–Tween 20

1 liter 1X PBS
1 ml Tween 20

5. 50 m$M$ Glycine-HCl (pH 2.0)

1 $M$ stock solution:
111.6 g Glycine
1 liter $ddH_2O$

Adjust pH to 2.0 with HCl
Autoclave
Dilute to 50 m$M$ with sterile $ddH_2O$.

6. 200 m$M$ Na $PO_4$ buffer (pH 7.5)

1 $M$ stock solution:
44.16 g $NaH_2PO_4 \cdot H_2O$
450.66 g $Na_2HPO_4 \cdot 7H_2O$
1 liter $ddH_2O$

Autoclave
Dilute to 200 m$M$ with sterile $ddH_2O$.

7. 2x YT media, 2x YT top agar, and 2x YT bottom agar

10 g tryptone
10 g yeast extract
5 g NaCl
1 liter $ddH_2O$

+15 g Bacto-agar for 2x YT bottom agar
+8 g Bacto-agar for 2x YT top agar
+12 g Bacto-agar for 2x YT pronging agar
Autoclave.

---

**PROTOCOL 8    Identification of binding clones by filter lifts**

### Introduction

The characterization of individual clones by titering and phage ELISA is useful for surveying a modest number of isolates from an enriched pool of phage. However, a filter lift may be necessary in cases where the ratio of enrichment is low and a greater number of isolates must be screened to identify specific binding clones. Filter lifts can also be used to identify rare high-affinity binding phage present in an excess of low-affinity clones, which have presumably been enriched on the basis of avidity rather than affinity. Although signal intensity is related to affinity, one should be aware that the signal strength may be influenced by other factors, including varied expression levels and protease susceptibility among different peptide sequences.

This protocol describes the screening of phage extruded from bacterial colonies by absorption onto nitrocellulose followed by probing with the target of interest (Blond-Elguindi et al., 1993). Screening filter lifts of phage plaques is identical except that more time is required to transfer a sufficient amount of phage to the nitrocellulose filter.

## Procedure

1. Dilute phage stock and infect DH5αF' bacterial cells. Plate on the appropriate antibiotic-containing medium and incubate at 37°C overnight. Typically 200 colonies per 100-mm plate (low density) or 5000 colonies per 150-mm plate (high density) are screened.

2. Place a nitrocellulose filter onto each plate and mark their orientation with a syringe needle dipped into India ink. After approximately 1 min, remove the filters and rinse with PBS–Tween 20 to remove any remaining adherent cells.

3. Block the filter in PBS–Tween 20 + 1.0% BSA at 4°C for 1 hr. Return the plate to 37°C for 4–8 hr to allow colonies to regenerate. For plaque lifts, place the filter on the plate for 4–8 hr during plaque development at 37°C and then treat as above.

4. Wash the blocked filter three times for 1 min each with PBS–Tween 20. Dilute the target protein in PBS–Tween 20 + 0.1% BSA and incubate the filters with target at 4°C from 1 hr to overnight on a rotating platform. Typical target protein concentrations are 0.5–5.0 µg/ml, or approximately $10^6$ cpm/ml for radiolabeled proteins.

5. Wash the filters as described above. Detect bound target protein by incubating the filter with an anti-target antibody (diluted to 1 mg/ml in PBS–Tween 20 + 0.1% BSA) at 4°C for 1 hr. Alternatively, use alkaline phosphatase-conjugated streptavidin to detect biotinylated target proteins. When using radiolabeled target protein, wash the filters and blot them dry prior to autoradiography.

6. Wash the filters as described above. Detect bound secondary reagent by incubating the filters with an alkaline phosphatase-conjugated secondary antibody (diluted 1:2000 in PBS–Tween 20 + 0.1.% BSA) at 4°C for 1 hr.

7. Wash the filters as described above. Add 3 ml substrate solution per filter. Incubate at 25°C until color develops and stop the reaction by rinsing the filters in ddH₂O.

8. Align the developed filters or film with the corresponding petri plates and pick colonies (or plaques) corresponding to positive spots. Replate and screen again at low density to isolate pure clones.

## Solutions

1. 1X PBS: 137 m$M$ NaCl, 3 m$M$ KCl, 8 m$M$ $Na_2HPO_4$, 1.5 m$M$ $KH_2PO_4$

10x stock solution:
  80.0 g NaCl
  2.0 g KCl
  11.5 g $Na_2HPO_4 \cdot 7H_2O$
  2.0 g $KH_2PO_4$
  1 liter $ddH_2O$

Confirm pH 7.5
Autoclave
Dilute 1:10 with sterile $ddH_2O$.

2. PBS–Tween 20

  1 liter 1X PBS
  0.5 ml Tween 20

3. PBS–Tween 20 + 0.1 or 1.0% BSA

  10% BSA stock solution:
    10 g BSA Fraction V
    100 ml PBS–Tween 20

Store at 4°C short-term or at -20°C long-term.
Dilute in PBS–Tween 20 to appropriate concentration.

4. Alkaline phosphatase substrate solution

  0.1 $M$ Tris·HCl, pH 9.5
  0.1 $M$ NaCl
  50 m$M$ $MgCl_2$

Immediately prior to use, add (per 10 ml)
  44 µl of 75 mg nitroblue tetrazolium chloride/ ml DMF (70%)
  33 µl of 50 mg 5-bromo-4-chloro-3-indoyl-phosphate-p-toluidine/ ml DMF.

## References

Blond-Elguindi, S., Cwirla, S., Dower, W., Lipsschutz, R., Sprang, S., Sambrook, J., and Gething, M.-J. (1993). Affinity panning of a library of peptides displayed on bacteriophages reveals the binding specificity of BiP. *Cell* **75,** 717–728.

Clackson, T., Hoogenboom, H. R., Griffiths, A. D., and Winter, G. (1991). Making antibody fragments using phage display libraries. *Nature* **352,** 624–628.

Clackson, T., and Wells, J. A. (1994). In vitro selection from protein and peptide libraries. *TIBS* **12,** 173–184.

Dedman, J. R., Kaetzel, M. A., Chan, H. C., Nelson, D. J., and Jamieson, J. G A. (1993). Selection of target biological modifiers from a bacteriophage library of random peptides: The identification of novel calmodulin regulatory peptides. *J. Biol. Chem.* **268,** 23025–23030.

Hammer, J., Takacs, B., and Sinigaglia, F. (1992). Identification of a motif for HLA-DR1 binding peptides using M13 display libraries. *J. Exp. Med.* **176,** 1007–1013.

Kay, B. K., Adey, N. B., He, Y.-S., Manfredi, J. P., Mataragnon, A. H., and Fowlkes, D. M. (1993). An M13 library displaying 38-amino-acid peptides as a source of novel sequences with affinity to selected targets. *Gene* **128,** 59–65.

Lowman, H. B., and Wells, J. A. (1993). Selecting high affinity binding proteins by monovalent phage display. *J. Mol. Biol.* **234,** 564–578.

McConnell, S., Uveges, A., and Spinella, D. (1995). Comparison of plate versus liquid amplification of M13 phage display libraries. *Biotechniques* **18,** 803–804.

Scott, J. K., and Smith, G. P. (1990). Searching for peptide ligands with an epitope library. *Science* **249,** 386.

Smith, G. P. (1985). Filamentous fusion phage: Novel expression vectors that display cloned antigens on the virion surface. *Science* **228,** 1315–1317.

Sparks, A. B., Quilliam, L. A., Thorn, J. M., Der, C. D., and Kay, B. K. (1994). Identification and characterization of Src SH3 ligands from phage-displayed random peptide libraries. *J. Biol. Chem.* **269,** 23853–23856.

Sparks, A. B., Rider, J. E., Hoffman, N. G., Fowlkes, DM., Quilliam, L. A., and Kay, B. K. (1996). Distinct ligand preferences of SH3 domains from Src, Yes, Abl, cortactin, p53BP2, PLCγ, Crk, and Grb2. *Proc. Natl. Acad. Sci. U.S.A.* **93,** 1540–1544.

# *14*

# Substrate Phage

## *David J. Matthews*

Substrate phage is a powerful adaptation of phage display which makes it possible to determine substrate specificity by screening peptide phage libraries. The method has been successfully applied to determining the substrate specificity of several different proteases (Table 1) but could also be applied to determining peptide sequence specificity for a wide variety of peptide post-translational modifications. The requirements for a substrate phage experiment are described below, illustrated by reference to protease substrate phage.

## A SUITABLE LIBRARY OF RANDOM PEPTIDES DISPLAYED ON THE SURFACE OF PHAGE PARTICLES

In the case of protease substrate phage, a randomized peptide coding sequence is inserted between an "affinity domain" (in Fig. 1, a variant of human growth hormone [hGH]) and a truncated form of M13 gene III. Other forms of affinity domain could also be used for protein immobilization, such as "epitope tag" peptides which bind to anti-peptide antibodies (for example, the Flag-tag (Hopp *et al.,* 1988; Brizzard *et al.,* 1994) and the Myc-tag (Munro and Pelham, 1984; Ward *et al.,* 1989). The phage library is ideally expressed in a phagemid system (Wells and Lowman, 1992), so that only a small percentage of phage particles display the

**TABLE I**    Proteases Screened Using Substrate Phage Display

| Protease | Reference |
|---|---|
| Subtilisin BPN′ mutant S24C/H64A/ E156S/G166A/G169A/Y217L | Matthews and Wells (1993) |
| Human factor Xa | Matthews and Wells (1993) |
| HIV protease (preliminary data only) | Matthews and Wells (1993) |
| Human furin | Matthews *et al.* (1994) |
| Human matrilysin | Smith *et al.* (1995) |
| Human stromelysin | Smith *et al.* (1995) |
| Human tissue-type plasminogen activator | Ding *et al.* (1995) |
| Subtilisin BPN′ mutant N62D/G166D | Ballinger *et al.* (1996) |

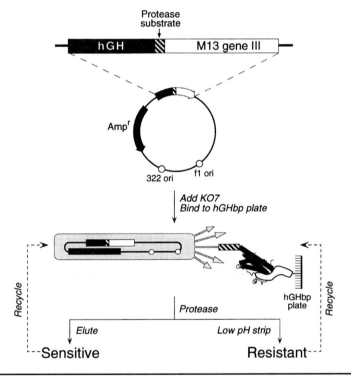

**FIGURE I**    Scheme for protease substrate phage selection. Reprinted with permission from Matthews, D. J., and Wells, J. A. (1993) Substrate phage: Selection of Protease substrates by monovalent phage display. *Science* **260**, 1113–7, © 1993 AAAS.

growth hormone–peptide–gene III fusion and the majority of these contain only one copy per phage particle. However, it should be noted that Smith *et al.* (1995) have reported the use of a polyvalent phage display library for discovery of substrates of the metalloproteinases matrilysin and stromelysin.

## A METHOD FOR *IN VITRO* MODIFICATION OF PEPTIDE SUBSTRATES

For protease substrates, *in vitro* modification involves simply cleaving the peptide substrates by incubation with protease under appropriate conditions. This selection may be employed to generate information on both protease-sensitive and protease-resistant sequences. The amount of protease used will depend on its specificity and catalytic efficiency, and usually must be determined empirically. If possible, it is helpful to initially make two phage constructs, one bearing a known substrate and the other a known nonsubstrate sequence. Phage produced from these constructs may then be used to determine appropriate selection conditions as follows:

i. Immobilize equal numbers of substrate and nonsubstrate phage particles to separate wells of a microtiter plate via the affinity domain. Set up several wells for each phage construct.

ii. Wash extensively to remove unbound phage particles.

iii. Incubate with varying concentrations of protease for a fixed period of time (based on an estimate of the enzyme's turnover number if known). It is important to include a control incubation with buffer alone.

iv. Remove the eluted phage from the well and measure the phage titer. This should allow determination of the protease concentration which gives the best enrichment for substrate over nonsubstrate and for protease incubation over control incubation.

v. Proceed with phage selection using the protease concentration determined in the above steps.

## A METHOD FOR SEPARATING MODIFIED PEPTIDES FROM UNMODIFIED PEPTIDES

Substrate phage are immobilized on a solid support via the affinity domain (in this case, binding of the hGH variant to immobilized hGH binding protein). Techniques for protein immobilization, phage binding, and washing are essentially as discussed elsewhere in this volume; however, phage elution is performed by incubation with protease. On proteolytic cleavage, phage particles bearing substrates are released into solution whereas those containing nonsubstrates remain bound to the solid support. If desired, the protease-resistant phage may be subsequently eluted using a low-pH buffer (50 m$M$ glycine, pH 2.0). The pools of protease-sensitive and protease-resistant phage may be propagated by infection of *Escherichia coli* and the process repeated to further enrich for protease-sensitive and protease-resistant sequences. DNA sequencing of individual clones is performed to obtain detailed sequence information about substrate specificity. Phage titers may be measured after each round of selection, although this may be of limited benefit; it has been found that consensus sequences may emerge even when little or no enrichment is observed by phage titer (D. J. Matthews and J. A. Wells, unpublished observations).

Following phage sorting, it is useful to perform a rapid assay to confirm that sequences isolated from the phage library are indeed good substrates. Matthews and Wells (1993) have described such an assay for protease substrates. Sequences from phage sorting are assembled by PCR into a vector which directs expression of a tripartite fusion protein. The fusion protein comprises the substrate sequence fused between hGH at the N-terminus and alkaline phosphatase (AP) at the C-terminus. This protein is expressed, partially purified, and bound via hGH to immobilized hGH binding protein. On incubation with protease the AP moiety is released, thus allowing substrate cleavage to be monitored by measurement of AP activity in the supernatant. It is also possible to determine the N-terminal sequence of the cleaved AP component, thus allowing the scissile bond to be identified (Matthews and Wells, 1993; Matthews *et al.,* 1994; Ballinger *et al.* (1996).

## CAVEATS, TROUBLESHOOTING, AND HINTS

• In the case of protease substrate phage, it is important that cleavage occurs preferentially in the linker between the affinity domain and gene III. Flanking the substrate region with flexible linker peptides may help to ensure that cleavage occurs preferentially within the library cassette. Fortunately, filamentous phage are remarkably resistant to proteolysis; among commonly used proteases, only subtilisin is known to degrade phage coat proteins (Schwind *et al.,* 1992)

• *E. coli* periplasmic proteases may cleave some substrates before they even leave the cell. Nevertheless, substrates with trypsin-like motifs (which one might expect to be degraded in the periplasm) have been successfully screened using substrate phage (Matthews *et al.,* 1994).

• If experiments with a test system involving known substrate/nonsubstrate sequences are successful but no selection from a library is observed, it may be that the enzyme has broad specificity and thus the sorting process will not converge on any one family of substrate sequences. This is a limitation of any technique used to determine substrate specificity.

There are some 300 amino acid modifications which occur *in vivo* (Yan *et al.,* 1989), many of which could be studied using the substrate phage concept. For example, to study protein tyrosine phosphorylation one may construct a random peptide library with a fixed tyrosine residue surrounded by random codons. The selection procedure would involve incubating the phage library with a tyrosine kinase, and phosphorylated peptides could be separated from unphosphorylated peptides using an anti-phosphotyrosine antibody. In this case, care must be taken that the antibody recognizes phosphorylated tyrosine alone and is not sequence-specific; otherwise, this step introduces another selective pressure on the substrate phage screening process. Other post-translational modifications which could be studied with this technique include dephosphorylation, glycosylation, isoprenylation, methylation, acylation, and carboxylation. Such uses of phage display have not been reported to date. However, Westendorf *et al.* (1994) have recently reported the use of peptide phage libraries to determine the epitope of an anti-phosphopeptide antibody which binds to M-phase phosphoproteins, and Schatz (1993) has used a related technique—the so-called "peptides on plasmids" method—to determine the substrate specificity of *E. coli* biotin holoenzyme synthetase. A variation of the protease substrate phage technique has also been used to determine the nucleophile specificity of "subtiligase," a mutant of subtilisin which acts as a peptide ligase (Chang e*t al.,* 1994).

## References

Chang, T. K., Jackson, D. Y., Burnier, J. P. and Wells, J. A. (1994). Subtiligase: A tool for semisynthesis of proteins. *Proc. Natl. Acad. Sci. U.S.A.* **91,** 12544–12548.

Ballinger, M. D., Tom, J., and Wells, J. A. (1996). Designing Subtilisin BPN′ to cleave substrates containing dibasic residues. *Biochemistry* **34,** 13312–13319.

Ding, L., Coombs, G. S., Strandberg, L., Navre, M., Corey D. R., and Madison E. L. (1995). Origins of the substrate specificity of tissue-type plasminogen activator. *Proc. Natl. Acad. Sci. U.S.A.* **92,** 7627–7631.

Brizzard, B. L., Chubet, R. G., and Vizard, D. L. (1994). Immunoaffinity purification of Flag epitope-tagged bacterial alkaline phosphatase using a novel monoclonal antibody and peptide elution. *Biotechniques* **16,** 730–735.

Hopp, T. P., Prickett, K. S., Price, V. L., Libby, R. T., March, C. J., Ceretti, D. P., Urdal, D. L. and Conlon, P. J. (1988). A short polypeptide marker sequence useful for recombinant protein identification and purification. *Bio/Technology* **6,** 1204–1210.

Matthews, D. J., Goodman, L. J., Gorman, C. M., and Wells, J. A. (1994). A survey of furin substrate specificity using substrate phage display. *Prot. Sci.* **3,** 1197–1205.

Matthews, D. J., and Wells, J. A. (1993). Substrate phage: Selection of protease substrates by monovalent phage display. *Science* **260,** 1113–1117.

Munro, S., and Pelham, H. R. B. (1984). Use of peptide tagging to detect proteins expressed from cloned genes: Deletion mapping functional domains of *Drosophila* hsp 70. *EMBO J.* **3,** 3087–3093.

Schatz, P. (1993). Use of peptide library to map the substrate specificity of a peptide-modifying enzyme: A 13 residue consensus peptide specified biotinylation in *Escherichia coli. Bio/Technology* **11,** 1138–1142.

Schwind, P., Kramer, H., Kremser, A., Ramsberger, U., and Rasched, I. (1992). Subtilisin removes the surface layer of the phage fd coat. *Eur J Biochem* **210,** 431–436.

Smith, M., Shi, L., and Navre, M. (1995). Rapid identification of highly active and selective substrates for stromelysin and matrilysin using bacteriophage peptide display libraries. *J. Biol. Chem.* **270,** 6440–6449.

Ward, E. S., Gussow, D., Griffiths, A. D., Jones, P. T., and Winter, G. (1989). Binding activities of a repertoire of single immunoglobulin variable domains secreted from *Escherichia coli. Nature* **341,** 544–546.

Wells, J. A., and Lowman, H. B. (1992). Rapid evolution of peptide and protein binding properties *in vitro. Curr. Op. Struct. Biol.* 597–604.

Westendorf, J. M., Rao, P. N., and Gerace, L. (1994). Cloning of cDNAs for M-phase phosphoproteins recognized by the MPM2 monoclonal antibody and determination of the phosphorylated epitope. *Proc. Natl. Acad. Sci. U.S.A.* **91,** 714–718.

Yan, S. C. B., Grinnell, B. W., and Wold, F. (1989). Post-translational modifications of proteins: Some problems left to solve. *Trends Biochem. Sci.* **14,** 264–268.

# 15

# Phage Display: Factors Affecting Panning Efficiency

## *John McCafferty*

## INTRODUCTION

In 1988 Parmley and Smith demonstrated that peptides could be displayed on the surface of filamentous bacteriophage and that these fusion phage could be recognized by antibodies. This was followed in 1990 by the demonstration of the "reverse" situation where functional antibody fragments recognizing antigen were displayed on the surface of phage and specific antibody-phage enriched by panning on the immobilized antigen (McCafferty *et al.*, 1990).

With increasing use of this technology some general principles quickly emerged:

• It is possible to enrich specific binders from a background of irrelevant clones for a range of different affinities.

• The system permits the relative enrichment of clones encoding high-affinity binders over low-affinity clones.

This relative enrichment on the basis of affinity enables affinity maturation to be carried out. This requires mutagenesis of a clone followed by selection of the best of the resultant repertoire of variants.

An understanding of the parameters affecting panning efficiency will permit greater control over the affinity, diversity, or specificity of the isolated clones leading to faster delivery of useful antibodies. This control will become increasingly important as bigger repertoires representing a whole spectrum of affinities are constructed (e.g., Griffiths *et al.,* 1994 Vaughan *et al.,* 1996). This chapter will present experimental and theoretical discussion of the effects of display level, clonal stability, affinity, and antigen presentation on the efficiency of selection.

Model experiments were used to monitor the effect of these parameters on the performance of the system and this is expressed in terms of enrichment factors or numbers of phage binding. In these experiments and discussions, the interactions of displayed antibody fragments with their antigens will be used as examples, but most of the principles discussed apply to any protein or peptide displayed on phage.

## VARIATIONS IN DISPLAY LEVEL

### Display System

Protein and peptide display on phage has been demonstrated by fusion to either gene 3 or gene 8 of the filamentous phage. Discussions in this chapter will be restricted to gene 3 systems since this site is most commonly used and fulfills all the requirements with respect to selection and maturation of binding proteins and peptides. Proteins can be displayed as gene 3 fusions either by cloning directly within the phage genome (Parmley and Smith, 1988; Cwirla *et al.,* 1990; McCafferty *et al.,* 1990, 1991; Clackson *et al.,* 1991; Swimmer *et al.,* 1992) or by cloning into gene 3 present within a phagemid plasmid (Bass *et al.,* 1990; Barbas *et al.,* 1991; Brietling *et al.,* 1991; Hoogenboom *et al.,* 1991; Marks *et al.,* 1991). The latter "phagemid-display" system requires rescue with a helper virus, such as VCSM13 helper phage, to generate particles displaying gene 3 fusions. Phage particles derived in this way will display gene 3 protein (g3p) from both the wild-type gene 3 of the helper phage and the fusion gene 3 from the resident phagemid. This is in contrast to display of g3p fusions encoded within the phage genome itself, where all g3p molecules are originally present as fusions.

The level of display in the various systems can be assessed by Western blots by probing phage with anti-gene 3 antibodies. Figure 1 shows such an experiment where the same antibody clone is expressed in the different display systems. Intact fusion can be resolved on the gel from degraded fusion, which runs at the same position as wild-type g3p. Both species are observed on particles derived from phage display vectors. In this case only 20–30% of the fusion is intact (Fig. 1b).

With particles derived by rescue of phagemid clones, the majority of g3p molecules present are of wild-type g3p size (Fig. 1c). This is caused by incorporation of g3p from gene 3 of the helper phage and by degradation of the fusion. This result indicates that a large proportion of particles present are in fact "bald" with respect to display of fusions, particularly for the phagemid display system. Although there appears to be no intact fusion in Fig. 1c, intact fusion is visible on this preparation after longer exposure, and the sample is positive on ELISA (not shown).

In order to demonstrate that the observed difference in display level between the phage and phagemid systems was due to the presence of gene 3 protein (g3p) encoded by the helper virus, phagemid particles were rescued with a helper phage lacking

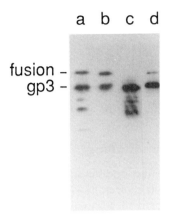

**FIGURE 1**     Yield of intact fusion using phage vectors or phagemid vectors with different helper viruses. A comparison was made of the display level of an antibody fragment expressed as a p3 fusion in either a phage-display vector (fd-DOG1, Clackson *et al.,* 1991) or in a phagemid-display vector (pHEN-1) rescued with M13 helper phage (e.g., VCSM13, Stratagene). An anti-oxazalone antibody (moOX16) in the form of a single-chain Fv was cloned in both vectors. Display level was determined by comparing the relative proportions of wild-type gene 3 protein (g3p) and fused g3p on Western blots probed with anti-g3p antiserum. The phagemid clone was also rescued with M13∂g3 (Griffiths *et al.,* 1993). This helper phage lacks gene 3 and results in phagemid particles having g3p encoded entirely from within the phagemid genome (i.e., as a fusion) rather than from the wild-type gene 3 of the helper phage. (The M13∂g3 helper was constructed by deleting most of gene 3 from the helper virus genome. The product of gene 3 is necessary for viral infection and so the new helper virus is cultured in a clone of *E. coli* cells which produce g3p *in trans* to give rise to phage particles which are themselves infectious but do not encode gene 3). Cultures were grown overnight and phage particles were concentrated by PEG precipitation. The equivalent of 1 ml of culture was loaded onto a 10% polyacrilamide gel, transferred to nitrocellulose, and detected with anti-gene 3 antibody (gift from I. Rasched). Samples are as follows: (a) clone moOX16 in phagemid vector pHEN1, rescued with M13∂g3; (b) clone pAbmoOX16 in phage vector fd-DOG1; (c) clone moOX16 in phagemid vector pHEN1, rescued with VCSM13; (d) clone pAbD1.3 in phage vector fd-DOG1.

gene 3 (M13∂g3, Griffiths *et al.,* 1993). Figure 1a shows that rescue with M13∂g3 restores the display level from phagemid vectors to that achieved from phage vectors, indicating that competing gene 3 from the helper phage is indeed the reason for the difference in display level. (The major hindrance to the generalized use of M13∂g3 for phagemid rescue is the fact that the titres produced are very low.)

## Clonal Variation

Regardless of which vector system is used, there can be clonal variations in display levels even among closely related sequences. These expression differences may be caused by differences in translation, transport, folding, or stability of the fusion. It is interesting to note in Fig. 1 that the better yield is from a clone selected by the phage system (pAbmoOX16, Fig. 1b) compared with a clone derived from hybridomas (pAbD1.3, Fig. 1d). In a library there will probably be a bias for selection of clones which have fusions which are well expressed since their progeny will make a greater contribution to the displayed repertoire (see following discussion on the effect of display level on enrichment).

For peptide libraries it is likely that there will be significant differences in the presentation of different peptide sequences. It is more difficult, however, to demonstrate differences in display level for short peptides than for proteins since the difference in size of intact fusion and native gene 3 protein on Western blots is less pronounced.

## Effect of Display Level on Binding

The effect of differences in display level on enrichment is discussed later in this chapter. It is clear, however, even from ELISA, that the improved display level in phage vectors leads to improved binding. Figure 2A compares the binding profile of an anti-oxazalone antibody displayed in either phage or phagemid vectors.

Figure 2B shows the ELISA binding profile of phagemid clones displaying antibodies of different affinity. In this experiment, phagemid clones were rescued with either M13∂g3 (which restores display level of phagemid clones to that occuring with phage-based vectors) or wild-type helper virus. It is clear that the enhancement in phage binding with the higher display system occurs across the whole range of affinities represented by these clones (10–18,000 n$M$).

## EFFECT OF AFFINITY ON PANNING EFFICIENCY

### Measuring the Performance of Phage Display Systems

Model experiments are often used as a means of measuring the performance of the phage display system. These are usually set up to mimic the situation occuring in the various stages of panning a genuine library. The proportion of binders to a particular target may range throughout a selection from 1–10 binders in $10^7$ clones at the begining to 1–10 binders for every 10 clones at the end of the selection. For model experiments, binding and nonbinding phage are grown and titred then mixed at the appropriate ratio. The output ratio, following selection, is then measured, e.g., by infecting the eluted population, into *Escherichia coli* and transferring the resultant colonies onto a membrane and probing with a labeled oligonucleotide specific for one of the clones. The enrichment factor is calculated from the input and output ratios.

In the experiment below, output ratio is calculated as the fraction of binding clones relative to the total output, i.e., binding and nonbinding clones (McCafferty *et al.,* 1990; Brietling *et al.,* 1991). Others have described the output as the fraction of binders relative only to nonbinders (Bass *et al.,* 1990; Barbas *et al.,* 1991). Whatever method is used to calculate enrichment, the message derived from the data is the same:

• There is improved enrichment with increasing affinity of the displayed binder (values up to 40–50,000-fold were calculated for a clone with an affinity constant, ($K_d$) of 10 n$M$).

• There is improved enrichment with higher display levels (discussed below).

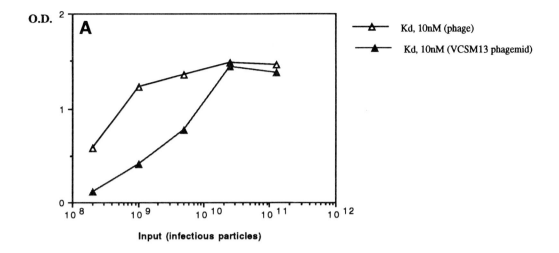

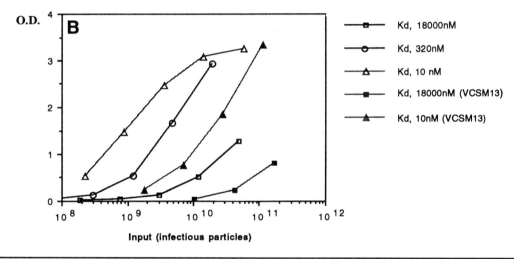

**FIGURE 2**   (A) Comparison of phage and phagemid display systems. Phage particles displaying the anti-oxazalone clone moOX16 ($K_d$, 10 n$M$) present in the phage vector fd-DOG1 or the phagemid vector pHEN-1 (rescued with VCSM13 helper phage) were prepared from fresh overnight culture supernatants. These were bound to Nunc Immunosorb ELISA plates coated with oxazalone–BSA and detected with anti-fd antiserum. VCSM13 rescue is represented by closed symbols and phage particles are represented by open symbols. (B) Effect of affinity and display level on binding in ELISA. A number of phagemid-display clones encoding antibody single-chain Fv fragments binding 2-phenyl-5-oxazolone were used to examine the effect of affinity on binding in ELISA (this figure) and during panning (Table 1, Fig. 3). moOX2 is a low-affinity mouse antibody with dissociation constant ($K_d$), 18,000 n$M$ moOX16 is a higher affinity mouse antibody with $K_d$, 10 n$M$. Both were isolated in the phage vector fd-DOG1 (Clackson *et al.,* 1991) and were transferred for this study to the phagemid vector pHEN1 (Hoogenboom *et al.,* 1991). αphOX15 is a human clone with $K_d$, 320 n$M$ (Marks *et al.,* 1992) and was isolated from a library derived from peripheral blood lymphocytes from nonimmunized human donors, cloned in the phagemid vector pHEN1. αphOX31-E (Tables 2 and 3) is a derivative of this clone which was improved by chain shuffling and by shuffling of V genes segments to generate an antibody fragment with $K_d$, 1.0 n$M$ (Marks *et al.,* 1992). All affinities were measured by fluorescence quench. These clones were rescued with VCSM13 or M13∂g3 and the phage allowed to bind Falcon PVC ELISA plates coated with oxazalone–BSA. Bound phage were detected with sheep anti-fd antiserum. VCSM13 rescue is represented by closed symbols and M13∂g3 rescue is represented by open symbols.

Although the ability to enrich specific clones over background or lower-affinity clones is the ultimate measure of the performance of the phage display system, it is also illuminating to independently examine the actual numbers of specific and non-specific phage which are introduced and recovered.

## Effect of Affinity and Display Level in a Model Selection

The effect of affinity on panning efficiency was determined in a model experiment by mixing a range of anti-oxazalone phagemid clones, at a range of ratios, with a fixed number of viral particles derived from a background clone (TEL9; Marks *et al.,* 1991). These mixes were panned against oxazalone conjugated to BSA, coated onto polystyrene tubes (Nunc Immunosorb). Following binding and washing, colonies derived from the eluted phage were probed with oligonucleotides specific for the oxazalone binder (as in McCafferty *et al.,* 1990). In this way the numbers of specific phage binding and recovered during a single round of panning were calculated for each clone. The data for specific phage input and output for the 10 and 18,000 n$M$ clones from Table 1 (columns B and C) were plotted on a log/log scale in Fig. 3. A linear relationship between log input and log output was observed.

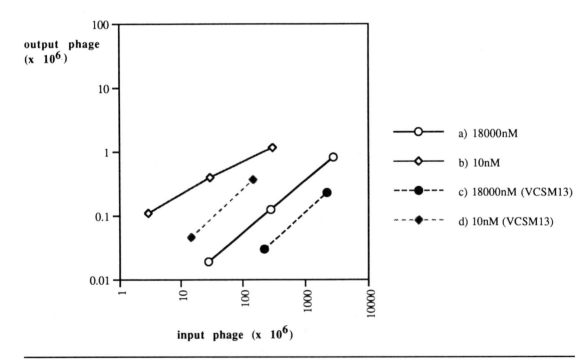

**FIGURE 3**        Effect of affinity and display level on panning efficiency. The input and output phage titers from Table 1 were plotted. The values for the 18,000 and 1 n$M$ affinity clones are shown (solid lines). Results from a parallel experiment where the same clones were rescued with VCSM13 are also shown (broken lines).

**TABLE I**   Effect of Displayed Affinity on Binding Capability of Phage[a]

| A. Sample | B. Sample Input numbers | | C. Specific phage recovered[b] | D. Nonspecific phage recovered | E. Enrichment over background |
| --- | --- | --- | --- | --- | --- |
| | No. | (c.f.u. x $10^6$) | | | |
| $K_d$ 18,000 nM | 1 | 28 | 19,000 | 311,000 | 954 |
| (moOX2) | 2 | 280 | 125,000 | 305,000 | 485 |
| | 3 | 2800 | 830,000 | 590,000 | 97 |
| $K_d$ 320 nM | 4 | 35 (with low background)[d] | 36,000 | 74,000 | 700 |
| ($\alpha$phOX15) | 5 | 35 | 22,000 | 214,000 | 1,996 |
| | 6 | 350 | 187,000 | 299,000 | 824 |
| $K_d$ 10 nM | 7 | 30 (with low background)[d] | 710,000 | 190,000 | 1,970 |
| (moOX16) | 8 | 3 | 108,000 | 521,000 | 42,925 |
| | 9 | 30 | 396,000 | 522,000 | 10,700 |
| | 10 | 300 | 1,170,000 | 230,000 | 2,075 |
| Background | 11 | 0 | N/A | 470,000 | N/A |
| only[e] | 12 | 0 (with low background)[d] | N/A | 147,000 | N/A |

[a]Panning tubes were coated with 1 mg/ml oxazalone coupled to BSA at a ratio of 14:1. Phagemid particles were prepared following rescue with the M13∂g3 helper phage, concentrated as described in McCafferty *et al.* (1990), and mixed at a range of ratios with $4.7$–$7.5 \times 10^{11}$ background phagemid particles (derived by rescue of TEL9, Marks *et al.,* (1991) in 0.5 ml PBS/2% dried milk powder. Preparation of panning tubes, binding, washing, and elution were essentially as in Marks *et al.* (1991).

[b]The proportion of OX-binding phage in the eluate was determined by transferring colonies to nitrocellulose filters and probing with radiolabeled oligonucleotides specific to each OX binding clone as in McCafferty *et al.* (1990).

[c]Enrichment is calculated by dividing output ratios by input ratios. For samples 1, 5 and 8 input ratios are 16,000:1, 21,400:1, 250,000:1, respectively.

[d]Samples labeled "low background" have a 10-fold reduction in background phage input.

[e]Samples 11 and 12 were carried out in a separate experiment using $2.36 \times 10^{11}$ and $2.36 \times 10^{10}$ background phage, respectively.

As expected the numbers of specific phage recovered for a given input are dependent on the affinity of interaction of the displayed antibody with its ligand.

The results of Table 1 were calculated on phagemid clones rescued with M13∂g3 (which restores the display level from phagemid vectors to that achieved from phage vectors). For comparison, the same two phagemid clones were rescued with the wild-type helper virus (VCSM13) in a parallel experiment and the results are also plotted in Fig. 3. A reduced proportion of binding phage were observed with the lower level display arising from wild-type helper phage. This is in line with the ELISA results of Fig. 2B. Thus for a given input of specific phage the proportion recovered after panning is dependent on the display level. The use of phage vectors (with a higher display level than phagemids) will give higher recoveries after panning.

Ultimately the calculated enrichment factor is the product of what occurs to both specific and nonspecific phage. Under the binding and washing conditions of Table 1, there is a significant recovery of background phage which appears to increase with increasing input of background. This is true whether or not specific phage are included (Table 1, column D, samples 4 versus 5, 7 versus 9, 11 versus 12). Thus the level of background phage in the eluate together with the numbers of specific phage binding ultimately affect the enrichment achieved at each step. Column E of Table 1 summarizes the enrichment factors calculated and illustrates the increases with improved affinities.

## Implications for Selection of Libraries

Model systems are a convenient way of monitoring the performance of phage display systems although they do not entirely show what goes on in a real library. It is likely that most libraries will present a spectrum of affinities for most antigens, ranging from high affinities, in some cases, through moderate and low affinities, to clones with no detectable antibody binding. In addition there may be clones which bind the panning matrix. Thus enrichment factors in libraries for high-affinity clones may not be as good as those measured in model experiments if there is an excess of phage exhibiting low-affinity interactions or reactivities with the matrix.

The actual number of rounds of selection is usually defined empirically for each situation by screening the output phage from different rounds of selection for binders. We usually work with the first two pannings which give 10% or greater positives. In practice this means that two to five rounds of selection are usually employed.

Although the experiment in Fig. 3 represents a comparison of different systems for display, it also illustrates the expected effect of variation in the expression level/stability of different clones on phage. If two clones in a library have the same affinity but different expression levels, the more highly expressed one will be preferentially selected. Differences from the two display systems shown in Fig. 3 can be used to illustrate this effect (compare a with c, b with d in Fig. 3).

Clonal variation may extend not just to stability of the fusion but to the yield of phage from different clones. If two clones of equal affinity have different growth rates, they will give rise to different numbers of progeny phage in the final population. There will be a bias toward the clone with greatest yield. If one clone yields more phage than another of equal affinity, it will make a greater contribution to the output population; the magnitude can be seen by comparing the different inputs of phage in Fig. 3.

Table 1 and Fig. 3 show that not all of the input phage are recovered in the eluate. This type of analysis has led some workers to conclude that only a low proportion of phage in a population are capable of binding. Table 2, however, shows that in some circumstances, where low inputs of phage are used, significant proportions of the input phage bind. Under these "optimal" conditions, up to 34% of particles are capable of being bound and recovered. As the phage input is increased, however, the proportion binding is reduced as illustrated in Table 1. For example, a 100-fold increase of input for the 10 n$M$ (moOX16) clone results in only a 10-fold increase in recovered phage, i.e., there is a 10-fold reduction in the proportions which are binding. Thus phage which have the capacity to bind are not being bound.

One possible reason for this effect is that the matrix is being saturated with phage. A phage particle lying flat on the matrix would occupy an area of only 0.006 $\mu m^2$ (1 x 0.006 $\mu m$). Theoretically 1.5 x $10^{10}$ phage could fit onto the surface of a 100-$\mu l$ well (assuming a surface area of 90 $mm^2$ per well). In practice we have measured

**TABLE 2**    Measurement of Binding Capability of Fusion Phage

| Sample | Input | A. M13∂g3 Output | Binding fraction (%) |
|---|---|---|---|
| $K_d$ 320 nM (αphOX15) | $10^6$ | 91,000 | 9.1 |
| $K_d$ 10 nM (moOX16) | $10^6$ | 342,000 | 34 |
| $K_d$ 1 nM (OX31-E) | $10^6$ | 309,000 | 31 |

| Sample | Input | B. VCSM13 Output | Binding fraction (%) |
|---|---|---|---|
| $K_d$ 320 nM (αphOX15) | $10^7$ | 265,000 | 2.6 |
| $K_d$ 10 nM (moOX16) | $10^7$ | 236,000 | 2.4 |
| $K_d$ 1 nM (OX31-E) | $10^7$ | 738,000 | 7.4 |

*Note.* Oxazalone-binding phagemid clones were rescued with the appropriate helper phage as before and mixed with $5 \times 10^{11}$ phage particles derived from pAbD1.3 which confers tetracycline resistance (McCafferty *et al.,* 1990) in a volume of 0.5 ml PBS/ 2% dried milk powder. After washing, binding, and elution, the numbers of OX-binding phage were determined by infecting TG1 cells and plating on ampicillin. The number of tetracycline clones in the eluate was between 5 and 20,000.

binding and recovery of $1.4–2.6 \times 10^8$ phage from an ELISA well (from the experiment in Fig. 2A). Thus, from a theoretical and experimental point of view, saturation of the surface of the well is unlikely to be the reason for the reduced proportion of binders with increasing input.

Another proposal to explain this is that phage form aggregates after binding, or bind as aggregates, resulting in steric inhibition of incoming phage by bound aggregates. Thus, as the binding numbers increase, the rate of association of incoming phage with the surface is reduced. Nonspecific interactions between phage have been observed directly by electron microscopy (Day *et al.,* 1988). The suggestion of aggregation is in line with observations made during physical measurements on phage particles where aggregation and large excluded volumes were observed (see Day *et al.,* 1988). As well as reducing the effective capacity of the surface, aggregations could result in "trapping" of specifically bound phage and a resultant "avidity-like" effect at the surface. This could explain why proteins with relatively fast off-rates can successfully be isolated by panning within phage display systems, despite extended washing times.

Whatever the reasons for the reduced proportion of binders, the results discussed above have important implications if one is trying to deselect using immobilized antigens. With the phage display system one has the option of deselecting unwanted cross-reactivities, i.e., exposing the repertoire to the unwanted antigen to remove binders before selection on the desired antigen. The results of Table 1 suggest that a high proportion of cross-reactive phage which are capable of binding will not be removed in a single step. Perhaps a better way to carry out deselection is to use a large excess of the "deselection antigen" in solution where surface capacity and accessibility to the competing antigen will be less of a problem (e.g., Parsons *et al.*, 1996).

## EFFECT OF AFFINITY AND DISPLAY LEVEL ON DISCRIMINATION BETWEEN BINDERS

It is clear from the experiments above that greater enrichment factors are achieved with higher-affinity clones, and as a result these would be expected to be preferentially selected in mixtures of high- and low-affinity clones. The effect of affinity and display level on the ability of the phage system to discriminate between clones was compared. Clones of different affinities were mixed together and the relative enrichment (selectivity) was measured directly (Table 3). Phagemid particles from a clone with $K_d$, 1 n$M$ (OX31-E) were mixed with a clone of $K_d$, 10 n$M$ (moOX16). In addition, phagemid particles from the 10 n$M$ clone were mixed with the lower-affinity clone ($K_d$, 320 n$M$, pho-OX15). The experiment was carried out by mixing the phagemid particles, rescued with either VCSM13 or M13∂g3, in equal proportion with an excess of a nonbinding background phage (pAbD1.3) and panning on immobilized antigen.

As before, increased affinity gives increased recovery regardless of which helper phage is used (Table 3). The absolute numbers recovered with the low-level expres-

**TABLE 3**  Selectivity for Higher-Affinity Clones

| Input phage (200 x 10⁶ each) | Recovery | Selectivity |
|---|---|---|
| High-level gene 3 display (M13∂g3) | | |
| $K_d$, 320 n$M$ | 2.7 x 10⁶ | |
| $K_d$, 10 n$M$ | 11.3 x 10⁶ | 4.2 |
| High-level gene 3 display (M13∂g3) | | |
| $K_d$, 10 n$M$ | 6.4 x 10⁶ | |
| $K_d$, 1 n$M$ | 17.6 x 10⁶ | 2.75 |
| Low-level gene 3 display (VCSM13) | | |
| $K_d$, 320 n$M$ | 0.24 x 10⁶ | |
| $K_d$, 10 n$M$ | 1.0 x 10⁶ | 4.2 |
| Low-level gene 3 display (VCSM13) | | |
| $K_d$, 10 n$M$ | 0.98 x 10⁶ | |
| $K_d$, 1 n$M$ | 1.83 x 10⁶ | 1.9 |

*Note.* The various oxazalone-binding phagemid clones, rescued with either VCSM13 (low-level gene 3 display) or M13∂g3 (high-level gene 3 display), were mixed in equal proportion (200 x 10⁶ input each) with a 2000-fold excess of background phage derived from pAbD1.3 in 0.5 ml PBS/2% dried milk powder. The eluates were used to infect TG1 cells and plated onto ampicillin plates to select only the phagemid clones, i.e. the oxazalone binders. The proportion of each OX binding phagemid clone present on these plates was determined by probing with an oligonucleotide specific to the 10 n$M$ (moOX16) clone. Approximately 10³–10⁴ background phage were recovered from an input of 4x10¹¹.

sion system (VCSM13 rescue) are smaller than with the high-level expression system (M13∂g3 rescue).

Since equal inputs of the different specific phage were used, we can examine the selectivity between affinities directly by looking at the numbers recovered. There is approximately a fourfold enrichment of the 10 n*M* clone over the 320 n*M* clone regardless of which helper phage is used. Similarly there is an enrichment of 1.9- and 2.7-fold for the 1 n*M* versus the 10 n*M* clone using the two different helper phage. Thus, increasing the antibody display level of phagemid clones to that achieved by fusing directly to gene 3 within the phage genome does not reduce the selectivity factor. In one example it is improved. This is despite the use of hapten coupled to carrier at a ratio of 14:1, where the potential for avidity effects from polyvalent phage is high.

It is often assumed that display of fusions from a system with only one gene 3, as occurs with phage-display vectors (or rescue with a helper deficient in gene 3), will give poorer discrimination between affinities than phagemid systems using one fused and one unfused gene 3 (Bass *et al.,* 1990; Barbas *et al.,* 1991; Garrard *et al.,* 1991; Lowman *et al.,* 1991). The assumption is based on the belief that polyvalent phage will be produced from phage-display vectors. This is expected to prevent separation of high-affinity antibodies from lower-affinity antibodies due to the apparent increased affinity of low-affinity clones binding cooperatively, i.e., avidity. While there is a greater likelihood of polyvalency when there is a single gene 3 present as a fusion, the fact is that significant degradation occurs and the majority of particles are likely to be monovalent or "bald" (Fig. 1). The main benefit of the phage vector system, therefore, is that a greater number of particles will have at least one copy and so are capable of participating in binding.

The original assumption that polyvalency occurs and leads to poor discrimination in phage-display vectors, is based on the suggestion of a "chelate effect," offered by Cwirla *et al.* (1990). This was used to explain the absence of the expected nanomolar affinity clones after three rounds of panning a peptide-display library with an anti-β-endorphin antibody. Peptides on phage may well be more stable than antibody fusions giving polyvalent phage which lead to the "chelate effect" they propose. An alternative explanation, however, is that for every nanomolar affinity clone there are clones with lower affinity in greater number or with greater expression level. The discrimination values shown in Table 3 (1.9- to 4-fold) suggest that three rounds of selection on immobilized hapten may not be enough to isolate the high-affinity clones from an excess of lower-affinity clones.

# SELECTION WITH SOLUBLE ANTIGEN: THEORETICAL ASPECTS

The interaction of proteins with their ligands in solution, particularly antibodies, has been the subject of much study and sophisticated methods have been developed to quantify the kinetic and equilibrium behavior of these binding interactions. At first glance one might expect that this knowledge might be transposed to predict the interactions of proteins displayed on phage and to calculate the relative proportion of high- and low-affinity phage interactions which will occur during panning. The fact that one partner of the interaction is immobilized is a fundamental difference making extrapolation from solution-phase kinetic calculations difficult for both native and phage displayed proteins.

There are a number of other factors specifically related to the phage display systems which have a direct effect on the performance of the system. One obvious difference between antibodies displayed on phage and native antibodies is the physical difference in size of the binding entity. This could affect both the accessibility and the diffusion of the antibody to immobilized ligand. As we have seen using immobilized antigen there are associated limitations, such as reduced proportions binding with increasing inputs.

The use of soluble, recoverable target, e.g., biotinylated antigens, means that the binding step can be carried out in solution (Parmley and Smith, 1988; Hawkins et al., 1992). Following binding, the complex between biotinylated antigen and phage antibody is recovered using immobilized streptavidin or streptavidin-coated magnetic beads. Hawkins et al. (1992) mutated an antibody with $(K_d)$ of 42 nM (for lysozyme) and selected an improved form with affinity of 9.4 nM by using concentrations of biotinylated antigen below the desired $K_d$ value. With a similar approach Thompson et al. (1996) improved an anti-gp 120 antibody. Thus in the initial solution phase binding step, the effective concentration of the antigen may be more closely controlled and the outcome may be more predictable by the laws of mass action.

The following discussion describes antibody:antigen interactions but relates equally to the interaction of any protein or peptide interacting with target. If two interacting molecules, e.g., monovalent antibody (Ab) and antigen (Ag) are mixed, they can form a complex (Ab:Ag) and eventually reach equilibrium. The rate at which equilibrium is reached depends on the on- and off-rate constants for that particular interaction ($k_{on}$ and $k_{off}$). This can be represented as:

$$Ab + Ag \underset{k_{off}}{\overset{k_{on}}{\rightleftharpoons}} Ab:Ag.$$

The rate of complex formation is defined as $[Ab][Ag]k_{on}$ and the rate of dissociation is $[Ab:Ag]k_{off}$. At equilibrium the rates of association and dissociation are equal. This can be written as follows (the subscript eq denotes that the concentration of the species at equilibrium is being described):

$$[Ab_{eq}][Ag_{eq}]k_{on} = [Ab:Ag_{eq}]k_{off}$$

$$K_d = \frac{k_{off}}{k_{on}} = \frac{[Ab_{eq}][Ag_{eq}]}{[Ab:Ag_{eq}]}.$$

Thus the equilibrium constant, $K_d$, is related to the on- and off-rate constants, but can be measured independently of them. Knowing the dissociation constant and the input concentration of antibody, one can predict the concentration of complex $x$ which will be formed.

$$K_d = \frac{[\text{Ab}_{\text{input}}\text{-}x][\text{Ag}_{\text{input}}\text{-}x]}{[x]}$$

Concentrations of individual binding phage are likely to be picomolar, or less, during the early stages of panning a library, and so antigen will in most cases be in excess of phage. (For example, a library with 1 binder in every 10,000 clones and a total titre of $6 \times 10^{12}$ phage particles/ml will have $6 \times 10^8$ binders/ml, i.e., 1 p$M$ assuming all phage particles have an antibody.) In this situation of antigen excess, there will be little or no difference in antigen concentration at equilibrium, i.e., $\text{Ag}_{\text{eq}} = \text{Ag}_{\text{input}}$. The above equation can therefore be rewritten:

$$K_d = \frac{[\text{Ab}_{\text{input}}\text{-}x][\text{Ag}_{\text{input}}]}{[x]}. \qquad [1]$$

This can be rearranged to the form below:

$$x = \frac{\text{Ab}_{\text{input}}\text{Ag}_{\text{input}}}{K_d + \text{Ag}_{\text{input}}} \qquad [2]$$

$$\text{proportion bound} = \frac{x}{\text{Ab}_{\text{input}}} = \frac{\text{Ag}_{\text{input}}}{K_d + \text{Ag}_{\text{input}}}. \qquad [3]$$

Using Eq. [3], the graph of Fig. 4, showing proportion of antibody bound at different antigen concentrations, can be derived for clones with $K_d$ of 1 and 10 n$M$ (i.e., the same affinities as the clones used in the model experiment of Table 3). This profile is true whatever antibody concentration is used (provided the concentration of Ag remains greater than the concentration of Ab). The profiles also describe what will happen with different clones in a mixture. Since antigen is in excess over both antibodies, there will be no competition between the clones for antigen. From these curves one can predict the degree of selectivity that can be achieved in a mixture. At higher antigen concentrations, e.g., 10 n$M$, the selectivity is 1.8-fold. This is the same as was found experimentally during panning (Table 3). As the antigen concentration is reduced the selectivity increases to a maximum of 10, which is the ratio of the $K_d$ values of the two clones (Fig. 4).

The value for the equilibrium constant is determined by the relative on-rate and off-rate for the interaction. Hawkins *et al.* (1992) have increased the selectivity between clones by selecting on the basis of off-rate. This is done by diluting bound phage-antibody in an excess of nonbiotinylated antigen. Phage-displaying antibodies with the slowest off-rates are retained for longer on the biotinylated antigen. In this way off-rates can be used to create a large difference in the relative recovery of clones with relatively small differences in their off-rates.

In summary, selection with soluble, biotinylated antigen provides one means to circumvent some of the problems associated with immobilized antigen and provides an opportunity to increase the degree of discrimination between populations of clones with closely related affinities, e.g., during the latter stages of panning and during affinity maturation of a binder. Notwithstanding this, immobilized antigens

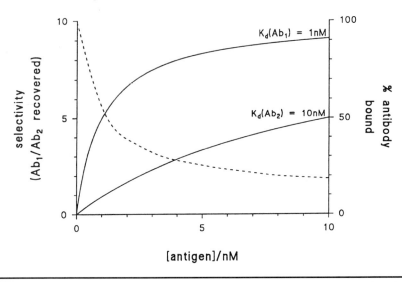

Antigen concentration determines selectivity

**FIGURE 4**    Calculating binding and selectivity with soluble antigen. Using Eq. [3] (see text) the proportion of phage bound at a range of antigen concentrations was calculated for clones with 1 and 10 n$M$ affinity (solid lines, right axis). From these graphs the selectivity of one clone over the other was calculated (hatched lines, left axis). The calculations assume that antigen is in excess.

have proved, and continue to prove, useful as a convenient means of preparing panning matrices and for isolating rare binders present in repertoires.

## CONCLUSION

Phage display is proving to be a very powerful means of selecting specific binders in a library from nonbinders or low-affinity binding clones. The efficiency of the system in recovering rare binders from repertoires is affected by a number of parameters including the affinity of the displayed clone and the display level which is achieved on phage particles. Thus, clones which are well expressed have a selective advantage. Phage vectors give greater display levels and binding characteristics than phagemid vectors rescued with wild-type helper virus such as VCSM13 helper phage. Regardless of which system is used, there can be a significant decrease in display level by proteolysis of the fusion.

Incubation of low amounts of binding phage with immobilized antigen reveals that up to 34% of these particles are capable of binding and infecting upon recovery. As the input is increased, however, there is a reduction in the proportion binding. One consequence of the reduced binding with richer mixes is that there will be a significant proportion of binders which have been unable to associate during the binding step. This means that depletion of unwanted binders by prior deselection using immobilized antigen can be an inefficient process.

The effect of affinity was studied in a series of model experiments using anti-oxazalone antibodies with affinities in the range $1-18,000$ n$M$. The numbers binding and the enrichment of specific clones over background increases with affinity across the range. The selective enrichment of high- over low-affinity clones was unaffected by increased display level and in the experiment here was 4-fold for a 10 n$M$ over a 320 n$M$ clone and was 2- to 3-fold for a 1 n$M$ over a 10n $M$ clone. Thus, while enrichment over nonbinder can be 40,000-fold or greater for a 10 n$M$ clone, the enrichment relative to other binding clones is more modest.

It is likely that interactions between phage and soluble ligand will more closely mirror the predicted behavior of two interacting molecules. In theory and in practice the use of soluble, biotinylated ligand enables better discrimination of clones with closely related affinities than selections on immobilized antigen and it is likely that the usage of this route to enrichment will increase in coming years.

## Acknowledgements

I acknowledge the contribution of Ron Jackson and Kevin Pritchard to theoretical and mathematical discussions and Kevin Johnson and Alex Duncan for comments on the manuscript.

## References

Barbas, C. F., Bain, J. D., Hoekstra, D. M., and Lerner, R. A. (1991) Assembly of combinatorial antibody libraries on phage surfaces: The gene III site. *Proc. Natl. Acad. Sci. U.S.A.* **88,** 7978–7982.

Bass, S. H., Greene, R., and Wells, J. A. (1990). Hormone phage: An enrichment method for variant proteins with altered binding properties. *Proteins* **8,** 308–314.

Brietling, F., Dübel, S., Seehaus, T., Klewinghaus, I., and Little, M. (1991). A surface expression vector for antibody screening. *Gene* **104,** 147–153.

Clackson, T., Hoogenboom, H. R., Griffiths, A. D., and Winter, G. P. (1991). Making antibody fragment using phage display libraries. *Nature (London)* **352,** 624–628.

Cwirla, S. E., Peters, E. A., Barrett, R. W., and Dower, W. J. (1990). Peptides on phage: A vast library of peptides for identifying ligands. *Proc. Natl. Acad. Sci. U.S.A.* **87,** 6378–6382.

Day, L. A., Marzec J., Reisberg, S. A., and Casadevall (1988). DNA packing in filamentous phage. *Annu. Rev. Biophys. Chem.* **17,** 509–539.

Garrard, L. J., Yang, M., O'Connell, M. P., Kelley, R. F., and Henner, D. J. (1991). $F_{Ab}$ assembly and enrichment in a monovalent phage display system. *Bio/Technology* **9,** 1373–1377.

Griffiths, A. D., Malmqvist, M., Marks, J. D., Bye, J. M., Embleton, M. J., McCafferty, J., Gorick, B. D., Hughes-Jones, N. C., Hoogenboom, H. R., and Winter, G. P. (1993). Human anti-self antibodies with high specificity from phage display libraries. *EMBO J.* **12,** 725–734.

Griffiths, A. D., Williams, S. C., Hartley, O., Tomlinson, I. T., Waterhouse, P., Crosby, W. L., Kontermann, R. E., Jones, P. T., Low, N. M., Allison, T. J., Prospero, D., Hoogenboom, H. R., Nissim, A., Cox, J. P. L., Harrison, J. L., Zaccolo, M., Gherardi, E., and Winter, G. P. (1994). Isolation of high affinity antibodies directly from large synthetic repertoires. *EMBO J.* **13,** 3245–3260.

Hawkins, R. E., Russell, S. J., and Winter, G. P. (1992). Selection of phage antibodies by binding affinity: mimicking affinity maturation. *J. Mol. Biol.* **226,** 889–896.

Hoogenboom, H. R., Griffiths, A. D., Johnson, K. S., Chiswell, D. J., Hudson, P., and Winter, G. P. (1991). Multi subunit proteins on the surface of filamentous phage: Methodologies for displaying antibody (Fab) heavy and light chains. *Nucleic. Acids Res.* **19,** 4133–4137.

Lowman, H. B., Bass, S. H., Simpson, N., and Wells, J. A. (1991). Selecting high-affinity binding proteins by monovalent phage display. *Biochemistry* **30,** 10832–10838.

Marks, J. D., Hoogenboom, H. R., Bonnert, T. P. , McCafferty, J., Griffiths, A. D., and Winter, G. P. (1991). By-passing immunization: Human antibodies from V gene libraries displayed on phage. *J. Mol. Biol.* **222,** 581–597.

Marks, J. D., Griffiths, A. D., Malmqvist, M., Clackson, T. P., Bye, J. M., and Winter, G. P. (1992). By-passing immunization: Building high affinity human antibodies by chain shuffling. *Bio/Technology* **10,** 779–783.

Parsons, H. L., Earnshaw, J. C., Wilton, J., Johnson, K. S., Schueler, P. A., Mahoney, W., and McCafferty, J. (1996). Directing phage selections towards specific epitopes. *Protein Eng.* (in press).

McCafferty, J., Griffiths, A. D., Winter, G. P., and Chiswell, D. J. (1990). Phage antibodies: filamentous phage displaying antibody variable domains. *Nature (London)* **348,** 552–554.

McCafferty, J., Jackson, R. H., and Chiswell, D. J. (1991). Phage-enzymes: Expression and affinity chromatography of functional alkaline phosphatase on the surface of bacteriophage. *Protein Eng.* **4,** 955–961.

Parmley S. F., and Smith, G. P. (1988). Antibody-selectable filamentous fd phage vectors: Affinity purification of target genes. *Gene* **73,** 305–318.

Swimmer, C., Lehar, S. M., McCafferty, J., Chiswell, D. J., Blättler, W. A., and Guild, B. C. (1992). Phage display of ricin B chain and its single binding domains: System for screening galactose-binding mutants. *Proc. Natl. Acad. Sci. U.S.A.* **89,** 3756–3760.

Thompson, J., Pope, T., Tung, J. S., Chan, C., Hollis, G., Mark, G., and Johnson, K. S. (1996). Affinity maturation of a high affinity monoclonal antibody against the third hypervariable loop of human immunodeficiency virus: Use of phage display to improve affinity and broaden strain reactivity. *J. Mol. Biol.* **256,** 77–88.

Vaughan, T. J., Williams, A. J., Pritchard, A. J., Osbourn, J. K., Pope, A. R., Earnshaw, J. C., McCafferty, J., Wilton, J., and Johnson, K. S. (1996). Isolation of human antibodies with sub-nanomolar affinities direct from a large non-immunised phage display library. *Nature Biotechnol.* **14,** 309–314.

# *16*

# Preparation of Second-Generation Phage Libraries

## *Nils B. Adey, Willem P. C. Stemmer, and Brian K. Kay*

Once recombinants have been characterized from a phage display library, it is often useful to construct and screen a second-generation library that displays variants of the original sequence. This methodology is much like one cycle of Darwinian evolution, where additional mutations are generated and selected to yield optimal properties. It is often warranted because the number of combinations for proteins or peptides longer than seven residues is so great that all permutations cannot exist in the primary library. Furthermore, by mutating sequences, the "sequence landscape" around the isolated sequence can be examined to find local optima (Kauffman, 1992).

There are several methods available to the experimenter for the purposes of mutagenesis. They are site-directed mutagenesis, cassette mutagenesis, error-prone PCR, and DNA shuffling of single recombinants (Fig. 1). It is also possible to do applied evolution on populations of isolates (Fig. 2) with DNA shuffling. Coupled with phage display, these methods have been quite effective in protein engineering, such as creating a form of bovine pancreas trypsin inhibitors that can inhibit elastase (Roberts *et al.,* 1992), improving antibody affinities (Barbas *et al.,* 1994), generating a high-affinity inhibitor of urokinase-type plasminogen activator by phage display of mutagenized ecotin (Wang *et al.,* 1995), and altering the substrate specificity of glutathione S-transferase (Widersten and Mannervik, 1995).

In making these second-generation libraries, it is helpful to mark the vector differently from the original isolated recombinant. This is for two reasons. First, trace

*Phage Display of Peptides and Proteins*

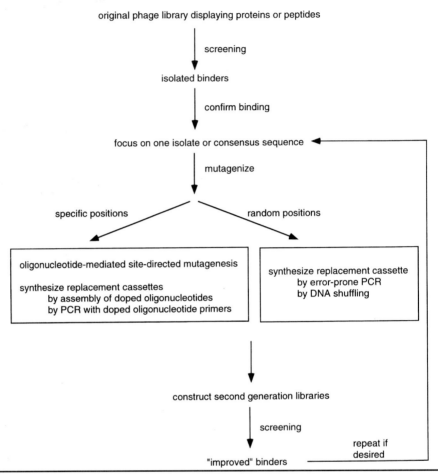

**FIGURE I**          Flow chart of one cycle of molecular evolution. Boxes contain the various methods of mutagenizing phage inserts (see text) for the purpose of selecting displayed proteins or peptides that have altered or "improved" binding properties.

amounts of the original recombinant virus usually contaminate the laboratory and it is useful to have a "marker" to distinguish the original virus from mutagenized viruses easily and quickly. Second, the original virus isolated from the primary library and an aliquot of the second generation library can be intentionally mixed and screened together; a better recovery of some of the engineered viruses compared to the original virus would suggest that they have improved binding characteristics (Fig. 3). Convenient markers are *lac*Z (Fowlkes *et al.,* 1992), drug resistance, restriction enzyme site polymorphism, PCR product length polymorphism, and epitope tags.

From these second-generation libraries, very often "better" or "stronger" binders can be isolated. Selective enrichment of such phage can be accomplished by screening with lower target concentrations immobilized on a microtiter plate or in solu-

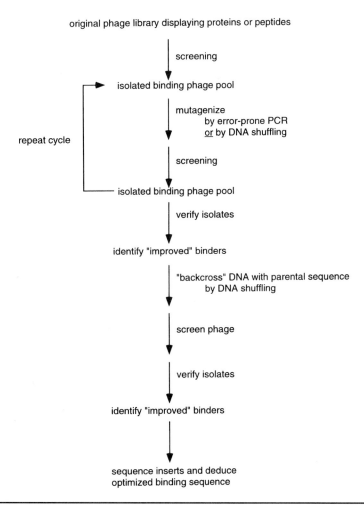

original phage library displaying proteins or peptides

↓ screening

isolated binding phage pool

↓ mutagenize
   by error-prone PCR
   or by DNA shuffling

↓ screening

isolated binding phage pool

repeat cycle

↓ verify isolates

identify "improved" binders

↓ "backcross" DNA with parental sequence
   by DNA shuffling

↓ screen phage

↓ verify isolates

identify "improved" binders

↓

sequence inserts and deduce
optimized binding sequence

**FIGURE 2**    Flow chart of evolution of mixed populations. The process of adaptive or forced evolution can also be applied to populations of recombinants displaying either proteins or peptides. However, due to the short nature of peptides, this process is better suited to evolving proteins. Once recombinants have undergone several rounds of mutagenesis and selection, it is possible to eliminate silent mutations through backcrossing isolated DNA with parental sequences during DNA shuffling (see Fig. 6).

tion, combined with extensive washing (Barrett *et al.*, 1992). Another option is to display the mutagenized population of molecules at a lower valency on phagemids to select for peptides with better $K_d$'s. Finally, it is possible to include a low concentration of binding competitor (i.e., target, ligand) in the wash buffer to select for phage that have the slowest off-rate.

The various methods are described in detail below. Each technique offers different extents and locations of mutations; they are summarized in the following table.

**Petri Plates & Bacterial Lawns**

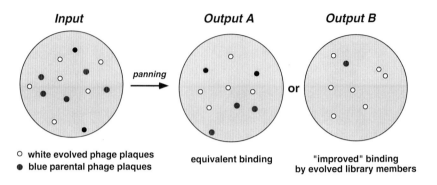

○ white evolved phage plaques
● blue parental phage plaques

equivalent binding

"improved" binding
by evolved library members

| | |
|---|---|
| **FIGURE 3** | Hypothetical results of affinity selection with a mixture of parental and mutagenized phage. Input is shown along with two potential output results. Plaques are shown as dark or clear on the surface of bacterial lawns in Petri plates (large circles); dark or clear plaques are formed by phage which carry or lack the *lacZ* gene, respectively, when grown in the presence of IPTG and XGal. The input is an artificial mixture of evolved and original (parental) plaque-forming units. The relative numbers of plaques for the output and input samples are shown to be comparable for illustration purposes, except that the number of recovered pfu in the output is much lower (i.e., $10^3$–$10^6$) than the input. Selection for "improved" binding phage is typically accomplished by screening with decreased concentrations of target, shorter incubation times, longer or more extensive washes, or elevated temperatures. |

| Technique | Size of mutated region (nt) | Position of mutations |
|---|---|---|
| Site-directed | 1–10 | Fixed |
| Cassette | 5–40 | Random or fixed |
| Error-prone PCR | 1 | Scattered |
| Shuffling | 1 | Scattered |

## OLIGONUCLEOTIDE-MEDIATED SITE-DIRECTED MUTAGENESIS

One of the most widely used forms of recombinant DNA-based mutagenesis is known as oligonucleotide-mediated site-directed mutagenesis. An oligonucleotide is designed such that it can base-pair to a target DNA, while differing in one or more bases near the center of the oligonucleotide. When this oligonucleotide is base-paired to the single-stranded template DNA, the heteroduplex is converted into double-stranded DNA in *vitro;* in this manner one strand of the product will carry the nucleotide sequence specified by the mutagenic oligonucleotide (Fig. 4). These DNA molecules are then propagated *in vivo* and the desired recombinant is ultimately identified among the population of transformants.

A protocol for single-stranded mutagenesis is described below. Readers are directed to the literature (Hutchison *et al.,* 1978; Kunkel *et al.,* 1991) for additional details

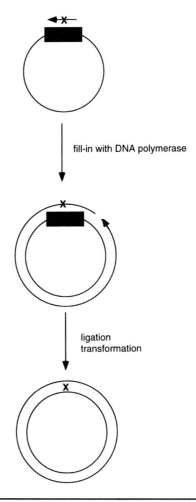

**FIGURE 4**    Oligonucleotide-mediated site-directed mutagenesis. The region of interest is shown as a solid box on the single-stranded genome of M13 phage or secreted phagemid particles. An oligonucleotide, with one or several mutated bases (specified or at random), is annealed to the single-stranded DNA to form a template for DNA polymerase. The second strand is synthesized and closed by ligation. The double-stranded DNA is then introduced into bacteria via electroporation, yielding mutated and the orginal genomes. The two genomes can be distinguished by a variety of means (i.e., restriction enzyme digestion).

before beginning this method. Kits for oligonucleotide-directed mutagenesis of phage and phagemids can be obtained from Bio-Rad [Catalog No. 170-3571 (phage), No. 170-3576 (phagemid)].

1. Prepare single-stranded DNA from M13 phage or phagemids (see Appendix). Isolate ~2 μg of DNA. The DNA can be isolated from a *dut⁻ung⁻* bacterial host (i.e., CJ236) so that the recovered DNA contains uracil in place of many thymine residues. (The incorporation of uracil in DNA is not itself mutagenic.)

## Cassette Mutagenesis

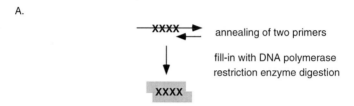

segment of DNA with the box corresponding to an area targeted for mutation

A.

annealing of two primers

fill-in with DNA polymerase
restriction enzyme digestion

B.

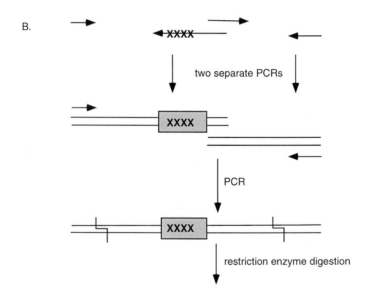

two separate PCRs

PCR

restriction enzyme digestion

clone cassette back into phage-display vector

---

**FIGURE 5**    Four forms of cassette mutagenesis. Double-stranded DNA is shown with the black and stippled boxes corresponding to the area targeted for mutagenesis. **X** refers to sites of mutation. (A) Resynthesis of a cassette with two oligonucleotides. One long oligonucleotide is synthesized with fixed bases at the two termini and randomized (i.e., NNN, NNK, NNS, spiked, triplets, doped) codons in the middle. This oligonucleotide is annealed to a shorter one that serves to prime DNA synthesis *in vitro* with DNA polymerase. The resulting double-stranded DNA is then cleaved with restriction enzymes to generate "sticky ends" for cloning of the cassette. This method is useful when restriction enzyme sites immediately flank the region targeted for mutation. (B) Synthesis of a cassette through PCR. Two oligonucleotides, one with perfect and the other with randomized codons (see **A**), are used to amplify a coding region in two pieces. The resulting PCR product is then extended through a second round of "pull-through" PCR to generate the final fragment which is then cleaved with restriction enzymes to generate "sticky ends" for cloning. This method is useful when convenient restriction enzyme sites do not directly flank the region targeted for mutation. (C) Synthesis of

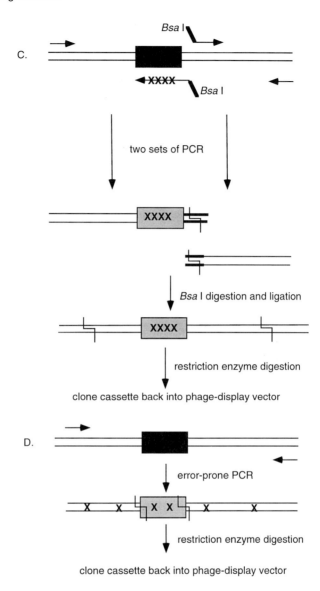

the cassette by PCR and ligation of two DNA segments. Two different PCR are performed with oligonucleotides such that one contains randomized codons and 3′ tails that carry the *Bsa*I restriction site. This class 2s enzyme cleaves the following sequence to generate 5′ overlapping ends of NNNN:

$$GGTCTCN^{\downarrow}$$
$$CCAGAGNNNNN_{\uparrow}.$$

Thus the overlapping ends can be any sequence desired and randomization can occur throughout the cassette. Other class 2s enzymes are *Bbs*I, *Esp*3I, *Ear*I, *Bsm*I, *Bsp*MI, *Bsg*I, and *Bpm*I. This method is very similar to that of (B), except that the type 2s restriction sites are used to rejoin the two segments to create a cassette for recloning. (D) Error-prone PCR. This method is intended for the mutagenesis of, i.e., cassettes >100–1000 nt in length. PCR is accomplished under conditions that lead to a high rate (0.7 %) of misincorporation in the amplified sequence. The resulting double-stranded DNA is then cleaved with restriction enzymes, positioned at the end of the area targeted for mutagenesis, to generate a cassette for cloning.

2. Design an oligonucleotide that has at least 15 residues of complementarity on either side of the site to be mutated. In the oligonucleotide, the region to be randomized can be represented degenerate codons (see below). If the noncomplementary region is large (i.e., >12 nucleotides), then the flanking regions should be extended to ensure proper base-pairing. The oligonucleotide should be synthesized with a 5′ PO$_4$ group, as it improves the efficiency of the mutagenesis procedure; this group can also be added enzymatically with T4 polynucleotide kinase. (In an Eppendorf tube, incubate 100 ng of oligonucleotide with 2 units of T4 polynucleotide kinase in 50 m*M* Tris·HCl (pH 7.5), 10 m*M* MgCl2, 5 m*M* DTT, and 0.4 m*M* ATP for 30 min at 37°C.)

3. Anneal the oligonucleotide with the single-stranded DNA in a 500-μl Eppendorf tube containing:

> 1 μg single-stranded DNA
> 10 ng oligonucleotide
> 20 m*M* Tris·HCl (pH 7.4)
> 2 m*M* MgCl$_2$
> 50 m*M* NaCl

4. Mix the solutions together and centrifuge the tube for a few seconds to recollect the liquid. Heat the tube in a flask containing water heated to 70°C. After 5 min, transfer the flask to the lab bench and let it cool to room temperature slowly.

5. Take the tube out of the water bath and put it on ice. Add the following reagents to the tube, for a total volume of 100 μl.

> 20 m*M* Tris·HCl (pH 7.4)
> 2 m*M* DTT
> 0.5 m*M* dATP, dCTP, dGTP and dTTP
> 0.4 m*M* ATP
> 1 unit T7 DNA polymerase
> 2 units T4 DNA ligase

6. After 1 hr, add EDTA to 10 m*M* final concentration.

7. Take 20 μl from the sample and run on an agarose gel. Most of the single-stranded DNA should be converted to covalently, closed circular DNA. Electrophorese some controls in adjacent lanes (i.e., template, template reaction without oligonucleotide). Add T4 DNA ligase to close the double-stranded circular DNA.

8. Extract the remainder of the DNA (~80 μl) by phenol extraction and recover by ethanol precipitation.

9. Electroporate into *dut*$^+$*ung*$^+$ bacteria (i.e., JM101).

10. Harvest the second-generation phage by PEG precipitation (see Chapter 4).

**Notes**: (a) One systematic method of testing different amino acids at a particular residue is to introduce an amber stop codon, TAG, by site-directed mutagenesis. When the recombinant is transferred into bacteria carrying different suppressor tRNAs, different amino acids are incorporated in that position. Such bacteria can be obtained from Promega Corporation (Catalog No. Q5080); each strain will incorporate 1 of 12 different amino acids (i.e., C, E, F, G, H, K, L, Q, P, R, S, Y) during translation of the amber codon. (b) Typically, the products of *in vitro* mutagenesis are a mixture of recombinant and parental molecules. One powerful means of selecting for the mutagenized molecules is to electroporate the double-stranded DNA into bacteria deficient for dUTP pyrophosphatase *(dut)* and uracil-N-glycosylase *(ung)*. Only the synthetic strand, the one incorporating the mutagenic oligonucleotide, will be propagated efficiently (i.e., >80%). Other methods of selection are loss of a restriction site at the site randomized or correction of an accompanying stop codon; thus, the parental DNA sequences can be destroyed by restriction enzyme digestion or propagation in a suppressor-minus bacterial host, respectively. (c) Double-stranded DNA can be used as a template for site-directed mutagenesis as well. Promega (Cat. No. Q6080) has a system, Altered States II, in which the double-stranded DNA is alkaline-denatured to yield single-stranded DNA for oligonucleotide-mediated mutagenesis. In this kit, two additional oligonucleotides are supplied to repair and knock-out the ampicillin and tetracycline resistance genes, respectively, in the pALTER-1 vector. Transformants carrying the replicated strand can be selected on ampicillin-containing plates; furthermore, the resulting colonies should be sensitive to tetracycline when replica plated.

## CASSETTE MUTAGENESIS

A convenient means of introducing mutations at a particular site within a coding region is by cassette mutagenesis. As seen in Fig. 5, the "cassette" can be generated several different ways: (A) by annealing two oligonucleotides together and converting them into double-stranded DNA (Oliphant *et al.,* 1986; Hutchison *et al.,* 1991; Reidhaar-Olson *et al.,* 1991), (B and C) by PCR, and (D) by error-prone PCR. The cassettes formed by these four procedures are fixed in length and coding frame, but have codons which are mutated; thus, cloning and expression of the cassettes will generate a plurality of peptides or proteins that vary by one or more residues. (For the sake of space we have highlighted only these four schemes; there are many possible schemes, and we apologize for omitting anyone's favorite.)

When oligonucleotides are used as the method of introducing mutations into a cassette, the oligonucleotides can be designed several ways. The codons at the randomized positions can be synthesized as NNK or NNS, with N (all four bases), K (G+T), and S (C+G) representing different combinations of incorporated bases to encode all 20 residues. They can also be synthesized as preformed triplets (Virnekäs *et al.,* 1994) or by mixing oligonucleotides synthesized by the split-resin method which together cover all 20 codons at each desired position (Glaser *et al.,* 1992). Conversely, a subset of codons could also be used to favor certain amino acids and exclude others at certain positions (Arkin and Youvan, 1992). Finally, all or some of the codons in the cassette can have some low probability of being mutated. This is accomplished by synthesized oligonucleotides with bottles "spiked" with the other three bases (Hutchison *et al.,* 1986) or by altering the ratio of oligonucleotides mixed together by the split-resin method (Huse *et al.,* 1992). The doping ratio can also differ based on the average amino acid use in natural globular proteins (LaBean and Kauff-man, 1993) or other algorithms (Arkin and Youvan, 1992; Delagrave and Youvan, 1993; Siderovski and Mak, 1993). There is a commercially available computer program, CyberDope, which can be used to aid in determining the base mixtures for synthesizing oligonucleotides with particular doping schemes. A demonstration copy of the CyberDope program for the PC can be obtained by sending an e-mail request to cyberdope@aol.com.

For mutagenesis of short regions, cassette mutagenesis with synthetic oligonucleotides is ideal. Furthermore, several cassettes can be used at a time (Crameri and Stemmer, 1995) to alter several regions simultaneously. This approach is ideal when creating a library of mutant antibodies, where all six complementarity determining regions (CDRs) are altered concurrently.

## ERROR-PRONE PCR

There are several protocols based on altering standard PCR conditions (Saiki *et al.*, 1988) to elevate the level of mutations during amplification. Usually, the rate of mutation during PCR is one nucleotide per 10 kb replicated (Keohavong and Thilly, 1989). However, when the concentration of deoxynucleoside triphosphate is increased and $Mn^{2+}$ is added, the rate of mutation increases significantly to $\sim 7 \times 10^{-3}$ per nucleotide (Cadwell and Joyce, 1992) for *Taq* DNA polymerase. Since the mutations are introduced at random without significant sequence bias, this is a convenient mechanism for generating populations of novel proteins (Cadwell and Joyce, 1994). On the other hand, error-prone PCR is not ideal for altering short peptide sequences because the number of mutations is low; this can be overcome somewhat by recursive rounds of error-prone PCR (Bartel and Szostak, 1993).

1. Design oligonucleotide primers that flank the coding region of interest in the phage. They should be ~21 nucleotides in length and flank the region to be mutagenized. The fragment to be amplified should also carry restriction sites within it to permit easy subcloning in the appropriate vector.

2. Set up the following reaction:

> 30 pmol of each primer
> 20 fmol of the DNA template
> 50 m*M* KCl
> 10 m*M* Tris (pH 8. 3)
> 7 m*M* MgCl$_2$
> 1 m*M* DTT
> 0.2 m*M* dATP, 0.2 m*M* dGTP, 1 m*M* dCTP, 1 m*M* TTP
> Bring the final volume to 88 μl

3. Add 10 μl of 5 m*M* MnCl$_2$. Mix well.

4. Add 5 units of *Taq* DNA polymerase. Cover the liquid with mineral oil. Alternatively, use AmpliWax PCR GEM tubes (Cat. No. N808-0100) by Perkin Elmer

which use a wax barrier to eliminate evaporation and that can be easily penetrated by a pipet tip to withdraw the PCR.

5. Cycle 24 times between 10 sec at 94°C, 30 sec 45°C, and 30 sec at 72°C to amplify fragments up to 1 kb. For longer fragments, lengthen the 72°C step by 30 sec for each kb.

6. The success of the PCR can be verified by gel electrophoresis.

7. Digest the PCR product with the appropriate restriction enzyme(s) to generate sticky ends. It may be advisable to gel purify the restriction fragment. Alternatively, use biotinylated oligonucleotide primers in step 1 for PCR; excess primers and the restriction enzyme generated terminal fragments can then be removed with strepta-vidin–agarose (Gibco; Cat. No. 15942-014). All of the primer molecules need to be biotinylated. Excess primers and terminal fragments can also be removed with Quick Spin columns (Boehringer-Mannheim) or a PCR cleanup kit (Qiagen).

8. Clone the DNA segment into the appropriate vector by ligation.

9. Remove the salts from the ligating DNA and electroporate into bacterial cells.

## DNA SHUFFLING

A newly introduced method for generating mutations *in vitro* is known as DNA shuffling (Fig. 6). This method has been applied to interleukin 1β (IL-1β) (Stemmer, 1994a), and β-lactamase (Stemmer, 1994b). In DNA shuffling, genes are broken into small, random fragments with DNase I, and then reassembled in a PCR-like reaction, but without any primers. The process of reassembling can be mutagenic in the absence of a proofreading polymerase, generating up to 0.7% error when 10- to 50-bp fragments are used. These mutations consist of both transitions and transversion, randomly distributed over the length of the reassembled segment.

1. PCR-amplify the fragment to be shuffled. Often it is convenient to PCR from a bacterial colony or plaque. Touch the colony or plaque with a sterile toothpick and swirl in a standard PCR reaction mix (10 m$M$ Tris·HCl, 50 m$M$ KCl, 1.5 m$M$ MgCl$_2$, 0.2 m$M$ each of dATP, dCTP, dGTP, and dTTP, 0.005% Brij 35, 0.1 to 1 μ$M$ of each primer). Remove the toothpick and heat the reaction for 10 min at 99°C. Cool the reaction to 72°C, add 1–2 units of *Taq* DNA polymerase, cycle the reaction 35 times for 30 sec at 94°C, 30 sec at 45°C, 30 sec at 72°C, and finally heat the sample for 5 min at 72°C. (All the conditions here are for a 1-kb gene.)

2. Remove the free primers, preferably by purification of the DNA with Wizard PCR Preps (Promega, Madison, WI), or by gel purification. Complete primer removal is essential.

3. Fragment about 2–4 μg of the DNA with 0.15 units of DNase I (Sigma, St. Louis, MO) in 100 μl of 50 m$M$ Tris·HCl (pH 7.4), 1 m$M$ MgCl$_2$, for 5–10 min at room temperature. Freeze on dry ice, check size range of fragments on 2% low-melting-point agarose gel or equivalent, and thaw to continue digestion until desired size range is obtained. The desired size range depends on the application; for shuffling of a 1-kb gene, fragments of 100–300 bases are adequate.

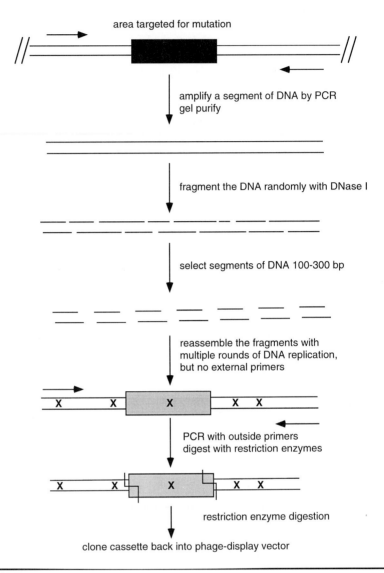

**FIGURE 6**    Generation of mutants through DNA shuffling. A segment of DNA is amplified by PCR and purified free of primers. This segment is then cleaved randomly with DNase I and size selected. The fragments are then denatured, annealed, and extended with DNA polymerase in the absence of any primers. After several cycles, the restored fragment is amplified by PCR. The resulting double-stranded DNA is then cleaved with restriction enzymes, positioned at the end of the area targeted for mutagenesis, to generate a cassette for cloning.

4. Gel purification. Gel-purify the desired DNA fragment size range (100–300 bp) from a 2% low-melting-point agarose gel or equivalent. One easy method is to insert a small piece of Whatman DE-81 ion-exchange paper just in front of the DNA, run the DNA into the paper, put the paper in 0.5 ml 1.2 $M$ NaCl in TE, vortex 30

sec, then carefully spin out all the paper, transfer the supernatant, and add 2 vol of 100% ethanol to precipitate the DNA; no cooling of the sample should be necessary. The DNA pellet is then washed with 70% ethanol to remove traces of salt.

5. DNA reassembly. The DNA pellet is resuspended in PCR mix (Promega) containing 0.2 m$M$ each dNTP, 2.2 m$M$ MgCl$_2$, 50 m$M$ KCl, 10 m$M$ Tris·HCl, pH 9.0, 0.1% Triton X-100, at a high concentration of 10–30 ng of fragments per microliter of PCR mix (typically 100–600 ng per 10–20 μl PCR reaction). No primers are added in this PCR reaction. *Taq* DNA polymerase (Promega) alone can be used if a substantial rate of mutagenesis (up to 0.7% with 10- to 50-bp DNA fragments) is desired. The inclusion of a proofreading polymerase, such as a 1/30 (vol/vol) mixture of *Taq* and *Pfu* DNA polymerase (Stratagene, San Diego, CA) is expected to yield a lower error rate (Barnes, 1994) and allows the PCR of very long sequences. A program of 30–45 cycles of 30 sec 94°C, 30 sec 45–50°C, 30 sec 72°C in an MJ Research PTC-150 minicycler (Cambridge, MA) is run. The progress of the assembly can be checked by gel analysis but this is normally not necessary. The PCR product at this point should contain the correct size product in a smear of larger and smaller sizes.

6. Amplification. The correctly reassembled product of this first PCR is amplified in a second PCR reaction which contains the outside primers. Aliquots of 2.5 μl of the PCR assembly are diluted 40x with PCR mix containing 0.8 μ$M$ of each primer. A PCR program of 20 cycles of 30 sec 94°C, 30 sec 50°C, and 30–45 sec at 72°C is run, with 5 min at 72°C at the end. This amplification results in a large amount of PCR product of the correct size.

7. Cloning. The best PCR product is then digested with terminal restriction enzymes, gel-purified, and cloned back into a phage or phagemid genome.

**NOTES**: (a) Site-specific recombination can also be used, for example, to shuffle heavy and light antibody chains inside infected bacterial cells as a means of increasing the binding affinity and specificity of antibody molecules. It is possible to use the *Cre/lox* system (Waterhouse *et al.,* 1993; Griffiths *et al.,* 1994) and the *int* system (See Chapter 12). Such approaches also have the attraction of increasing the diversity of phage-displayed antibody libraries. (b) It is possible to take recombinants and to shuffle them together to combine advantageous mutations that occur on different DNA molecules (Stemmer, 1994b). Figure 7 demonstrates this process. It is also possible to take a recombinant displayed insert and to "backcross" with parental sequences by DNA shuffling to remove any mutations that do not contribute to the desired traits (Stemmer, 1994a, b). Recombination is an efficient means of searching sequence space (Stemmer, 1995).

## CONCLUSION

Once one has isolated a phage-displayed recombinant with desirable properties, it is generally appropriate to improve or alter the binding properties through a round of molecular evolution. We have generated second-generation libraries of displayed peptides and antibodies, as described in this chapter, and isolated phage with improved

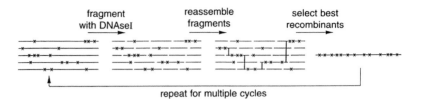

**FIGURE 7**      DNA shuffling of populations of molecules. One important aspect of DNA shuffling is the capacity to have recombination occur among populations of DNA molecules. A population of DNA inserts from isolated phage displaying proteins can be prepared by PCR and fragmented with DNase I. By the process of DNA shuffling, the inserts are reassembled, creating new combinations of preexisting mutations in the protein coding region. A library of phage can be generated with newly formed DNA molecules and screened for the best phage recombinants. This process of adaptive molecular evolution can be repeated several times. It can also occur with an excess of parental DNA molecules to "backcross" out silent mutations.

(i.e., 3- to 1000-fold) apparent binding strength. Thus, through repeated rounds of library generation and selection it is possible to "hill-climb" through sequence space to optimal affinity.

## References

Arkin, A., and Youvan, D. (1992). Optimizing nucleotide mixtures to encode specific subsets of amino acids for semi-random mutagenesis. *Bio/Technology* **10**, 297–300.

Barbas, C., III, Hu, D., Dunlop, N., Sawyer, L., Cababa, D., Hendry, R. M., Nara, P. L., and Burton, D. R. (1994). *In vitro* evolution of a neutralizing human antibody to human immunodeficiency virus type 1 to enhance affinity and broaden strain cross-reactivity. *Proc. Natl. Acad. Sci. U.S.A.* **91**, 3809–3813.

Barnes, W. M. (1994). PCR amplification of up to 35-kb DNA with high fidelity and high yield from lambda bacteriophage templates. *Proc. Natl. Acad. Sci. U.S.A.* **91**, 2216–2220.

Barrett, R. W., Cwirla, S. E., Ackerman, M. S., Olson, A. M., Peters, E. A., and Dower, W. J. (1992). Selective enrichment and characterization of high affinity ligands from collections of random peptides on filamentous phage. *Anal. Biochem.* **204**, 357–364.

Bartel, D., and Szostak, J. (1993). Isolation of new ribozymes from a large pool of random sequences. *Science* **261**, 1411–1418.

Cadwell, R., and Joyce, G. F. (1992). Randomization of genes by PCR mutagenesis. *PCR Methods Appl.* **2**, 28–33.

Cadwell, R., and Joyce, G. (1994). Mutagenic PCR. *PCR Methods/Appl.* **3**, S136–S140.

Crameri, A., and Stemmer, W. P. C. (1995). Combinatorial multiple cassette mutagenesis creates all the permutations of mutant and wildtype sequences. *BioTechniques* **18**, 194–196.

Delagrave, S., and Youvan, D. C. (1993). Searching sequence space to engineer proteins: Exponential ensemble mutagenesis. *Bio/Technology* **11**, 1548–1552.

Fowlkes, D., Adams, M., Fowler, V., and Kay, B. (1992). Multipurpose vectors for peptide expression on the M13 viral surface. *BioTechniques* **13**, 422–427.

Glaser, S. M., Yelton, D. E., and Huse, W. D. (1992). Antibody engineering by codon-based mutagenesis in a filamentous phage vector system. *J. Immunol.* **149**, 3903–3913.

Griffiths, A. D., Williams, S. C., Hartley, O., Tomlinson, I. M., Waterhouse, P., Crosby, W. L., Kontermann, R. E., Jones, P. T., Low, N. M., Allison, T. J., Prospero, T. D., Hoggenboom, H. R., Nissim, A., Cox, J. P. L., Harrison, J. L., Zaccolo, M., Gheradi, E., and Winter, G. (1994). Isolation of high affinity human antibodies directly from large synthetic repertoires. *EMBO J.* **13**, 3245–3260.

Huse, W. D., Stinchcombe, T. J., Glaser, S. M., Starr, L., MacLean, M., Hellström, K. E., and Yelton, D. E. (1992). Application of a filamentous phage pVIII fusion protein system suitable for efficient production, screening, and mutagenesis of F(ab) antibody fragments. *J. Immunol.* **149,** 3914–3920.

Hutchison, C. A., Phillips, S., Edgell, M. H., Gillam, S., Jahnke, P., and Smith, M. (1978). Mutagenesis at a specific position in a DNA sequence. *J. Biol. Chem.* **253,** 6551–6560.

Hutchison, C. A., Nordeen, S. K., Vogt, K., and Edgell, M. H. (1986). A complete library of point substitution mutations in the glucocorticoid response element of mouse mammary tumor virus. *Proc. Natl. Acad. Sci. U.S.A.* **83,** 710–714.

Hutchison, C. A., Swanstrom, R., and Loeb, D. (1991). Complete mutagenesis of protein coding regions. *In* "Methods in Enzymology" (J. J. Langone, ed.), vol. 202, pp. 356–390. Academic Press, San Diego, CA.

Kauffman, S. A. (1992). Applied molecular evolution. *J. Theor. Biol.* **157,** 1–7.

Keohavong, P., and Thilly, W. G. (1989). Fidelity of DNA polymerases in DNA amplification. *Proc. Natl. Acad. Sci. U.S.A.* **86,** 9253–9257.

Kunkel, T. A. Bebenek, K., and McClary, J. (1991). Efficient site-deirected mutagenesis using uracil-containing DNA. *Methods Enzymol.* **204,** 125–139.

LaBean, T. H., and Kauffman, S. A. (1993). Design of synthetic gene libraries encoding random sequence proteins with desired ensemble characteristics. *Protein Sci.* **2,** 12449–1254.

Oliphant, A., Nussbaum, A., and Struhl, K. (1986). Cloning of random sequence oligodeoxynucleotides. *Gene* **44,** 177–183.

Reidhaar-Olson, J., Bowie, J., Breyer, R., Hu, J., Knight, K., Lim, W., Mossing, M., Parsell, D., Shoemaker, K., and Sauer, R. (1991). Random mutagenesis of protein sequences using oligonucleotide cassettes. *In* "Methods in Enzymology" (R. T. Sauer, ed.), vol. 208, pp. 564–587. Academic Press, San Diego, CA.

Roberts, B., Markland, W., Ley, A., Kent, R., White, D., Guterman, S., and Ladner, R. (1992). Directed evolution of a protein: Selection of potent neutrophil elastase inhibitors displayed on M13 fusion phage. *Proc. Natl. Acad. Sci. U.S.A.* **89,** 2429–2433.

Saiki, R. K., Gelfand, D. H., Stoffel, S., Scharf, S. J., Higuchi, R., Horn, G. T., Mullis, K. B., and Erlich, H. A. (1988). Primer-directed enzymatic amplification of DNA with a thermostable DNA polymerase. *Science* **239,** 487–491.

Siderovski, D. P., and Mak, T. M. (1993). RAMHA: A PC-based Monte-Carlo simulation of random saturation mutagenesis. *Comput. Biol. Med.* **23,** 463–474.

Stemmer, W. P. C. (1994a). Rapid evolution of a protein *in vitro* by DNA shuffling. *Nature (London)* **370,** 389–391.

Stemmer, W. P. C. (1994b). DNA shuffling by random fragmentation and reassembly: *In vitro* recombination for molecular evolution. *Proc. Natl. Acad. Sci. U.S.A.* **91,** 10747–10751.

Stemmer, W. P. C. (1995). Searching sequence space. *Bio/Technology* **13,** 549–553.

Virnekäs, B., Ge, L., Plückthun, A., Schneider, K. C., Wellnhofer, G., and Moroney, S. E. (1994). Trinucleotide phosphoramidites: Ideal reagents for the synthesis of mixed oligonucleotides for random mutagenesis. *Nucleic Acids Res.* **22,** 5600–5607.

Wang, L.-F., du Plessis, D. H., White, J. R., Hyatt, A. D., and Eaton, B. T. (1995). Use of a gene-targeted phage display random epitope library to map an antigenic determinant on the bluetongue virus outer capsid protein VP5. *J. Immunol. Methods* **178,** 1–12.

Waterhouse, P., Griffiths, A. D., Johnson, K. S., and Winter, G. (1993). Combinatorial infection and *in vivo* recombination: A strategy for making large phage antibody repertoires. *Nucleic Acids Res.* **21,** 2265–2266.

Widersten, M., and Mannervik, B. (1995). Glutathione transferases with novel active sites isolated by phage display from a library of random mutants. *J. Mol. Biol.* **250,** 115–122.

# *17*

# Nonradioactive Sequencing of Random Peptide Recombinant Phage

## *Barbara L. Masecar, Christopher A. Dadd, George A. Baumbach, and David J. Hammond*

## INTRODUCTION

Issues of cost, disposal, and safety have made nonradioactive detection methods for DNA sequencing and other probe hybridization techniques the method of choice for many laboratories. Sanger di-deoxynucleotide chain-termination sequencing with a biotinylated oligonucleotide primer and subsequent chemiluminescent detection results in autoradiograms that rival the sensitivity of standard $^{32}$P and $^{35}$S sequencing methodology. Chemiluminescent detection methods utilize the dioxetane conjugated phosphoester family of luminescent substrates such as CSPD (disodium 3-{4-methoxyspiro[1,2-dioxetane-3,2′-(5′-chloro)tri-cyclo{3.3.1.1$^{3,7}$} decan]-4-yl} phenyl phosphate) or AMPPD (disodium 3-{methoxyspiro[1,2-dioxetane-3,2′-tricyclo-{3.3.1.1.$^{3,7}$}decan]-4-yl}phenyl phosphate), which are substrates for alkaline phosphatase. Dephosphorylation of these dioxetane substrates results in light emission at approximately 460 nm which is used to create exposures on X-ray film. CSPD is thought to be more sensitive with higher-quality band resolution than AMPPD.

Our standard protocol consists of 3 major parts: (A) asymmetric amplification of single-stranded phage DNA; (B) cycle sequencing with di-deoxynucleotides and a biotinylated primer; and (C) electroblotting of the sequencing gel onto a nylon membrane with detection of the transferred DNA by chemiluminescence. A general

*Phage Display of Peptides and Proteins*

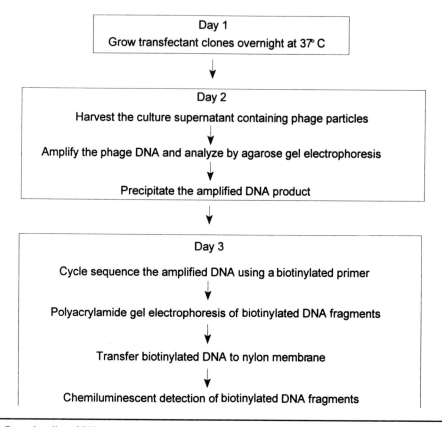

**FIGURE I**      General outline of DNA sequencing of recombinant phage with chemiluminescent detection.

outline is given in Fig. 1. By following this protocol, we typically find that greater than 90% of selected clones result in readable sequences at the first attempt.

The primers flanking the random insert region in the pIII gene of fUSE 5 phage libraries (generous gift of G. Smith, University of Missouri) that we employ to amplify inserts correspond to bases 1504–1523 (upstream, plus-strand primer) (see below) and the complementary sequence of bases 1708–1727 (minus-strand primer) (see Reagents for source of primers). The sequence of the upstream primer is identical in M13, f1, and fd native sequences at the same base positions; the minus-strand downstream primer corresponds to the complementary sequence of bases 1684 to 1704 in these closely related phage (Hill and Petersen, 1982).

We use the upstream primer in a five-fold molar excess over the downstream primer for asymmetric amplification favoring synthesis of the plus-strand, and subsequently sequence with biotinylated downstream primer. Our largest library contains a 15-amino acid residue random insert region; however, this protocol should be readily applicable to sequencing inserts of at least 60 base pairs without modification. Sequence verification is provided by sequencing with biotinylated internal primer corresponding to bases 1596 to 1616 in fUSE 5 virions (also identical in

<u>Upstream primer</u>  (G + C = 50%; T$_m$ = 60°C)

```
   1504                        1523
    |                           |
5' CAC CTC GAA AGC AAG CTG AT 3'
```

<u>Downstream primer</u>  (G + C = 50%; T$_m$ = 60°C)

```
   1727                        1708
    |                           |
5' GTC GTC TTT CCA GAC GTT AG 3'
```

<u>Sequencing primer</u>  (G + C = 50%; T$_m$ = 60°C)

```
       1727                            1708
        |                               |
5' BIOTINYL-GTC GTC TTT CCA GAC GTT AG 3'
```

M13, f1, and fd native sequences). Alternatively, the primer ratios can be inverted and sequencing performed with biotinylated upstream primer.

fUSE 5 phage carry the tetracycline resistance determinant from Tn10 (Zacher *et al.*, 1980). When using the replication-defective fUSE phage it is necessary to amplify the phage from a single transfected colony to obtain enough single-stranded phage DNA for sequencing. The protocol for this is given below. However, when using nonreplicative deficient phage, direct linear amplification sequencing from single plaques or bacterial colonies is possible (Krishnan *et al.*, 1991), and the CircumVent protocol should successfully sequence the nanogram quantities of DNA template in these preparations.

## Equipment and Reagents

1. Thermal cycler, standard sequencing reagents and apparatus, Hoefer Instruments Gene-Sweep (Model TE90, Hoefer Scientific Instruments, San Francisco, CA) or other large format electroblotter (substitution of electroblotting with capillary transfer is possible; see below), Spectrolinker UV crosslinker (Model XL-1500, Spectronics Corp., Westbury NY; or transilluminator can be substituted), rolling bottle hybridization incubator (Model 310, Robbins Scientific, Sunnyvale, CA) (or equivalent), fiberglass window screen (cut to approximately 35 x 60 cm; available at most hardware stores), polyvinyl-chloride plastic wrap (Fisherbrand all-purpose laboratory wrap, Cat. No. 15-1618, Fisher Scientific, Pittsburgh, PA), benchtop platform rotator, standard autoradiogram development reagents and equipment.

2. Terrific broth (LB can substitute): In 900 ml distilled deionized water (ddH$_2$O), dissolve:

> 12 g Bacto-tryptone
> 24 g yeast extract
> 4 ml (5.04 g) glycerol

Autoclave 90-ml aliquots in 125-ml glass bottles for 20 min at 120°C; when cool, add to each bottle 10 ml of separately autoclaved potassium phosphate buffer (0.17 $M$ KH$_2$PO$_4$, 0.72 $M$ K$_2$HPO$_4$, unadjusted pH approximately 7.4). The potassium phosphate buffer is made by dissolving 2.31 g KH$_2$PO$_4$ (anhydrous) and 12.54 g K$_2$HPO$_4$ (anhydrous) in 90 ml ddH$_2$O, adjusting volume to 100 ml, and autoclaving as above.

3. CircumVent Sequencing Kit [New England BioLabs (NEBL), Beverly, MA; Cat. No. 430]: includes Vent$_R$ (exo$^-$) DNA polymerase, 10X CircumVent buffer, Triton X-100, di-deoxynucleotide triphosphates (ddNTPs), and bromphenol blue/xylene cyanol stop dye.

4. SeqLight chemiluminescence DNA detection kit and Tropilon-Plus nylon membrane (Cat. Nos. SQ100D and XQ100M, respectively, Tropix Inc., Bedford, MA).

5. Deoxynucleotide triphosphates (dNTPs) for DNA amplification: dNTPs (dATP, dGTP, dCTP, and dTTP) from NEBL (Cat. No. 446, not included in sequencing kit) are 100 m$M$. Make a 1:10 dilution of combined dNTPs; add 1 μl of this 1:10 working solution to each amplification reaction (equivalent to 0.25 μl of each individual dNTP at 1:10, in a final reaction volume of 50 μl). Final concentration of each dNTP is 50 μ$M$.

6. 2% agarose in 1X Tris-phosphate buffer (TPE). Molecular biology grade agarose. 10X TPE: 108 g Tris base, 15.5 ml 85% phosphoric acid (1.679 g/ml) 40 ml 0.5 $M$ EDTA (pH 8.0). Agarose gel loading buffer: 0.25% bromphenol blue 40% (w/v) sucrose in ddH$_2$0.

7. Oligonucleotide primers: Sequences are given above. All oligonucleotides were purchased from Midland Certified Reagents (Midland, TX) at ≥95% purity (HPLC and gel filtration purified). Sequencing primers are purified at the source after reaction with phosphoramidite-biotin at the end of oligonucleotide synthesis. These are received as OD units from Midland, where each unit is equivalent to 30 to 35 μg of primer. For a 20-mer, resuspend each 1 OD unit in 500 μl of ddH2O (equivalent to 10 pmol/μl), and store in 50-μl aliquots at -70°C.

8. Miscellaneous solutions: 10 $M$ ammonium acetate; Pierce Blocker Casein (Cat. No. 37528, Pierce Chemical Co., Rockford, IL), chloroform/isoamyl alcohol (24:1), 10X SSC (1.5 $M$ sodium chloride, 0.15 $M$ trisodium citrate dihydrate), pBR322-*Msp*I digest (New England Bio Labs, Cat. No. 303-2S), 95% ethanol (molecular biology grade).

9. Bacterial strains: *Escherichia coli* k91 Kan (see Appendix for genotype).

## AMPLIFICATION OF PHAGE DNA

1. Pick individual tet$^r$ *E. coli* k91 Kan transfectant clones and transfer into 2 ml of Terrific broth with 100 μg/ml kanamycin and 40 μg/ml tetracycline. Select as many clones as will fit in the thermocycler in step 3, one clone per tube. Grow for 6 hr to overnight at 37°C with vigorous aeration. Centrifuge 1 ml of culture in microfuge tube, 5 min, 13,000 rpm. Transfer culture supernatant to clean tube.

2. Combine the following in 0.5-ml microfuge tubes, in order, per reaction:

> 5 µl CircumVent buffer
> 1.7 µl Triton X-100 (3%)
> 1 µl dNTP working solution (see Reagents, above)
> 50 pmol forward primer (5 µl)
> 10 pmol reverse primer (1 µl)
> 2 µl Terrific Broth- or LB-grown phage supernatant
> 34.3 µl sterile $ddH_2O$
>
> 50 µl total volume

Include a (-) template control.

3. Hot start the mixture for 5 min at 95°C; place on ice for 1 min; spin briefly at 13,000 rpm.

4. Add 1 µl $Vent_R$ (exo-) enzyme per reaction, mix (avoid bubbling), and add 2 drops of mineral oil.

5. Place tubes into thermocycler. Our program with the primers specified above is 30 cycles of 95°C denaturation (20 sec), 55°C annealing (20 sec), and 72°C elongation (20 sec).

6. Analyze 5 µl of the aqueous phase (with 0.5 µl DNA sample buffer) by electrophoresis in 2% agarose/TPE minigel, containing 0.5 µg/ml ethidium bromide (100 V, 20 min). Include pBR322-*Msp*I digest as molecular weight standard. Using the fUSE 5 libraries and primers specified, the amplified DNA typically appears as a single band of approximately 200 base pairs.

## Precipitation of Amplified DNA

1. Add 50 µl chloroform/isoamyl alcohol (24/1) to invert the oil overlay and aqueous reaction phases. Vortex and spin briefly at 13,000 rpm at room temperature.

2. Remove 40 µl of the upper phase to a clean tube and add 10 µl 10 *M* ammonium acetate; mix. Add 100 µl cold (-20°C) 95% ethanol, mix, and place at -20°C for at least 30 min. Spin at 13,000 rpm for 15 min at -5°C.

3. Remove the supernatant and dry the pellets (in a centrifugal evaporator or by passing dry compressed air over the tops of the tubes). Resuspend the pellet in 20 µl sterile $ddH_2O$.

## SEQUENCING OF AMPLIFIED DNA

Sequence the amplified DNA according to NEBL CircumVent Thermal Cycle Dideoxy DNA Sequencing Kit, according to the manufacturer's recommended protocol, with the following modifications.

1. Prepare a sequencing Master Mix, in the following order (volumes are per reaction):

> 1.5 µl 10X Circumvent buffer
> 1 µl Triton X-100 (3%)
> 0.75 µl of biotinylated sequencing primer at 10 pmol/µl (7.5 pmol)

2. Aliquot 3.25 μl of Master Mix to each template tube. Add 3 μl precipitated DNA and ddH$_2$O for a total volume of 14 μl.

3. Hot start reaction mixture at 95°C for 5 min in thermocycler. Place reactions on ice for 2 min and spin briefly.

4. Add 1 μl Vent$_R$ (exo-) polymerase, mix (avoid bubbles), and immediately dispense 3.2 μl to each of four ddNTP tubes containing 3 μl of G mix, A mix, T mix, or C mix, according to manufacturer's protocol. Mix (avoid bubbles), overlay with mineral oil, and load tubes into thermocycler.

5. Cycle as above [30 cycles of 95°C denaturation (20 sec), 55°C annealing (20 sec), and 72°C elongation (20 sec)]. Add 4 μl of CircumVent stop dye to each tube at the conclusion of cycle sequencing.

## Polyacrylamide Gel Electrophoresis

Electrophorese 3 to 5 μl of each reaction tube in a standard 8% acrylamide–urea sequencing gel, as described elsewhere in this manual.

# CHEMILUMINESCENT DETECTION

## Transfer of DNA to Positively Charged Nylon Membrane

Chemiluminescent detection of sequencing fragments requires transfer of the biotinylated DNA to a positively charged nylon-66 membrane (0.45 μm), such as Tropilon Plus (Tropix, Bedford, MA). We transfer the DNA either with a Hoefer Transblotter (Hoefer Instruments, San Francisco, CA ) or by capillary action (both described below). The DNA is then covalently crosslinked to the membrane by UV irradiation. Wear a fresh pair of latex gloves throughout this entire procedure to reduce phosphatase contamination of the nylon membrane.

### Electrotransfer of DNA to nylon membrane

DNA fragments are electrophoretically transferred from the sequencing gel to a positively charged nylon-66 membrane using the Hoefer GeneSweep Sequencing Gel Transfer Unit as follows:

1. Prepare three sheets of 3MM blotting paper (Whatman LabSales, Hillsboro, OR, or equivalent): 1 sheet 34 x 59 cm and two sheets 2 cm larger than the gel on all sides. Also cut a piece of positively charged nylon membrane (Tropilon Plus, Tropix) 2 cm larger than the gel (or the region of interest on the gel, see Note below) on all sides.

2. Disassemble the sequencing gel plates with the gel remaining on the back (smaller) plate.

3. Wet the 34 x 59-cm piece of blotting paper in 1X TBE, place it on the transfer platform, and remove any air bubbles.

NOTE: When electrophoresed for 1.5 hr, the insert region of the pIII gene in the fUSE 5 6- and 15-mer libraries typically lies approximately 10 cm down from the top edge of the gel, and 10 cm up from the bottom edge. Nylon use can be minimized by cutting blotting paper to include only the region of interest on the gel, then carefully cutting the gel along the edge of the blotting paper with a scalpel or razor blade and discarding the unwanted gel. The gel/blotting paper can then be separated from the short plate as described below.

4. Wet a piece of the blotting paper cut 2 cm larger than the gel in 1X TBE and blot off excess buffer with another sheet of blotting paper. Starting at the bottom of the gel, lay the damp blotting paper on the gel moving slowly toward the top, being careful not to introduce air bubbles. It may be helpful to roll a 10-ml pipette along the blotting paper as it is laid down.

5. Turn the glass plate over so that the gel is facing up and is on top of the blotting paper. Starting at the bottom of the gel, slowly slide the gel and glass plate over the edge of the lab bench, allowing the blotting paper and gel to pull off of the glass plate. Support the gel as it pulls away from the glass plate.

6. Lay the gel face up on the electrotransfer platform, avoiding air bubbles between the sheets of blotting paper.

7. Wet the precut nylon membrane in 1X TBE. Starting at one end of the gel, lay the nylon membrane on the gel, rolling a pipette to avoid trapping air bubbles.

8. Wet the last piece of blotting paper and lay it onto the nylon using the above methods to avoid air bubbles.

9. With the transfer arm (anode) at the far left position, lower it onto the gel stack and turn the control switch to "run." The transfer should run between 1.8 and 2.0 A. If the current is less than 1.0 A, the gel stack may be too dry and should be sprayed lightly with 1X TBE.

10. When the arm reaches the right end of the platform, it will stop and the power will shut off automatically. Lift the arm to the up position, and set the control switch to "rev." The arm will return to the left starting position.

11. Disassemble the stack and peel the nylon membrane slowly from the gel. If the gel sticks to the membrane, spray the membrane with 1X TBE to remove the gel.

## Capillary transfer of DNA to nylon membrane

1. Prepare four sheets of 3MM blotting paper, one that is 2 cm wider and 20 cm longer than the sequencing gel and three that are the same size as the gel. Also cut a piece of positively charged nylon membrane (Tropilon Plus, Tropix) that is the same size as the gel.

NOTE: When electrophoresed for 1.5 hr the insert region of the pIII gene in the fUSE 5 6- and 15-mer libraries typically lies approximately 10 cm down from the top ed3ge of the gel, and 10 cm up from the bottom edge. Nylon use can be minimized by cutting blotting paper to include only the region of interest on the gel, then carefully cutting the gel along the edge of the blotting paper with a scalpel or razor blade and discarding the unwanted gel. The gel/blotting paper can then be separated from the short plate as described below.

2. Disassemble the sequencing gel plates with the gel remaining on the back (smaller) plate.

3. Lay the large piece of blotting paper on top of the gel, leaving 10 cm excess at both ends to act as wicks.

4. Turn the glass plate over, with the gel now facing up and on top of the blotting paper. Press the glass plate against the gel to ensure that it will adhere to the blotting paper. Starting at the bottom of the gel, slowly slide the gel and glass plate over the edge of the lab bench, allowing the blotting paper and gel to pull off of the glass plate. Support the gel as it separates from the glass plate.

5. Elevate a sequencing gel glass plate approximately 5 cm off the lab bench (on top of a test tube rack) and lay the blotting paper and gel face up with the excess paper hanging over the ends of the glass plate. Place each wick end into a dish containing 10X SSC buffer (Fig. 2).

6. Cut thin strips of plastic wrap and place along the edge of the gel to keep the lower and upper stacks from coming in contact and short-circuiting the transfer (do not cover areas to be transferred).

7. Starting at one end of the gel, lay the precut dry nylon membrane on the gel using a pipette as an aid to roll out air bubbles.

8. Wet one piece of the remaining blotting paper in 10X SSC and lay it on top of the nylon membrane, using the above method to avoid air bubbles.

9. Place the last two pieces of dry blotting paper on the stack followed by a 5-cm stack of paper towels.

10. Lay another glass plate on the stack with a 1-kg weight on top. Allow the transfer to proceed for 2 hr at room temperature.

11. Upon completion of the transfer, disassemble the stack and peel the nylon membrane slowly from the gel.

## DNA to Nylon Crosslinking

Crosslink the DNA fragments to the nylon by exposing the membrane to 144 mJ/cm$^2$; this is programmable on the Spectronics Crosslinker. Alternatively, a

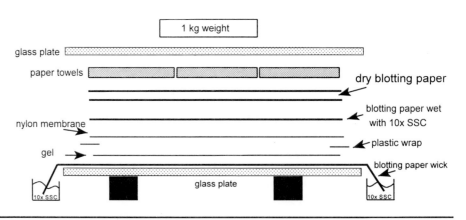

**FIGURE 2**          Assembly of capillary transfer of DNA fragments from sequencing gel to nylon membrane.

transilluminator can be used by exposing slightly overlapping quadrants of the nylon to the UV source. Ten minutes per section of cling-wrap-enclosed membrane has worked well for us in the past.

## Chemiluminescent DNA Detection

Wear a fresh pair of latex gloves throughout this procedure and use sterile water and vessels to make and store reagents. Contaminating bacterial phosphatases will interfere with the chemiluminescent reaction.

1. Lay the membrane onto a fiberglass screen of similar size and roll the screen and membrane up together (this prevents nylon-to-nylon overlap). Slide into a Rollins rolling bottle and add 200 ml of Pierce Blocker Casein. Roll at 65°C 1 hr to overnight.

NOTE: It is best to orient the nylon so that the DNA side is facing the outside of the rolled membrane/screen.

2. Unroll the membrane and lay into plastic wrap-lined tray containing 300 ml of conjugate solution, with DNA side facing up. Conjugate solution: 300 ml of 0.2% I-Block reagent in 1X PBS (or 1X Pierce Blocker Casein) with approximately 50 ng/ml of streptavidin–alkaline phosphatase conjugate (for example, 60–120 μl of Avidx-AP conjugate; Tropix). 10X PBS: 0.58 $M$ disodium phosphate, 0.17 $M$ monosodium phosphate, 0.68 $M$ NaCl, pH 7.4. Incubate 30 min at room temperature, rotating at 100 rpm. Do NOT extend incubation, as background will increase.

NOTE: For each new solution, use a tray lined with a clean piece of plastic wrap. Fisher wrap is good here because it is wider than SaranWrap. Pick up the blot by two corners with forceps, let drain, and lay into the next solution. Avoid trapping air bubbles under the blot. Rotate all washes at 100 rpm.

3. Wash twice with 300 ml each 0.2% I-Block reagent in PBS with 0.5% SDS, 5 min each wash.
4. Wash three times with 300 ml each PBS with 0.5% SDS, 10 min each wash.
5. Wash twice with 200 ml each Assay buffer, 2–5 min each wash. Assay buffer: 0.1 $M$ diethanolamine, 1 m$M$ magnesium chloride, pH 10.
6. Incubate with 50 ml CSPD solution (0.25 m$M$ in Assay buffer), hand-rotating the tray to ensure even distribution of the substrate, for 8 to 10 min. Drain the blot and lay it onto an exposed sheet of X-ray film (for backing), avoiding air bubbles. If the blot begins to dry out, add back a few drops of CSPD solution between the blot and the backing. Immediately cover with plastic wrap (do not let the membrane dry out); gently roll out any air bubbles with a pipet. Ensure that the plastic wrap covers the membrane and backing totally; leaking CSPD solution will create black areas on the X-ray film.

NOTE: CSPD is very stable and leftover solution can be decanted into a storage bottle, frozen at -20°C, and reused two to three times.

Barbara L. Masecar *et al.*

7. Assemble into exposure cassette against Kodak X-OMAT XAR film; expose 1 hr to overnight. The film can be cut to the same size as the membrane to economize use of the film. Enhancing screens are not recommended. The X-ray film can be layered to compensate for very strong chemiluminescent signals, which may be better seen on a more distant film. Develop film according to standard autoradiogram development techniques. A typical autoradiogram is shown in Fig. 3.

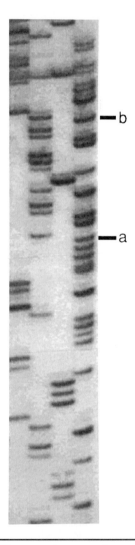

**FIGURE 3**        Example of a section of a typical DNA sequencing autoradiogram (2-hr exposure) using chemiluminescent detection. Order of lanes: G, A, T, C. DNA template was recombinant fUSE 5 phage with first base of insert (a) and last base of insert (b) as indicated. Transfer of DNA to nylon membrane was accomplished by capillary transfer.

# TROUBLESHOOTING GUIDE

## Weak Signal on Autoradiogram

Increase exposure time.

Repeat with fresh CSPD solution (if the solution was reused).

Ensure that all other reagents (including X-ray film developer) are fresh and that contaminating phosphatases are avoided during the detection. Allow the membrane to contact CSPD solution for at least 8 to 10 min to decrease required exposure times.

Verify that amplification of phage DNA gives visible band on agarose gel analysis.

Aliquot all primers and store at -70°C. Do not store biotinylated primers more than 6 months.

Verify that capillary or electrotransfer occurred; transfer of sequencing stop dye is one indicator. Follow above instructions for transfer exactly. Most problems arise from insufficient transfer of DNA onto the nylon membrane; the above transfer protocols have proven reliable and reproducible.

If detection reagents are in question, the membrane can be stripped by overnight incubation in 5% SDS in PBS at 65°C, and the detection steps repeated.

The concentrations of ddNTPs during sequencing can be increased to favor generation of readable sequence close to the biotinylated primer.

## High background

We have found Pierce Blocker Casein at 65°C to be superior to nonfat dry milk, 5% gelatin, and 5% human serum albumin (all with and without 0.1% Tween 20) in blocking nonspecific sites on the nylon membrane. The I-Block provided with the Tropix kit was also effective and could be substituted. Ensure that the fiberglass screen is preventing overlap of the nylon membrane.

### References

Hill, D. F., and Petersen, G. B. (1982). Nucleotide sequence of bacteriophage f1 DNA. *J. Virol.* **44,** 32–46.

Krishnan, B. R., Blakesly, R. W., and Berg, D. E. (1991). Linear amplification DNA sequencing directly from single phage plaques and bacterial colonies. *Nucleic Acids Res.* **19,** 1153.

Zacher, A. N., Stock, C. A., Golden, J. W., II, and Smith, G. P. (1980). A new filamentous phage cloning vector: fd-tet. *Gene* **9,** 127–140.

# Measurement of Peptide Binding Affinities Using Fluorescence Polarization

## *Thomas Burke, Randall Bolger, William Checovich, and Robert Lowery*

## INTRODUCTION TO FLUORESCENCE POLARIZATION

Fluorescence polarization was first described almost 70 years ago (Perrin, 1926) and is now a versatile laboratory technique for measuring equilibrium binding and enzyme catalysis. Fluorescence polarization assays are homogeneous in that they do not require a separation step such as centrifugation, filtration, chromatography, precipitation, or electrophoresis. These assays are done in real time, directly in solution, and do not require an immobilized phase. Polarization values can be measured repeatedly and after the sequential addition of reagents since measuring polarization is rapid and does not destroy the sample. Generally, this technique can be used to measure polarization values of molecules with concentrations from low picomolar to micromolar levels. The purpose of this chapter is to describe how fluorescence polarization can be used in a simple and quantitative way to measure equilibrium binding affinities between peptides and their target molecules. The binding between an antibody and its cognate peptide is described below but any large molecule (protein, DNA, cell surface, etc.) could be used in a binding study.

When a fluorescently labeled molecule is excited with plane polarized light, it emits light that has a degree of polarization that is inversely proportional to its molecular rotation. Large fluorescently labeled molecules remain relatively stationary during the excited state (4 nsec in the case of fluorescein) and the polarization of the

*Phage Display of Peptides and Proteins*
Copyright © 1996 by Academic Press, Inc. All rights of reproduction in any form reserved.

light remains relatively constant between excitation and emission. Small fluorescently labeled molecules rotate rapidly during the excited state and the polarization changes significantly between excitation and emission. Therefore, small molecules have low polarization values and large molecules have high polarization values. For example, a fluorescein-labeled peptide has a low polarization value but, when bound to an antibody, has a high polarization value.

Fluorescence polarization ($P$) is defined as:

$$P = \frac{\text{Int}_{\parallel} - \text{Int}_{\perp}}{\text{Int}_{\parallel} + \text{Int}_{\perp}},$$

where $\text{Int}_{\parallel}$ is the intensity of the emission light parallel to the excitation light plane and $\text{Int}_{\perp}$ is the intensity of the emission light perpendicular to the excitation light plane. $P$, being a ratio of light intensities, is a dimensionless number. The Beacon System used in these experiments expresses polarization in millipolarization units (1 Polarization Unit = 1000 mP Units).

Fluorescence anisotropy ($A$) is another term commonly used to describe this phenomenon of measuring molecular rotation and size by monitoring the change in emission polarization. In this chapter, we will use only the term polarization instead of anisotropy, but the literature uses the terms somewhat interchangeably. Polarization and anisotropy are related in the following way.

$$A = \frac{2P}{3 - P}$$

The relationship between molecular rotation and size is described by the Perrin equation but it is beyond the scope of this chapter to explicate the Perrin equation. The reader is referred to Jolley (1981), which gives a more thorough explanation. Summarily, the Perrin equation states that polarization is directly proportional to the **correlation time,** the time that it takes a molecule to rotate through an angle of approximately 68.5°. Correlation time is related to viscosity $\eta$, absolute temperature ($T$), molecular volume ($V$), and the gas constant ($R$) by the following equation:

$$\text{Correlation time} = \frac{3V\eta}{RT}.$$

The correlation time is small ($\approx 1$ nsec) for small molecules (e.g., fluorescein) and large ($\approx 100$ nsec) for large molecules (e.g., immunoglobulins) (Jolley, 1981). If viscosity and temperature are held constant, correlation time, and therefore polarization, are related directly to the molecular volume. Changes in molecular volume may be due to interactions with other molecules, dissociation, polymerization, degradation, or conformational changes of the fluorescently labeled molecule. For example, fluorescence polarization has been used to measure enzymatic cleavage of large fluorescein-labeled polymers by proteases, DNases, and RNases (Bolger and Checovich, 1994; Bolger and Thompson, 1994; Yonemura and Maeda, 1982). It also has been used to measure equilibrium binding for protein/protein interactions, antibody/antigen binding, and protein/DNA binding (LeTilly and Royer, 1993;

Dandliker *et al.,* 1981; Radek *et al.,* 1993; Wei and Herron, 1993). In this article, we will show that fluorescence polarization is a simple and affordable way to measure binding affinities, and by way of example, we will use fluorescence polarization to determine the dissociation constant for a cdc2Hs peptide and a monoclonal antibody. Readers requiring more details on fluorescence polarization applications are referred to the "Beacon Fluorescence Polarization Applications Guide," available from PanVera Corporation.

## SELECTION OF A FLUORESCENCE POLARIZATION INSTRUMENT

The selection of a reliable and sensitive instrument is the most important factor in using fluorescence polarization to measure binding affinities. One of the most commonly asked questions is, "Can I use the fluorometer in the lab next door to do these experiments?" The answer depends on how sensitive the instrument is and whether it can be operated in a polarization mode. The Beacon line of instruments used in these experiments can measure polarization values on concentrations as low as 10–20 p$M$ fluorescein, but it is much more sensitive in the intensity mode. Even though an instrument is very sensitive in measuring fluorescence intensity, it may not be sensitive measuring polarization values. Instruments which have monochrometers and mirrors often cannot measure polarization values in the p$M$ or n$M$ ranges. Therefore, the types of experiments described here would be impossible to perform on less sensitive instruments. To assess the sensitivity of an instrument, serially dilute fluorescein or a fluorescein-labeled molecule from 500 n$M$ to 10 p$M$ and obtain polarization measurements for each dilution. While the intensity should decrease linearly over this entire range, the polarization values should not change significantly (± 2 mP). If polarization values vary with intensity in the equilibrium binding range that you want to measure, you will need to use another instrument, one with the sensitivity of the Beacon instrument. Characteristic data demonstrating the sensitivity of the Beacon instrument are shown in Fig. 1.

Fluorescein was serially diluted into 0.1 $M$ potassium phosphate, pH 7.4, and 100 µg/ml bovine gammaglobulin (PanVera Cat. No. P2013). The typical dynamic range in polarization measurement is from 20–25 mP for free fluorescein to >400 mP for a fluorescein/antibody complex. For purposes of showing the real sensitivity of this instrument, only a brief part of the dynamic range of the instrument is shown. Note that polarization values remain constant and intensity decreases, as fluorescein concentrations drop several orders of magnitude.

## TIME REQUIRED TO DETERMINE PEPTIDE BINDING CONSTANTS

Typically, equilibrium binding experiments using fluoresceinated peptides are set up over a 2-day period. Peptide labeling and purification is done on the first day and the

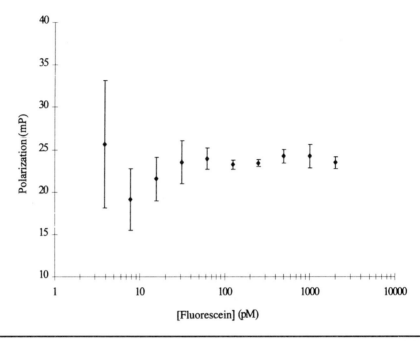

Fluorescein concentration vs polarization (mP). Fluorescein concentration was plotted vs. polarization (mean ± 1 SD, n = 5). Measurements were done on a Beacon fluorescence polarization instrument.

equilibrium binding measurements are done on the second day. A single labeling reaction will generate enough labeled peptide for several hundred binding experiments. This system seems ideally suited for laboratories which are handling large numbers of peptides identified by phage display systems.

*Day 1*

| | |
|---|---|
| Set up peptide labeling reactions | 30 min |
| Labeling reaction | 60 min |
| Quench reaction | 30 min |
| Set up thin layer chromatography | 15–30 min |
| Thin-layer chromatography | 3–6 hr to overnight |

*Day 2*

| | |
|---|---|
| Elution of peptides | 30–60 min |
| Preparation of binding buffers | 15–30 min |
| Diluting and mixing labeled peptide with binding protein | 15–30 min |
| Allow equilibrium to occur | Depending on the system from a few minutes to a few hours |
| Read polarization values | 15 sec per tube, typically 15–20 samples are analyzed |
| Analysis of data | 15–60 min |

# LABELING PEPTIDES WITH FLUORESCEIN

Three important considerations in labeling peptide tracers for fluorescence polarization are: (1) the peptide molecular weight and composition, (2) the labeling method, and (3) the purity of the labeled product. A general rule for fluorescence polarization is to label as small a molecule as possible which is still representative of the binding being studied. Proteins as large as 65 kDa have been used as fluorescent tracers but the first choice for a tracer should be a peptide, then a protein domain, followed by a complete protein. The system should be designed so that the polarization will be low for the free tracer and high for the bound tracer. If the tracer is too large, then the starting polarization value is too high and it is difficult to see changes in polarization upon binding. For example, many labeled, unbound peptides have polarization values of 45–100 mP. Upon binding, values typically increase by 50 to 250 mP.

## Choice of Labeling Methods

Peptides can be fluorescein-labeled during solid-phase synthesis or they can be chemically labeled in the laboratory. Labeling during synthesis is convenient but often expensive. In contrast, chemical labeling can provide plenty of labeled material at a minimum cost. There are a wide variety of reactive fluorescein derivatives that can be used for labeling peptides; some of which are listed in the following table. Most of these derivatives react primarily with thiol or amine groups. Hydroxyls and carboxyl groups can be labeled, with reagents other than those in Table 1, but these side groups are generally less reactive than thiols and amine groups.

It is difficult to predict how attachment of a fluorescein will affect the binding properties of a given peptide, but in most cases the fluorescein molecule will be less likely to interfere sterically if it is attached to the end of the peptide rather than one of the interior amino acids. In addition, many of the existing fluorescein conjugation methods are inefficient and yield unstable conjugates. In these experiments, the succinimidyl ester of fluorescein (FS) was chosen for three reasons: (1) The succinimidyl ester preferentially forms amide bonds, thus the conjugates formed with these compounds are more stable than those formed with FITC or the thiol-reactive fluorophores. (2) Fluorescein is available either as a direct conjugate to the succinimidyl ester or with a six-carbon spacer that may minimize steric interference by the bulky fluorescein moiety. (3) The succinimidyl esters react more rapidly at neutral pH than does FITC, thus exploiting the selective labeling of N-terminal amino groups at neutral pH. This may be desirable in order to minimize modification of amine groups at other sites on the molecule. For proteins in which the N-terminus is not exposed, the efficiency of sidegroup amine labeling may be increased by executing the reaction at pH 8.2 instead of 7.0.

The reactive fluorescein derivatives are very easy to use: the target peptide is mixed with an excess of the fluorescein derivative and the reaction is quenched after a suitable incubation time. Generally, a mixture of fluoresceinated peptide products

**TABLE I**    Reactive Fluorescein Derivatives

| Compound | Selectivity | Compound stability | Conjugate stability | Notes |
|---|---|---|---|---|
| Fluorescein-maleimide | SH (some -NH$_2$ at high pH); most selective SH reagent | Autohydrolysis, light sensitive in solution | Moderate, some ring opening | Requires prior –S–S– reduction in target peptide |
| Fluorescein-iodoacetamide | SH (Tyr, His, Met) | Light sensitive | Moderate | Requires prior S–S reduction in target peptide, long reaction time (18 hr) |
| 5-Bromomethyl-fluorescein | SH, (NH$_2$ Tyr, His, COOH) | Light sensitive in solution | High, most stable SH conjugate | Requires prior S–S reduction |
| Fluorescein isothiocyanate (FITC) | NH$_2$, (SH, Tyr) | Poor, unstable in air, light, water | Moderate | Commonly used for labeling antibodies |
| Fluorescein-succinimidyl ester (FS) | NH$_2$—more selective than FITC | Light sensitive in solution, autohydrolysis at high pH | High—amide bond is as stable as peptide | Lower pH optimum allows selectivity for N-terminus |
| Fluorescein-C6 succinimidyl ester (FXS) | NH$_2$—more selective than FITC | Autohydrolysis at high pH | High—same as FS | Lower pH optimum allows selectivity for N-terminus; C6 spacer minimizes steric hindrance by fluorescein |
| Fluorescein-dichlorotriazine (DTAF) | NH$_2$—OH at high temp. | Light sens. in solution | Moderate | Good reactivity |

result. If one of the fluorescein succinimidyl esters is used, for example, some of the peptide will be unlabeled, some will be labeled only at the N-terminus, some at one or more side chain lysines or histidines, and some at multiple positions. An abbreviated procedure for labeling the N-terminal of peptides is described below and is taken from the Fluorescein Amine Labeling Kit (PanVera Cat. No. 2058).

## Purity of the Tracer

One cannot underestimate the importance of tracer purity in fluorescence polarization. The higher the specific activity of the fluoresceinated peptide, the better the results with fluorescence polarization. The use of a widely heterogeneous or contaminated mixture of labeled peptides often yields poor results.

We have optimized conditions for selective labeling of the N-terminus using the succinimidyl esters, but there is no way to avoid some reaction with amine-containing side chains. The purity of the small fluoresceinated-peptide product can usually be assessed using thin-layer chromatography (TLC). If one chooses, subsequent purification can be carried out by any number of chromatographic methods. Reverse-phase HPLC is the most common approach, but we have found that the same TLC system used for analytical analysis offers a simple and inexpensive alternative to HPLC. In addition, TLC allows multiple samples to be chromatographed simultaneously, making it the method of choice if several different peptides or labeling conditions are to be analyzed. Because fluorescence polarization uses such a small amount of tracer, a spot eluted from an analytical TLC system can provide enough fluorescent peptide for hundreds of assays.

## Design of the Peptide

In designing peptides for use in fluorescence polarization, be sure that the selected chemical groups are available for labeling. For example, do not acetylate peptides if amino groups are to be labeled and do not block the carboxyl groups with amides if the C-terminus is to be labeled. If N-terminal labeling is preferred, internal lysine residues will often label significantly and this must be kept in mind when designing the peptide. Even if multiple amino acids are labeled in the peptide, it is possible to separate the different labeled species by using thin-layer-chromatography or high-performance liquid chromatography.

## Materials for Peptide Labeling

The following materials are needed for peptide labeling:

A. Fluorescein succinimidyl ester (FS) or fluorescein-C6-succinimidyl ester (FXS). These should be reconstituted in DMSO immediately prior to use in the vial provided. We recommend that this step be done in a chemical safety hood to avoid contamination of laboratory equipment with fluorescent particles (Pan-Vera Cat. No. 2058).
B. Dimethylsulfoxide (DMSO)
C. Coupling Buffer, 10x, 1 $M$ $KH_2PO_4$, pH 7.0
D. Quench Buffer, 10x, 1 $M$ Tris-HCl, pH 8.0
E. Distilled water
F. Microcentrifuge tubes
G. Pipetting devices for accurate delivery of microliter volumes
H. A water bath or heating block set at 37°C
I. A peptide to be labeled

## Experimental Design

1. Choosing the proper succinimidyl ester derivative. When labeling a peptide or protein for the first time, we suggest using the FS derivative. Some peptides or proteins may lose biological activity or affinity for receptors after coupling to the FS derivative. In these cases, we recommend repeating the coupling with FXS, a fluorescein succinimidyl ester with a six-carbon spacer arm (also available from Pan-Vera Corporation). FXS may prevent loss of biological activity for some peptides; however, the increased mobility of the fluorophore conferred by the larger spacer arm may result in a smaller polarization shift upon binding.

2. Using the proper labeling ratio. When labeling a peptide for the first time, we suggest labeling at molar ratios of 2:1 and 5:1, label to protein. Many protocols suggest higher labeling ratios, but these procedures are aimed at larger proteins (e.g., antibodies). If your peptide does not label efficiently at 2:1 or 5:1, or if you use this kit to label large proteins or antibodies, it may be appropriate to increase the ratio to 10:1, or greater. Once optimal labeling conditions have been identified for the peptide or protein of interest, a single labeling reaction with the appropriate molar ratio is usually sufficient.

3. Preparing the peptide or protein. The peptide or protein being labeled should be reconstituted in distilled water to a concentration of at least 2 mg/ml (for a final concentration of at least 1 mg/ml, in the labeling reaction). If the final concentration is below 1 mg/ml, labeling will proceed more slowly, and longer incubation times (see step 6, below) may be needed.

If it is necessary to use a solvent other than water, either to maintain biological activity or for solubility reasons, there are two constraints. First, a nonamine containing buffer must be used in order to avoid reaction of the buffer molecules with the fluorescein succinimidyl esters. Tris and most Good buffers (HEPES, MOPS, etc.) contain amines and should be avoided. Second, if the sample substantially changes the pH of the labeling reaction, the labeling efficiency and selectivity for the N-terminus will be adversely affected.

4. Determining FS concentration and volume. Table 2 should be used to determine the concentration of FS that should be prepared (20 or 10 m$M$) and the amount of FS that should be added to a typical reaction mixture containing 100 µg protein. For molecular weights other than those listed, interpolation should be made to determine the required volumes or calculate the actual nanomoles of FS required. (Table 3 provides a sample reaction.)

For pilot experiments using less than 100 µg peptide or preparative studies using more than 100 µg, the volumes of each component should be either reduced or increased, respectively. The only limitation on this adjustment is that the coupling buffer should account for 10% of the final volume (see Table 3).

**TABLE 2**    Determination of FS Concentration and Volume

| Molecular weight (Da) | nmole in 100 µg | µl of stock FS 2:1 | µl of stock FS 5:1 | FS stock concentration (mM) |
|---|---|---|---|---|
| 500 | 200 | 20 | 50 | 20 |
| 1,000 | 100 | 10 | 25 | 20 |
| 2,000 | 50 | 5 | 12.5 | 20 |
| 5,000 | 20 | 2 | 5 | 20 |
| 10,000 | 10 | 2 | 5 | 10 |
| 20,000 | 5 | 1 | 2.5 | 10 |

**TABLE 3**   Reaction Components

| Reagent | Reaction blank | Reaction |
|---|---|---|
| Coupling buffer, 10x | 10 μl | 10 μl |
| Peptide + deionized water | 90 μl - $X$ (no peptide) | 90 μl - $X$ |
| FS | $X$ μl (from Table 2) | $X$ μl (from Table 2) |
| Total volume | 100 μl | 100 μl |

5. Preparing FS stock solution. Prepare 20 m$M$ solutions of the fluorescein succinimidyl ester by adding 200 μl of DMSO directly to a single 2-mg vial. The fluorescein succinimidyl esters are light sensitive and are hydrolyzed rapidly in solution. They should be made up fresh for each day's experiment, kept on ice, and used as soon as possible. If required, the 10 m$M$ stock should be prepared from the 20 m$M$ stock using DMSO as the diluent.

## Precautions

Normal precautions exercised in handling laboratory reagents should be followed. Refer to the Material Safety Data Sheet(s) (MSDS) for any updated risk, hazard, or safety information. It is recommended that gloves, lab coats, and eye protection be used when working with any chemical reagents.

## Procedure

1. Prepare three microcentrifuge tubes (one Reaction Blank at the highest labeling ratio and containing no protein; two reactions at different molar ratios) for each peptide to be labeled, according to Table 3.
2. Add the coupling buffer, peptide, and water, then chill the tubes to 4° C.
3. Add FS to the ice-cold peptide solutions according to the calculations above and vortex immediately. Place the tubes at 37°C for 60 min. Note the two examples shown below in Tables 4 and 5.

**TABLE 4**   Hypothetical FS Labeling Reactions[a]

| | FS blank | 2:1 FS rxn | 5:1 FS rxn |
|---|---|---|---|
| Coupling buffer, 10X | 5 μl | 5 μl | 5 μl |
| Peptide solution, 10 mg/ml | None | 5 μl (50 nmol) | 5 μl (50 nmol) |
| Deionized water | 32.5 μl | 35 μl | 27.5 μl |
| FS, 20 m$M$ | 12.5 μl | 5 μl of 20 m$M$ (100 nmol) | 12.5 μl of 20 m$M$ (250 nmol) |
| Total volume | 50 μl | 50 μl | 50 μl |

[a]Contains 50 μg of a 1000-Da peptide.

**TABLE 5**    Hypothetical FS labeling reactions[a]

| Reagent | FS blank | 2:1 FS rxn | 5:1 FS rxn |
|---|---|---|---|
| Coupling buffer, 10X | 10 µl | 10 µl | 10 µl |
| Peptide solution, 10 mg/ml | None | 10 µl (5 nmol) | 10 µl (5 nmol) |
| Deionized water | 87.5 µl | 79 µl | 77.5 µl |
| FS, 10mM | 2.5 µl | 1 µl of 10 mM (10 nmol) | 2.5 µl of 10 mM (25 nmol) |
| Total volume | 100 µl | 100 µl | 100 µl |

[a]Contains 100 µg of a 20,000-Da protein.

4. Quench all tubes with 10 µl 1 $M$ Tris-HCl, pH 8.0, per 100 µl reaction volume, vortex, and leave at room temperature for an additional 30 min. Tris contains amine groups and will react with the remaining unreacted fluorescein succinimidyl esters.

5. Store fluoresceinated peptide (F-peptide or protein) wrapped in foil at -20° to -80°C. A total of 1–2 µl should be sufficient for purity analysis.

## ANALYSIS AND PURIFICATION OF LABELED PEPTIDES USING THIN-LAYER CHROMATOGRAPHY

Thin layer chromatography is a convenient way to remove unwanted hydrolysis products and to resolve the multiple fluoresceinated products that result when peptides that contain amino acids with amine side chains are labeled. The maximum size limit for purification of fluoresceinated peptides by TLC is between 13 and 30 amino acids, and will depend on the composition of the individual peptide. Because of the small amounts of fluorescent tracer used in fluorescence polarization experiments, the methods described below for analysis of reaction products should be sufficient also to provide a reasonable quantity of fluoresceinated peptide for binding studies. For instance, if 2 µl of a labeling reaction containing 1 mg/ml of a 1000-Da peptide is used for analysis and the yield of useful product is 20%, there would be about 0.4 nmol of fluoresceinated peptide available for the binding experiments. This reagent would be sufficient for 400 1-ml assays containing 1 n$M$ fluoresceinated peptide.

## Materials

A. Silica TLC plates, 5 x 20 cm (PanVera Cat. No. P2108).
B. *n*-Butanol:acetic acid:water, 4:1:1 (v:v).
C. TLC strip solvent (optional; PanVera Cat. No. P2088).
D. TLC MiniChamber (PanVera Cat. No. P2109)..
E. Elution buffer (50 m$M$ Tris-HCl, pH 8.0, qualified as low fluorescence; made from PanVera Cat. No. 2104, Tris-HCl, 1$M$, pH 8.0).

## TLC Procedure

1. Draw a light pencil line 2.5 cm from the narrow end of a TLC plate to mark the origin line without etching the chromatographic medium. Mark up to five light crosshatch lines at no less than 1-cm intervals and no closer than 0.5 cm to the long edge of the plate. These marks identify the origins where samples should be applied. Using a razor or thin spatula edge, score the plate lengthwise by etching the medium in order to isolate the lanes from one another. Be sure to score lines along the outer edge of the plate.

2. Apply, by means of a micropipetting device or syringe, 1–5 µl of sample, in 1- to 2-µl aliquots, on the origins. Allow each spot to dry completely between applications. A blow dryer may be used to accelerate drying and ensure that spots remain as small as possible. Avoid gouging the silica coating on the TLC plate when spotting the sample. Overloading the plate can cause smearing during the chromatography. Loading different amounts of sample to different lanes can alleviate this problem. Be sure to include a sample from the mock labeling reaction where no peptide was added. This will show where fluorescein, its hydrolysis products, and fluoresceinated Tris migrate in TLC.

3. When the spots have dried completely, place the TLC plate into a TLC chamber or jar containing solvent to a depth of approximately 1 cm. The origin line should be well above the level of the solvent.

4. Place the lid on the TLC chamber and allow the solvent front to climb to within 1 or 2 cm of the top of the plate. This usually takes about 5–6 hr. The scored lane markers will prevent cross-contamination of one lane into the other. Remove the plates from the TLC chamber and allow them to dry completely. A fan or blow dryer will accelerate drying. Be careful of blowing dried silica into the air, which may contaminate laboratory instruments. Often it is convenient to allow the system to proceed overnight. If this is the case, simple score the plates across the top (1 cm from edge) to stop the solvent front.

After the plates have dried, the fluoresceinated compounds can be visualized by placing the plate on an ultraviolet transilluminator or by using a hand-held ultraviolet lamp (long wavelength). Be careful in limiting your exposure to the ultraviolet light. Wear certified protection for eyes, face, and any other exposed skin. Ultraviolet light can cause significant burns. The chromatograph can be documented using the type of equipment typically used for photographing ethidium bromide-stained DNA in agarose gels using the same red filter. An exposure of 0.5–1 sec, with the aperture set at f/11, is a good starting point when using Polaroid type 667 film.

The Tris-conjugated and autohydrolyzed fluorescein succinimidyl ester products run close to the solvent front. Generally, peptides that are fluoresceinated only at the N-terminus run closest to the origin and the various multisubstituted variants run increasingly further up the plate. Peptides with no lysine residues generally produce only one predominant labeled species; lysine-containing peptides will yield a more complex chromatogram. The mobilities above for fluoresceinated peptides are only guidelines. Depending on the amino acid side chains, peptide mobility may vary. In general, it is more important in fluorescence polarization applications to obtain a single species (i.e., single spot) than it is to identify the fluorescein substitution site.

# PEPTIDE ELUTION

It is recommended that a low fluorescence buffer be used for the elution (Pan-Vera Cat. No. 2104). While some lots of common laboratory reagents have low fluorescence, most do not. Even ionic salts such as sodium phosphate and sodium chloride may have very high fluorescent backgrounds. Be sure to check the fluorescence of the elution buffer before using it. It makes no sense to purify a product only to turn around and place it into a contaminated buffer. As an example, the Beacon Fluorescence Polarization Instrument measures the following intensities with the photomultiplier tube voltage set at 80 mV: ultrapure water, 3000 counts; low-fluorescence phosphate-buffered saline, 6500 counts; 100 p$M$ fluorescein, 25,000 counts; and 1 n$M$ fluorescein, 250,000 counts. If a 100 m$M$ solution of Tris-HCL contains 200,000 counts it will make it impossible to measure reliable polarization values. Even ultrapure grade reagents may have very high levels of contaminating fluorescence since fluorescence detection is not included in quality assurance specifications. Glycerol, detergents, ionic salts, and Tris buffers all typically contain significant amounts of background fluorescence.

After the plates are dry, the labeled peptides can be eluted from the silica with an aqueous buffer. A concentration of 50 m$M$ Tris, pH 8.0, will elute most peptides very efficiently, but use of other buffers may be desirable depending on the intended application and the solubility of the peptide or protein. The silica containing the peptide of interest is suspended in the eluting buffer, the suspension is allowed to incubate for approximately 0.5–2 hr at room temperature, and the silica is separated from the elution buffer by centrifugation. Some peptides may require longer elution times. Elution progress can be assessed visually after the centrifugation step. If a significant amount of the labeled protein is still adherent to the silica, mix the tube contents again and extend the incubation time. Vortexing samples or incubating the tube at 37°C may also enhance elution.

Traditionally, silica is removed from the glass plate by scraping with a razor blade or other sharp tool. This process is greatly simplified by using a TLC strip solvent (PanVera Corporation, Cat. No. P2088). Strip solvent detaches the selected silica from the glass plate without scraping. The treated silica can be removed from the glass plate in one piece and transferred to a tube containing elution buffer. Not only are the scraping and the associated silica dust eliminated, but also strip solvent does not contribute to the fluorescent background of the fluoresceinated peptide.

Removal of the silica matrix (containing fluorescent compounds) from the TLC plate should be done in a chemical fume hood or in a separate laboratory from that containing spectrophotometers, the Beacon fluorescence polarization instrument, or any fluorometers in order to avoid contamination of the instruments by airborne fluorescent particles.

## Elution Procedure Using TLC Strip Mix

1. Circle the desired spot lightly with pencil during visualization using ultraviolet light.

2. Apply enough TLC strip solvent to completely cover the circled area (about 5–20 µl).

3. Allow the TLC strip solvent to dry for a few minutes. As the TLC strip solvent dries, the edges of the spot will curl up, allowing removal of the entire spot with a pair of forceps. Transfer the spot containing the desired fluoresceinated peptide to a 1.5-ml microcentrifuge tube. Add 200–500 µl of elution buffer and vortex well. Allow 0.5–2 hr for the fluorescent peptide to completely elute from the silica particles.

4. Centrifuge the eluted silica matrix for 1 min in a microfuge and transfer the supernatant containing the fluorescent peptide to a second tube. Store at -20°C, in the dark.

5. The concentration and yield of small fluorescent peptides can be difficult to determine using any of the common protein assays (absorbance at 280 nm, Bradford assay, BCA assay, or Lowry assay) because of the wide variation in chromogenicity.

The concentration and yield can be estimated if one assumes that each peptide is labeled with only a single fluorescein. We suggest determining the total fluorescence intensity ($A_{492}$, $E_{530}$) of a small amount of the eluted F-peptide solution in a low-fluorescence buffer at pH 8.0 and then using the following calculations to estimate the peptide concentration and yield:

$$\text{Fluorescence yield} = [(A_{492})\,(E_{0.1\% \text{ protein}})(MW_{\text{protein g/mol}})]/ \{(68,000\ m^{-1})(A_{280} - [0.26 \times A_{492}])\},$$

where $A_{492}$ = absorbance at 492 nm, ($E_{0.1\% \text{ protein}}$) = extinction coefficient of the peptide or protein at 1 mg/ml, $MW_{\text{protein g/mol}}$ = molecular weight of the protein, $68,000\ m^{-1}$ = extinction coefficient of fluorescein, and $A_{280}$ = absorbance at 280 nm.

## QUENCHING

Proteins labeled to high specific activities may exhibit quenching of the fluorescein. The extent of quenching is determined by measuring the fluorescence intensity of the labeled protein in the presence and absence of 0.5% sodium dodecyl sulfate (SDS). SDS unfolds the protein and causes the fluorescence intensity to increase.

## BINDING OF cdc2Hs PEPTIDE TO MONOCLONAL ANTI-cdc2Hs

This section describes the characterization of the specific interaction between a monoclonal antibody and its antigen, the cdc2Hs peptide. The cdc2Hs peptide is a portion of p34 kinase, a component of Histone H1 kinase.

## Required Materials

A. cdc2Hs peptide, 28-mer corresponding to amino acids 76–104 of the cdc2Hs gene product (VAMKKIRLESEEEGVPSTAIREISLLKE)
B. Anti-cdc2Hs, propagated as mouse ascites; immunogen was cdc2Hs peptide coupled to KLH.
C. Fluorescein Amine Labeling Kit (PanVera Cat. No. P2058)
D. PBS-BSA, PBS, pH 7.4 (PanVera Cat. No. 2019), containing 100 µg/ml Ultralow fluorescence BSA (PanVera Cat. No. P2069)
E. TLC Fluorescein Peptide Purification Kit (PanVera Cat. No. P2067)

## Labeling cdc2Hs Peptide with Fluorescein

1. The cdc2Hs peptide was fluorescein labeled using the reagents and protocols contained in PanVera's Fluorescein Amine Labeling Kit. A 2:1 ratio of fluorescein-*n*-hydroxysuccinimide and 5 µg peptide was added into 50 µl 100 m$M$ phosphate buffer, pH 7.0, and incubated for 30 min at 37°C. The reaction was quenched by addition of 200 m$M$ Tris-HCl, pH 8.0.

2. The labeling reaction was purified by thin-layer chromatography using Pan-Vera's TLC Fluorescein Peptide Purification Kit. From the reaction, 3 µl was spotted on a TLC plate and allowed to air dry. The resolving solvent was *n*-butanol, glacial acetic acid, and water (4:1:1). The fluorescein-labeled peptide spots were visualized on an ultraviolet transilluminator.

3. Since the peptide contained a free N-terminal amine and internal lysine residues, many combinations of labeling could occur on the peptide. Eight fluorescent spots which were easily distinguished were removed from the TLC plate and eluted into 500 µl 50 m$M$ Tris, pH 8.0.

4. The eight fluorescent peptides were screened for relative anti-cdc2Hs binding activity. A 10 µl anti-cdc2Hs aliquot was added to 2 µl of each fluorescein-labeled peptide (F-peptide) in 1.2 ml PBS–BSA. Polarization and fluorescence intensity of the peptide solutions were measured with and without antibody present. The fluoresceinated cdc2Hs peptide that underwent the largest polarization shift in the presence of antibody was chosen for the antibody characterization study.

## Determination of the Time to Reach Equilibrium of F-Peptide/Anti-cdc2Hs Binding

To determine the equilibrium binding constant, the peptide concentration was held constant and the antibody concentration was varied. (The converse method, where the antibody concentration is held constant, can also be used, though this technique is not described here.)

1. 20 µl fluorescein-labeled cdc2Hs peptide (F-peptide) was added to 1.2 ml PBS–BSA in a Beacon test tube (10 p$M$ final concentration) and the polarization was measured using the Beacon instrument.

2. Approximately 100 n$M$ anti-cdc2Hs was added to the solution with gentle mixing, and it was quickly placed back in the Beacon instrument.

3. Using the instrument in the Kinetic mode, polarization was measured at 13-sec intervals for several minutes.

## Determination of the F-Peptide/Anti-cdc2Hs Equilibrium Binding Constant ($K_d$)

1. Anti-cdc2Hs was diluted serially in 16 Beacon tubes from 2.8 µ$M$ to 85 p$M$ in a final volume of 1.1 ml of PBS–BSA.

2. 20 µl of F-peptide was added to each tube and incubated at room temperature (approx. 22°C) for 30 min to reach equilibrium (10 p$M$ final peptide concentration).

3. Polarization values were measured for each sample.

## RESULTS

All data from these experiments are transferred directly from the Beacon instrument to a computer via a serial port as the data are generated. Data are then retrieved and processed as needed. Data transformation and analysis were performed on the polarization data as detailed below.

It is beyond the scope of this article to present mathematical details on fluorescence polarization and equilibrium binding. For a more detailed description, the reader is referred to the "Fluorescence Polarization Applications Guide" from PanVera Corporation (complimentary guide).

Fluorescein labeling resulted in the identification of eight fluorescent peptides (F-peptides) after separation by TLC (data not shown). Each labeled peptide was screened for anti-cdc2Hs binding activity by the ability of the antibody to increase the polarization values of the small F-peptides. The F-peptide that exhibited the greatest polarization shift in the presence of antibody was chosen for the antibody characterization because this F-peptide/antibody system would have the largest dynamic range in fluorescence polarization-based assays.

Before performing equilibrium binding experiments, it is important to determine the time it takes for a reaction to reach equilibrium. The time it takes to reach equilibrium can vary widely from one system to another and must be determined empirically. Figure 2 demonstrates graphically the association reaction of F-cdc2Hs and anti-cdc2Hs. Under the conditions used, the time required to reach equilibrium is less than 1 min. This may not be typical in that other systems may take several hours to reach equilibrium.

An equilibrium binding isotherm was constructed by titrating a fixed concentration of F-peptide (below the probable $K_d$) with increasing amounts of the anti-cdc2Hs (from below to at least 25 times above the $K_d$). The binding isotherm (mP vs n$M$ [antibody]) is shown in Fig. 3.

The same data were also used to construct a Klotz plot (mP vs log [anti-cdc2Hs]; Fig. 4.). The curve was fit by nonlinear regression using the Prizm curve-fitting

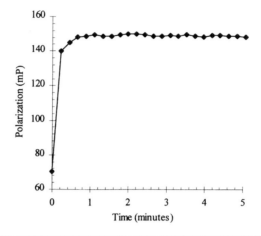

---

**FIGURE 2** Time course of association of fluoresceinated cdc2Hs peptide to anti-cdc2Hs antibody. F-cdc2Hs peptide (10 p*M*) was incubated with 100 n*M* of anti-cdc2Hs antibody at time 0 at 22°C. Polarization was measured for 5 min at 13-sec intervals.

software (Graphpad Software, San Diego, CA). The $K_d$ of F-cdc2Hs/anti-cdc2Hs binding was calculated from the fitted curve. Binding parameters and goodness of fit are listed below the graph. Note that only one inflection point is evident and that positive or negative cooperativity is not indicated. This is the most common method of reporting equilibrium binding data.

The binding isotherm was also analyzed by Scatchard analysis (Fig. 5.) [bound antibody]/[free antibody] was plotted vs [free antibody]. The Scatchard plot is sen-

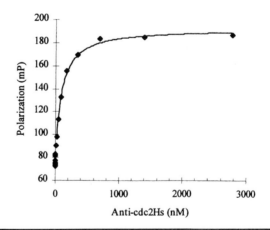

---

**FIGURE 3** Binding isotherm of fluoresceinated cdc2Hs peptide to anti-cdc2Hs antibody. Polarization values were measured after 10 p*M* F-cdc2Hs peptide was incubated with various concentrations of anti-cdc2Hs antibody for 30 min at 22°C.

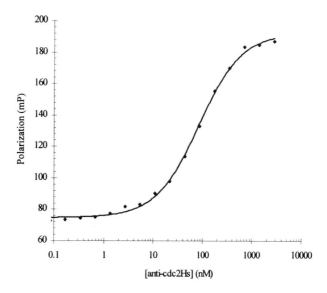

| Parameter | Value | 95% confidence Interval |
|---|---|---|
| Polarization of free F-cdc2Hs, (mP free) | 75 | 73.1, 76.4 |
| Polarization of bound F-cdc2Hs (mP bound) | 192 | 189.4, 195.2 |
| Equilibrium binding constant ($K_d$) | 83.4 nM | 74.4 nM, 93.5 nM |
| Correlation coefficient (R) | 0.999 | - |

**FIGURE 4**          Klotz plot. Plot of F-cdc2Hs/anti-cdc2Hs antibody binding data.

sitive to the presence of nonspecific binding, positive or negative cooperativity, and multiple classes of binding sites. The $K_d$, calculated from this plot (slope = $-1/K_d$), is 75 nM which agrees well with the Klotz plot $K_d$ determination. The data fit best to a linear function, indicating that there is a single class of binding sites whose concentration equals the $x$ intercept (approximately 10 pM).

The final analysis is the Hill plot (Fig. 6.). The slope of the Hill plot is approximately 1, which indicates that there is a single class of binding sites. The $K_d$ determined from this plot ($x$ intercept = log $K_d$) is 83.4 nM.

Normally, if the coefficient of the Hill Plot is a whole integer, the value indicates the number of identical interacting sites. For an antibody, this value is expected to be 2. Measuring the polarization value of the antigen, as we did, in order to construct the Hill Plot, we assumed that there is a 1:1 ratio of receptor to ligand. In reality, for an antibody, the ratio is 2:1. Therefore, the Hill Coefficient of 1, in this case, should be interpreted as 2.

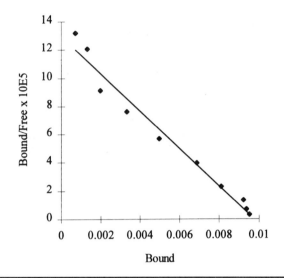

**FIGURE 5**     Scatchard plot. Plot of F-cdc2Hs/anti-cdc2Hs antibody binding data.

## DISCUSSION

These experiments with a fluoresceinated cdc2Hs peptide have demonstrated the steps for the analysis of equilibrium binding data. These steps involved: (1) Selection of an appropriate fluorescein-labeled peptide; (2) determination of the time required to reach equilibrium binding; (3) performance of the direct binding experiment; (4) analysis of the direct binding data by the four suggested transformations.

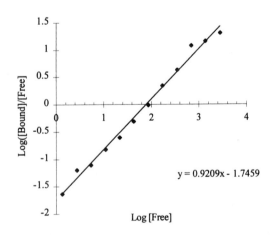

$$y = 0.9209x - 1.7459$$

**FIGURE 6**     Hill plot. Plot of cdc2Hs/anti-cdc2Hs antibody binding data.

This analysis indicated that the antibody preparation contained a single class of binding sites with an equilibrium dissociation constant ($K_d$) of ~80 n$M$ for the chosen fluoresceinated peptide.

## A COMMON PITFALL IN FLUORESCENCE POLARIZATION EQUILIBRIUM BINDING EXPERIMENTS

Many times a common mistake is made in designing equilibrium binding experiments. When troubleshooting experiments, many times the failed results were not from flawed reagents or instruments, but from experimental design. For an equilibrium reaction, the following equation can be used to calculate the percentage of bound molecules at various concentrations of ligand (tracer) and receptor (or other molecule which binds the tracer),

$$B = [(K_d + L + R) - \{(K_d + L + R)^2 - 4RL\}^{1/2}] / 2,$$

where $B$ = concentration of bound ligand, $K_d$ = the dissociation constant, $L$ = total concentration of ligand (tracer) added, and $R$ = receptor concentration. This equation was used to generate Table 6, which shows the percentage of the total labeled ligand bound at various receptor concentrations.

The percentage of labeled ligand bound varies widely depending on the concentration of both molecules. The widest range of polarization values occurs when the concentration of one molecule (e.g., the labeled ligand or tracer) is held constant at 10- to 100-fold below the $K_d$ and the unlabeled molecule (i.e., the receptor or antibody) is varied from at least 20-fold below the $K_d$ to 50-fold over the $K_d$ (see Table 7). For

**TABLE 6**   Percentage of Total Labeled Ligand Bound

| Receptor concentration (multiple of $K_d$) | Total ligand concentration (multiple of $K_d$) | % Ligand bound at equilibrium |
|---|---|---|
| 0.1 | 0.01 | 9 |
| 0.1 | 0.1 | 8 |
| 0.1 | 1 | 5 |
| 0.1 | 10 | 1 |
| 1 | 0.01 | 50 |
| 1 | 0.1 | 49 |
| 1 | 1 | 38 |
| 1 | 10 | 9 |
| 10 | 0.01 | 91 |
| 10 | 0.1 | 91 |
| 10 | 1 | 90 |
| 10 | 10 | 73 |

**TABLE 7**    Range of Percentages of Ligand Bound under Typical Experimental Conditions

| Receptor concentration (multiple of $K_d$) | Total ligand concentration (multiple of $K_d$) | % Ligand bound at equilibrium |
|:---:|:---:|:---:|
| 0.1 | 0.1 | 8 |
| 1 | 0.1 | 49 |
| 10 | 0.1 | 91 |
| 100 | 0.1 | 99 |

example, if a system was being studied that had a dissociation constant of 1 $\mu M$, and 100 n$M$ of ligand was added (0.1X $K_d$) to 100 n$M$ of receptor (0.1X $K_d$), then one should only expect 8% of the tracer to be bound at equilibrium. This means that 92% of tracer has a low polarization value and only 8% has a high polarization value.

Table 7 provides another example of binding, showing the range of 8–99% bound. For instance, in a model receptor-ligand system with a $K_d$ of 8 n$M$, the concentration of the labeled ligand should be less than 0.8 n$M$ and the receptor concentration should be varied from less than 0.4 n$M$ to at least 400 n$M$. Table 7 also shows that the suggested range of concentrations is identical to that required in any classical binding experiment and adequately covers >90% of the fractional occupancy of the labeled ligand. No matter what binding methodology is used, in order to correctly calculate equilibrium binding constants, one of the two binding molecules must be present at concentrations up to 50-fold over the $K_d$. This problem is not unique to fluorescence polarization.

## COMPETITION EXPERIMENTS

### General Considerations

Competition binding experiments are a second common method for analyzing binding affinities. They are generally used in one of the following ways: (1) to compare the relative binding affinities (IC$_{50}$) of two or more unlabeled ligands, or, when the $K_d$ of the labeled ligand is known, to calculate the dissociation constants $(K_d)$ of the unlabeled ligands; (2) to assess the effect of labeling on a ligand's binding affinity; or (3) to determine the concentration of an unlabeled antigen using a competitive assay. The great benefit of competition experiments is that they are a method for the determination of binding affinities that does not require labeling the ligand under study.

In the typical competition experiment, a small amount of a single labeled ligand is incubated with receptor and various concentrations of the unlabeled ligand. As the concentration of unlabeled ligand increases, it competes with labeled ligand for binding to the receptor. As the fraction of labeled ligand that is bound to the receptor drops, the bound signal drops (e.g., decrease in bound radioactivity or the polarization value).

Analysis of the competition binding curve yields $IC_{50}$ values, the concentration of unlabeled ligand necessary to displace 50% of the labeled ligand from the receptor. $IC_{50}$ values are very dependent on the experimental system, the concentration of labeled ligand, and the receptor concentration. $IC_{50}$ values, therefore, are not easily compared if these parameters vary between experiments. If the aim of the study is to compare the relative affinities of a series of ligands, comparison of $IC_{50}$ values, obtained under identical conditions, is an excellent approach. On the other hand, if a dissociation constant for each of the unlabeled ligands is desired, these will have to be calculated from the $IC_{50}$ values.

Competition experiments done using fluorescence polarization necessitate high receptor concentrations in order to bind significant amounts of the fluoresceinated ligand and therefore cause a change in the polarization value. If the [receptor]/$K_d$ ratio is $\leq 0.1$ according to Table 6, no more than 10% of labeled ligand will be bound, and the starting polarization value will still be only 10% of the maximum value. Fluorescence polarization competition experiments should be designed such that the [receptor]/$K_d$ ratio is at least 1, so that the starting polarization value will represent at least 50% of the maximal shift. Under these conditions, though, more inhibitor is required to see 50% inhibition compared to when less receptor is used, and the $IC_{50}$ will be an overestimation of the true $K_i$. If [receptor]/$K_d$ = 1, the error is in the range of 33%; when [receptor]/$K_d$ = 3, the error is approximately 4- to 5-fold, and if [receptor]/$K_d$ = 10, the error is nearly 10-fold. A mathematical correction of this error is possible, but it is complicated and beyond the scope of this discussion. More detailed information can be obtained by contacting PanVera Corporation.

## Receptor–Ligand Competition Experiments

The following example protocol is for a typical competition experiment.

1. Using a direct binding experiment (previously discussed), select a concentration of labeled ligand and receptor that results in 50–90% of the labeled ligand being bound into a complex in the mixture. Ideally, the labeled ligand and receptor can be premixed and added to assay tubes in a small volume.

2. Serially dilute the unlabeled ligands over a range of concentrations, with a total volume of 1.2 ml.

3. Measure the fluorescence background of each tube in the Beacon instrument.

4. Add identical aliquots of the labeled ligand/receptor mixture to each tube, mix, and allow to incubate until equilibrium is established.

5. Measure the polarization value of each tube.

The competition curves are plotted as mP vs unlabeled ligand concentration.

### Acknowledgments

The cdc2Hs antigen and antibody were generous gifts from Medical and Biological Laboratories Co., Ltd., Nagoya, Japan.

## References

Bolger, R., and Checovich, W. (1994). A new protease activity assay using fluorescence polarization. *BioTechniques* **17,** 585–589.

Bolger, R., and Thompson, D. (1994). A quantitative RNase assay using fluorescence polarization. *Am. Biotechnol. Lab.* **12,** 113–116.

Danliker, W. B., Hsu, M. L., Levin, J., and Rao, BR. (1981). Equilibrium and kinetic inhibition assays based upon fluorescence polarization. *In* "Methods in Enzymology" (J. J. Langone and H. Van Vunakis, eds.), vol. 74, pp. 3–28. Academic Press, New York.

Jolley, M. E. (1981). Fluorescence polarization immunoassay for the determination of therapeutic drug levels in human plasma. *J. Anal. Toxicol.* **5,** 236–240.

LeTilly, V., and Royer, C. A. (1993). Fluorescence anisotropy assays implicate protein-protein interactions in regulating trp repressor DNA binding. *Biochemistry,* **32,** 7753–7758.

Perrin, F. (1926). Polarisation de la lumière de fluorescence vie moyenne des molecules dans létat excité. *J. Phys. Radium* **7,** 390–401.

Radek, J. T., Jeong, J. M., Wilson, J., and Lorand, L. (1993). Association of the A subunits of recombinant placental factor XIII with the native carrier B subunits from human plasma. *Biochemistry* **32,** 3527–3534.

Wei, A. P., and Herron, J. N. (1993). Use of synthetic peptides as tracer antigens in fluorescence polarization immunoassays of high molecular weight analytes. *Anal. Chem.* **65,** 3372–3377.

Yonemura, K., and Maeda, H. (1982). A new assay method for DNase by fluorescence polarization and fluorescence intensity using DNA-ethidium bromide complex as a sensitive substrate. *J. Biochem. (Tokyo)* **92,** 1297–1303.

## I. General Reagents

| | | |
|---|---|---|
| Glycine | 200 m$M$ glycine·HCl (pH 2.5)<br>MW = 111.5 g/mol | |
| Na$_2$HCO$_3$ | 100 m$M$ Na$_2$HCO$_3$ (pH 8.5)<br>MW = 106 g/mol | |
| PEG/NaCl | 150 g polyethylene glycol 8000<br>146 g NaCl<br>Water to 1 liter | |
| 2x YT | 10 g tryptone, 5 g yeast extract, 5 g NaCl<br>1 liter water<br>  + 15 g Bacto-agar for bottom agar<br>  + 8 g Bacto-agar for top agar<br>autoclaved 25 min | |
| 2% IPTG | 0.2 g isopropyl-β-D-thiogalactopyranoside<br>  per 10.0 ml water<br>Filter sterilized<br>Stored at -20°C | |
| 2% XGal | 0.2 g 5-bromo-4-chloro-3-indoyl-β-D-galactoside<br>  per 10.0 ml water<br>Filter sterilized<br>Stored at -20°C; limit exposure to light | |

| | |
|---|---|
| 1x PBS | 137 mM NaCl, 3 mM KCl, 8 mM $Na_2HPO_4$, 1.5 mM $KH_2PO_4$ |
| 10x PBS | 80 g NaCl<br>2 g KCl<br>11.5 g $Na_2HPO_4$ ·7H20<br>2 g $KH_2PO_4$<br>Water to 1 liter |
| PBS–Tween 20 | PBS, 0.1% Tween 20 detergent |
| 1x TE | 10 mM Tris·HCl (pH 7.5), 1 mM EDTA |
| Phenol | Equilibrated by repeated extraction with 100 mM Tris·HCl (pH 9), 300 mM NaCl, 1 mM EDTA |
| 1xTAE | 50x stock solution (1 liter)<br>242 g Tris base<br>57.1 ml glacial acetic acid<br>37.2 g $Na_2EDTA·H_20$<br>Water to 1 liter |
| 1xTBE | 10x stock solution (1 liter)<br>108 g Tris base<br>55 g boric acid<br>40 ml 0.5M EDTA (pH 8) |

Alkaline phosphatase buffer
> 1M diethanolamine, pH 9.8
> 1.0 mM $MgCl_2$
> 0.04 mM $ZnSO_4$
> Paranitrophenylphosphate (pNPP)

## 2. Gel Recipes

| | |
|---|---|
| Agarose | FMC (Rockland, ME)<br>1x TAE or 1x TBE<br>GTG grade agarose<br>1% agarose resolves 0.6 to 15 kb dsDNA<br>2% agarose resolves 0.2 to 4 kb<br>3% NuSieve + 1% GTG agarose resolves 50 to 1000 bp<br>4% MetaPhor agarose resolves 50 to 600 bp |
| Polyacrylamide | 20% nondenaturing gel (50 ml)<br>16.7 ml 29% polycacrylamide/1% bisacrylamide<br>5 ml 10x TBE<br>28.3 ml $H_20$<br>25 µl TEMED |

250 µl 10% ammonium persulfate
Resolves 5 to 100 bp
XCFF runs at 12 bp

Sequencing      6% denaturing gel (60 ml)
25.2 g urea
9 ml 38% polyacrylamide/2% bisacrylamide
6 ml 10x TBE
24 ml $H_2O$
XCFF runs at ~30 nt

SDS–PAGE      Separating 15% polyacrylamide gel (15 ml)
7.5 ml 30% acrylamide/0.8% bisacrylamide
3.75 ml 4x Tris·HCl/SDS (pH 8.8)
3.75 ml $H_2O$
10 µl TEMED
50 µl 10% ammonium persulfate
Stacking gel (6 ml)
650 µl 30% acrylamide/0.8% bisacrylamide
1.25 ml 4x Tris·HCl/SDS (pH 6.8)
3.05 ml $H_2O$
5 µl TEMED
25 µl 10% ammonium persulfate

## 3. Gene III Coding Sequence (m663 recombinant)

```
ATT CAC CTC GAA AGC AAG CTG ATA AAC CGA TAC AAT TAA AGG CTC CTT TTG
TAA GTG GAG CTT TCG TTC GAC TAT TTG GCT ATG TTA ATT TCC GAG GAA AAC
-------ForwardSEQ--------->


GAG CCT TTT TTT TTG GAG ATT TTC AAC GTG AAA AAA TTA TTA TTC GCA ATT
CTC GGA AAA AAA AAC CTC TAA AAG TTG CAC TTT TTT AAT AAT AAG CGT TAA
                                V   K   K   L   L   F   A   I>
                               <---------SIGNALPEPTIDE------


CCT TTA GTT GTT CCT TTC TAT TCT CAC TCC TCG AGN NNK NNK NNK NNK NN
GGA AAT CAA CAA GGA AAG ATA AGA GTG AGG AGC TCN NNM NNM NNM NNM NN
 P   L   V   V   P   F   Y   S   H   S   S/T X   X   X   X   X :
------------SIGNALPEPTIDE----------->< XhoI-><-----Random-----


NNK NNK NNK NNK NNK NNK TCT AGA CCT TCG AGA ACT GTT GAA AGT TGT TT.
NNM NNM NNM NNM NNM NNM AGA TCT GGA AGC TCT TGA CAA CTT TCA ACA AA'
 X   X   X   X   X   X   S   R   P   S   R   T   V   E   S   C   L>
-----Peptide-------->< -Xba I->
```

```
GCA AAA CCC CAT ACA GAA AAT TCA TTT ACT AAC GTC TGG AAA GAC GAC AAA
CGT TTT GGG GTA TGT CTT TTA AGT AAA TGA TTG CAG ACC TTT CTG CTG TTT
 A   K   P   H   T   E   N   S   F   T   N   V   W   K   D   D   K>
```

```
ACT TTA GAT CGT TAC GCT AAC TAT GAG GGT TGT CTG TGG AAT GCT ACA GGC
TGA AAT CTA GCA ATG CGA TTG ATA CTC CCA ACA GAC ACC TTA CGA TGT CCG
 T   L   D   R   Y   A   N   Y   E   G   C   L   W   N   A   T   G>
                 <-------ReverseSEQ------->
```

```
GTT GTA GTT TGT ACT GGT GAC GAA ACT CAG TGT TAC GGT ACA TGG GTT CCT
CAA CAT CAA ACA TGA CCA CTG CTT TGA GTC ACA ATG CCA TGT ACC CAA GGA
 V   V   V   C   T   G   D   E   T   Q   C   Y   G   T   W   V   P>
```

```
ATT GGG CTT GCT ATC CCT GAA AAT GAG GGT GGT GGC TCT GAG GGT GGC GGT
TAA CCC GAA CGA TAG GGA CTT TTA CTC CCA CCA CCG AGA CTC CCA CCG CCA
 I   G   L   A   I   P   E   N   E   G   G   G   S   E   G   G   G>
```

```
TCT GAG GGT GGC GGT TCT GAG GGT GGC GGT ACT AAA CCT CCT GAG TAC GGT
AGA CTC CCA CCG CCA AGA CTC CCA CCG CCA TGA TTT GGA GGA CTC ATG CCA
 S   E   G   G   G   S   E   G   G   G   T   K   P   P   E   Y   G>
```

```
GAT ACA CCT ATT CCG GGC TAT ACT TAT ATC AAC CCT CTC GAC GGC ACT TAT
CTA TGT GGA TAA GGC CCG ATA TGA ATA TAG TTG GGA GAG CTG CCG TGA ATA
 D   T   P   I   P   G   Y   T   Y   I   N   P   L   D   G   T   Y>
```

```
CCG CCT GGT ACT GAG CAA AAC CCC GCT AAT CCT AAT CCT TCT CTT GAG GAG
GGC GGA CCA TGA CTC GTT TTG GGG CGA TTA GGA TTA GGA AGA GAA CTC CTC
 P   P   G   T   E   Q   N   P   A   N   P   N   P   S   L   E   E>
```

```
TCT CAG CCT CTT AAT ACT TTC ATG TTT CAG AAT AAT AGG TTC CGA AAT AGG
AGA GTC GGA GAA TTA TGA AAG TAC AAA GTC TTA TTA TCC AAG GCT TTA TCC
 S   Q   P   L   N   T   F   M   F   Q   N   N   R   F   R   N   R>
```

```
CAG GGG GCA TTA ACT GTT TAT ACG GGC ACT GTT ACT CAA GGC ACT GAC CCC
GTC CCC CGT AAT TGA CAA ATA TGC CCG TGA CAA TGA GTT CCG TGA CTG GGG
 Q   G   A   L   T   V   Y   T   G   T   V   T   Q   G   T   D   P>
```

```
GTT AAA ACT TAT TAC CAG TAC ACT CCT GTA TCA TCA AAA GCC ATG TAT GAC
CAA TTT TGA ATA ATG GTC ATG TGA GGA CAT AGT AGT TTT CGG TAC ATA CTG
 V   K   T   Y   Y   Q   Y   T   P   V   S   S   K   A   M   Y   D>
```

```
GCT TAC TGG AAC GGT AAA TTC AGA GAC TGC GCT TTC CAT TCT GGC TTT AAT
CGA ATG ACC TTG CCA TTT AAG TCT CTG ACG CGA AAG GTA AGA CCG AAA TTA
 A   Y   W   N   G   K   F   R   D   C   A   F   H   S   G   F   N>
```

```
GAA GAT CCA TTC GTT TGT GAA TAT CAA GGC CAA TCG TCT GAC CTG CCT CAA
CTT CTA GGT AAG CAA ACA CTT ATA GTT CCG GTT AGC AGA CTG GAC GGA GTT
 E   D   P   F   V   C   E   Y   Q   G   Q   S   S   D   L   P   Q>
```

```
CCT CCT GTC AAT GCT GGC GGC GGC TCT GGT GGT GGT TCT GGT GGC GGC TCT
GGA GGA CAG TTA CGA CCG CCG CCG AGA CCA CCA CCA AGA CCA CCG CCG AGA
 P   P   V   N   A   G   G   G   S   G   G   G   S   G   G   G   S>

GAG GGT GGT GGC TCT GAG GGT GGC GGT TCT GAG GGT GGC GGC TCT GAG GGA
CTC CCA CCA CCG AGA CTC CCA CCG CCA AGA CTC CCA CCG CCG AGA CTC CCT
 E   G   G   G   S   E   G   G   G   S   E   G   G   G   S   E   G>

GGC GGT TCC GGT GGT GGC TCT GGT TCC GGT GAT TTT GAT TAT GAA AAG ATG
CCG CCA AGG CCA CCA CCG AGA CCA AGG CCA CTA AAA CTA ATA CTT TTC TAC
 G   G   S   G   G   G   S   G   S   G   D   F   D   Y   E   K   M>

GCA AAC GCT AAT AAG GGG GCT ATG ACC GAA AAT GCC GAT GAA AAC GCG CTA
CGT TTG CGA TTA TTC CCC CGA TAC TGG CTT TTA CGG CTA CTT TTG CGC GAT
 A   N   A   N   K   G   A   M   T   E   N   A   D   E   N   A   L>

CAG TCT GAC GCT AAA GGC AAA CTT GAT TCT GTC GCT ACT GAT TAC GGT GCT
GTC AGA CTG CGA TTT CCG TTT GAA CTA AGA CAG CGA TGA CTA ATG CCA CGA
 Q   S   D   A   K   G   K   L   D   S   V   A   T   D   Y   G   A>

GCT ATC GAT GGT TTC ATT GGT GAC GTT TCC GGC CTT GCT AAT GGT AAT GGT
CGA TAG CTA CCA AAG TAA CCA CTG CAA AGG CCG GAA CGA TTA CCA TTA CCA
 A   I   D   G   F   I   G   D   V   S   G   L   A   N   G   N   G>

GCT ACT GGT GAT TTT GCT GGC TCT AAT TCC CAA ATG GCT CAA GTC GGT GAC
CGA TGA CCA CTA AAA CGA CCG AGA TTA AGG GTT TAC CGA GTT CAG CCA CTG
 A   T   G   D   F   A   G   S   N   S   Q   M   A   Q   V   G   D>

GGT GAT AAT TCA CCT TTA ATG AAT AAT TTC CGT CAA TAT TTA CCT TCC CTC
CCA CTA TTA AGT GGA AAT TAC TTA TTA AAG GCA GTT ATA AAT GGA AGG GAG
 G   D   N   S   P   L   M   N   N   F   R   Q   Y   L   P   S   L>

CCT CAA TCG GTT GAA TGT CGC CCT TTT GTC TTT AGC GCT GGT AAA CCA TAT
GGA GTT AGC CAA CTT ACA GCG GGA AAA CAG AAA TCG CGA CCA TTT GGT ATA
 P   Q   S   V   E   C   R   P   F   V   F   S   A   G   K   P   Y>

GAA TTT TCT ATT GAT TGT GAC AAA ATA AAC TTA TTC CGT GGT GTC TTT GCG
CTT AAA AGA TAA CTA ACA CTG TTT TAT TTG AAT AAG GCA CCA CAG AAA CGC
 E   F   S   I   D   C   D   K   I   N   L   F   R   G   V   F   A>

TTT CTT TTA TAT GTT GCC ACC TTT ATG TAT GTA TTT TCT ACG TTT GCT AAC
AAA GAA AAT ATA CAA CGG TGG AAA TAC ATA CAT AAA AGA TGC AAA CGA TTG
 F   L   L   Y   V   A   T   F   M   Y   V   F   S   T   F   A   N>

ATA CTG CGT AAT AAG GAG TCT TAA
TAT GAC GCA TTA TTC CTC AGA ATT
 I   L   R   N   K   E   S   *
```

## 4. Gene III Sequencing/PCR Primers

Forward: 5′-ATT CAC CTC GAA AGC AAG CTG-3′
MW = 6400.2
$T_m$ = 44°C

Reverse: 5′-GAG TAT CAA TCG CAT TGC-3′
MW = 5499.6
$T_m$ = 38°C

## 5. Gene VIII

```
CCT TCG TAG TGG CAT TAC GTA TTT TAC CCG TTT AAT GGA AAC TTC CTC
GGA AGC ATC ACC GTA ATG CAT AAA ATG GGC AAA TTA CCT TTG AAG GAG

ATG AAA AAG TCT TTA GTC CTC AAA GCC TCT GTA GCC GTT GCT ACC CTC
TAC TTT TTC AGA AAT CAG GAG TTT CGG AGA CAT CGG CAA CGA TGG GAG
 M   K   K   S   L   V   L   K   A   S   V   A   V   A   T   L>
<----------------------SIGNALPEPTIDE---------------------

GTT CCG ATG CTG TCT TTC GCT GCT GAG GGT GAC GAT CCC GCA AAA GCG
CAA GGC TAC GAC AGA AAG CGA CGA CTC CCA CTG CTA GGG CGT TTT CGC
 V   P   M   L   S   F   A   A   E   G   D   D   P   A   K   A>
   ----SIGNALPEPTIDE------->

GCC TTT AAC TCC CTG CAA GCC TCA GCG ACC GAA TAT ATC GGT TAT GCG
CGG AAA TTG AGG GAC GTT CGG AGT CGC TGG CTT ATA TAG CCA ATA CGC
 A   F   N   S   L   Q   A   S   A   T   E   Y   I   G   Y   A>

TGG GCG ATG GTT GTT GTC ATT GTC GGC GCA ACT ATC GGT ATC AAG CTG
ACC CGC TAC CAA CAA CAG TAA CAG CCG CGT TGA TAG CCA TAG TTC GAC
 W   A   M   V   V   V   I   V   G   A   T   I   G   I   K   L>

TTT AAG AAA TTC ACC TCG AAA GCA AGC TGA TAA ACC GAT ACA ATT AAA
AAA TTC TTT AAG TGG AGC TTT CGT TCG ACT ATT TGG CTA TGT TAA TTT
 F   K   K   F   T   S   K   A   S   *

GGC TCC
CCG AGG
```

## 6. Enzymes That Do Not Cut M13 mp18 DNA

This list is from New England BioLabs: *Aat*I, *Afl*II, *Age*I, *Apa*BI, *Apa*LI, *Avr*II, *Bbs*I, *Bcg*I, *Bcl*I, *Bsa*I, *Bsg*I, *Bsi*I, *Bsp*I, *Bss*HII, *Bst*1107I, *Bst*BI, *Bst*EII, *Bst*XI, *Eag*I, *Eam*1105I, *Eco*RV, *Fse*I, *Hpa*I, *Mlu*I, *Mun*I, *Nco*I, *Nhe*I, *Not*I, *Nru*I, *Xho*I, *Sac*II, *Sap*I, *Sca*I, *Sfi*I, *Spe*I, *Stu*I, *Sty*I, *Xcm*I

## 7. Map of Phagemid Vectors pCANTAB5-E and pCANTAB6

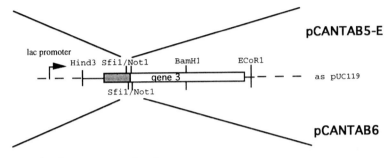

```
                signal sequence   -1 +1
    F   Y   A   A   Q   P   A   M   A   Q   V   Q   L   Q   E   L   E   I   K   R   A   A   A
   TTC TAT GCG GCC CAG CCG GCC ATG GCC CAG GTC CAA CTG CAG GAG CTC GAG ATC AAA CGG GCG GCC GCA
            SfiI          NcoI                      Pst I    (Xho I)                 Not
```

```
    G   A   P   V   P   Y   P   N   P   L   E   P   R   A   A amber
   ggt gcg ccg gtg ccg tat ccg aat ccg ctg gaa ccg cgt gcc gca tag-gene3
                          E tag
```

**pCANTAB5-E**

```
lac promoter
      Hind3 Sfi1/Not1      BamH1      ECoR1
                                                as pUC119
         gene 3
      Sfi1/Not1
```

**pCANTAB6**

```
                signal sequence   -1 +1
    F   Y   A   A   Q   P   A   M   A   Q   V   Q   L   Q   E   L   E   I   K   R   A   A   A
   TTC TAT GCG GCC CAG CCG GCC ATG GCC CAG GTC CAA CTG CAG GAG CTC GAG ATC AAA CGG GCG GCC GCA
            SfiI          NcoI                      Pst I    Xho I                   Not
```

```
    H   H   H   H   H   H   G   A   A   E   Q   K   L   I   S   E   E   D   L   N   G   A   A amber
   cat cat cat cac cat cac ggg gcc gca gaa caa aaa ctc atc tca gaa gag gat ctg aat ggg gcc gca tag-gene3
       polyhistidine                        myc tag
```

---

**FIGURE I**   The sequences around the cloning sites of pCANTAB5-E and pCANTAB6 are shown. pCANTAB5-E carries the E peptide tag which is recognized by the anti-E tag antibody (Pharmacia). pCANTAB6 carries the c-myc tag recognized by the 9E10 monoclonal antibody and a hexahistidine tag for affinity purification. 9E10 can be obtained from Cambridge Research Biochemicals (cat. # OM-11-908) or Oncogene Science (cat. # OP10), or the hybridoma can be obtained from ATCC (cat. # CR1729, MYC1-9E10.2).

Beyond the *Hin*d III and *Eco*R I sites, the plasmid is the same as pUC119 except that three *Apa*L I sites from pUC119 have been deleted. In pCANTAB5-E there is an additional Xho I site in the vector backbone which has been removed in pCANTAB6. Additional information regarding these vectors can be found in Pharmacia's Recombinant Phage Antibody System brochure and McCafferty *et al.,* (1994) *Appl. Biochem. Biotechnol.* **47,** 157–173.

## 8. Library Complexity

| Number of random aa residues | Number of permutations |
|:---:|:---:|
| 4 | $1.60 \times 10^5$ |
| 5 | $3.20 \times 10^6$ |
| 6 | $6.40 \times 10^7$ |
| 7 | $1.28 \times 10^9$ |
| 8 | $2.56 \times 10^{10}$ |
| 9 | $5.12 \times 10^{11}$ |
| 10 | $1.02 \times 10^{13}$ |
| 20 | $1.05 \times 10^{26}$ |

Assuming a Poisson distribution of sequences within the library, approximately 2.3 times more transformants are required to represent a given number of sequence permutations at a 90% confidence level.

Assuming a Poisson distribution of sequences within the library, approximately five times more transformants are required to represent a given number of sequence permutations at a 99% confidence level.

Because libraries composed of long peptides contain multiple overlapping short-peptide sequences, a long-peptide library may be considerably more complex than a short-peptide library with the same number of transformants. A library composed of $X$-mers contains $(X\text{-}N) + 1$ $N$-mers. This fact may have considerable practical consequences for identifying peptides from random libraries.

## 9. OD Determinations (Oligos and Vector DNA)

Concentration of double-stranded DNA: an $A_{260}$ of 1.0 = 50 µg/ml; concentration of single-stranded oligonucleotide: an $A_{260}$ of 1.0 = 37 µg/ml.

## 10. *Escherichia coli* Bacterial Strains

The genotypes of *E. coli* bacterial strains commonly used in phage display work are listed below.

• CJ236 (Bio-Rad, 800-424-6723): used for oligonucleotide-mediated site-directed mutagenesis, lacks a nonsense codon suppressor.

*dut-1 ung-1 thi-1 telA-1 pCJ105 (CmR)*

• DH5αF′ (Gibco BRL, 800-828-6686): useful host for many vectors.

*endA1 hsdR17 ($r_k^- m_k^-$) supE44 thi-1 recA1 gyrA (Nal$^r$)*
*relA1 Δ(lacZYA-argF)U169 deoR (Ø80dlacZΔM15) F′*

• JM110 (Stratagene, 800-424-5444): useful for "tight" control of *lac* promoter constructs; F factor selectable with proline-minus medium.

*rpsL (str$^r$) thr leu thi-1 lacY galK galT ara tonA tsx dam dcm*
*supE44 Δlac-prAB) (F′ traD36, proAB, lacI$^q$ZΔM15)*

• JS5 (Bio-Rad): lacks a non-sense codon suppressor.

*araD139 Δ(ara, leu)7697 (lac)$_{c74}$ galU galK, hsdR2 ($r_k^- m_k^-$)*
*mcrBC rpsL (Str$^r$), thi recA1 [F′ proAB, lacI$^q$ZΔM15, Tn10 (tet$^R$)]*

• MC1061: a household cloning strain; efficient transformation host; rec+.

*araD139, Δ(ara-leu)7696 galE15 galK16 Δ(lac)X75 rpsL (Str$^r$)*
*hsdR2 ($r_k^- m_k^+$) mcrA mcrB1F′*

• TG1: a household cloning strain; tight control of *lac* promoter constructs.

$$supE\ \Delta(hsdM\text{-}mcrB)5(r_k^- m_k^- McrB^-)thi\ \Delta(lac\text{-}proAB)$$
$$[F'\ traD36,\ LacI^q\ \Delta(lacZ)M15]$$

• XL1-Blue (Stratagene): commonly used strain; F factor is selectable with tetracycline.

$$recA1\ endA1\ gyrA96\ thi\text{-}1\ hsdR17\ supE44\ relA1\ lac$$
$$[F'\ proAB,\ lacI^q Z\Delta M15,\ Tn10\ (Tet^r),\ Cam^r]$$

## II. Anti-Phage Antibodies

• Pharmacia (800-526-3593) recombinant phage antibody system, detection module (Cat. No. 27-9402-01).
• Sigma (800-325-3010), rabbit anti-fd polyclonal antibody, Cat. No. B-7786
• Monoclonal antibodies to M13 coat proteins: Dente, L., Cesareni, G., Micheli, G., Felici, F., Folgori, A., Luzzago, A., Monaci, P., Nicosia, A., and Delmastro, P. (1994). Monoclonal antibodies that recognize filamentous phage: Tools for phage display technology. *Gene* **148**, 7–13. Bhardwaj, D., Singh, S. S., Abrol, S., and Chaudhary, V. K. (1995). Monoclonal antibodies against a minor and the major coat proteins of filamentous phage M13: their application in phage display. *J. Immunol. Methods* **179**, 165–175.

## 12. Codon Usage in *E. coli*

| Amino acid | Abbreviations | | MW | Codon | Usage[a] |
|---|---|---|---|---|---|
| Alanine | Ala | A | 89.1 | GCA | 0.28 |
| | | | | GCC | 0.10 |
| | | | | GCG | 0.26 |
| | | | | GCT | 0.35 |
| Arginine | Arg | R | 174.2 | AGA | 0.00 |
| | | | | AGG | 0.00 |
| | | | | CGA | 0.01 |
| | | | | CGC | 0.25 |
| | | | | CGG | 0.00 |
| | | | | CGT | 0.74 |
| Aspartic acid | Asp | D | 150.1 | GAC | 0.67 |
| | | | | GAT | 0.33 |
| Asparginine | Asn | N | 133.1 | AAC | 0.94 |
| | | | | AAT | 0.06 |
| Cysteine | Cys | C | 121.2 | TGC | 0.51 |
| | | | | TGT | 0.49 |

| Amino acid | Abbreviations | | MW | Codon | Usage[a] |
|---|---|---|---|---|---|
| Glycine | Gly | G | 75.1 | GGA | 0.00 |
| | | | | GGC | 0.38 |
| | | | | GGG | 0.02 |
| | | | | GGT | 0.59 |
| Glutamine | Gln | Q | 146.2 | CAA | 0.14 |
| | | | | CAG | 0.86 |
| Glutamic acid | Glu | E | 147.1 | GAA | 0.78 |
| | | | | GAG | 0.22 |
| Histidine | His | H | 155.2 | CAC | 0.83 |
| | | | | CAT | 0.17 |
| Isoleucine | Ile | I | 131.18 | ATA | 0.00 |
| | | | | ATC | 0.83 |
| | | | | ATT | 0.17 |
| Leucine | Leu | L | 131.2 | TTA | 0.02 |
| | | | | TTG | 0.03 |
| | | | | CTA | 0.00 |
| | | | | CTC | 0.07 |
| | | | | CTG | 0.83 |
| | | | | CTT | 0.04 |
| Lysine | Lys | K | 146.2 | AAA | 0.74 |
| | | | | AAG | 0.26 |
| Methionine | Met | M | 149.2 | ATG | 1.0 |
| Phenylanine | Phe | F | 165.2 | TTC | 0.76 |
| | | | | TTT | 0.24 |
| Proline | Pro | P | 115.1 | CCA | 0.15 |
| | | | | CCC | 0.00 |
| | | | | CCG | 0.77 |
| | | | | CCT | 0.08 |
| Serine | Ser | S | 105.1 | AGC | 0.20 |
| | | | | AGT | 0.03 |
| | | | | TCA | 0.02 |
| | | | | TCC | 0.37 |
| | | | | TCG | 0.04 |
| | | | | TCT | 0.34 |
| Threonine | Thr | T | 119.1 | ACA | 0.04 |
| | | | | ACC | 0.55 |
| | | | | ACG | 0.07 |
| | | | | ACT | 0.35 |
| Tryptophan | Trp | W | 204.2 | TGG | 1.0 |
| Tyrosine | Tyr | Y | 181.2 | TAC | 0.75 |
| | | | | TAT | 0.25 |

| Amino acid | Abbreviations | MW | Codon | Usage[a] |
|---|---|---|---|---|
| Valine | Val   V | 117.2 | GTA | 0.26 |
|  |  |  | GTC | 0.07 |
|  |  |  | GTG | 0.16 |
|  |  |  | GTT | 0.51 |

[a]Highly expressed genes (from GCG version 8).

## 13. Families of Amino Acids

Acidic side chains: D, E
Basic side chains: K, R, H
Nonpolar side chains: G, A, V, L, I, P, F, M, W
Uncharged polar side chains: N, Q, S, T, Y, C

## 14. Stop Codons and Suppression

| Stop codons | Nickname | Suppressor | Inserted residue |
|---|---|---|---|
| TAA, TGA | ochre, opal | supC | Y |
| TAG | amber | supE | Q |
|  |  | supF | Y |

## 15. M13 Phage Stability and Instability

| Stable[a] | Unstable |
|---|---|
| pH 2 | 1% SDS |
| pH 12 | Chloroform |
| 1% dimethylformamide | Phenol |
| 2 - 6 M urea |  |
| 10 mM DTT |  |
| 65°C (15 min) |  |

[a]Room temperature.